Introduction

The third BOC Priestley Conference was held at Imperial College London on 12-15 September 1983, thus celebrating the sestercentenary of Joseph Priestley's birth in 1733 and completing the first cycle of these triennial meetings held in turn in Leeds, Birmingham, and London, the cities most closely associated with Priestley's life and work in England. Like the previous meetings this one was generously sponsored by BOC PLC (Gases Division) and organised by the Royal Society of Chemistry.

This international meeting consisted of a scientific part with 15 lectures on the general theme of the role of oxygen in the conversion of present and future feedstocks, and an historical part, organised by the Historical Group of the Royal Society of Chemistry, with eight lectures on various aspects of Joseph Priestley's life and work. The very broad coverage of this significant and successful meeting can be seen from the titles of the lectures, all of which are collected in this volume.

The scientific part of the programme covered the production, purification, and conversion of synthesis gas, the gasification of oil, coal, and carbon, homogeneous and heterogeneous catalytic oxidation, the interconversion of heavy organic chemicals, Fischer-Tropsch chemistry, future liquid fuels, and biomass as a chemical raw material. The scene was aptly set for all of these contributions in an excellent review of the present and future use of oxygen in the conversion of hydrocarbons and synthesis gas into chemicals and transport fuels, and the pressing need for still more chemical and engineering elegance was clearly demonstrated.

The main scientific and historical programme was complemented by the Third Priestley Lecture which was delivered by the distinguished scientist and historian Dr. Joseph Needham. This lecture, entitled "The world's first chemical explosive - in China and the West", is also printed here. The Priestley Lectureship was established to commemorate the enormous breadth of Priestley's interests and influence, far beyond science, and we were fortunate to hear such a scholarly account of a subject of scientific and cultural interest which would surely have delighted Priestley as much as it did the large twentieth century audience containing embryo scientists from many schools and colleges.

The Programme Committee is most grateful to Dr. J.W. Barrett and the Organising Committee, Dr. B. Atkinson and the Local Committee, and Dr. J.F. Gibson and his staff for all their support, and particularly to the lecturers and participants who made this an excellent conference, the value of which is largely enshrined here.

C.W. Rees, Imperial College, London.
Chairman, Programme Committee.

Oxygen and the Conversion of Future Feedstocks

Special Publication No. 48

Oxygen and the Conversion of Future Feedstocks

The Proceedings of the Third BOC Priestley Conference, sponsored by the BOC Gases Division Trust and organised by the Royal Society of Chemistry in conjunction with the Imperial College of Science and Technology, London

London, September 12th—15th 1983

The Royal Society of Chemistry
Burlington House, London W1V 0BN

Copyright © 1984
The Royal Society of Chemistry

All Rights Reserved
No part of this book may be reproduced or transmitted in any form or by any means — graphic, electronic, including photocopying, recording, taping or information storage and retrieval systems — without written permission from The Royal Society of Chemistry

British Library Cataloguing in Publication Data

BOC Priestley Conference *(3rd : 1983 : London)*
Oxygen and the conversion of future feedstocks.
—(Special publication/Royal Society of
Chemistry, ISSN 0260-6291 ; 48)
1. Oxidation 2. Oxygen—Industrial applications
I. Title   II. BOC Gases Division Trust
III. Royal Society of Chemistry   IV. Series
661   TP156.09

ISBN 0-85186-915-7

Erratum

We regret that several pages of this publication appear out of sequence. The page numbered 356 should appear immediately after the page numbered 351. Consequently, pages 352—355 should be renumbered as 353—356. The present page 355 immediately precedes page 357.

Printed in Great Britain by
Whitstable Litho Ltd., Whitstable, Kent

Contents

Introductory Lecture: The Elegant Use of Oxygen *By R. Malpas*	1
The Catalysis of Synthesis Gas Production *By D.A. Dowden*	31
Development of the Shell Coal Gasification Process *By M.J. van der Burgt and J.E. Naber*	62
Development of the Fixed Bed Slagging Gasifier *By C.T. Brooks*	76
The Purification of Synthesis Gas *By S.P.S. Andrew*	95
Transition-metal Peroxides as Reactive Intermediates in Heterolytic and Homolytic Liquid-phase Catalytic Oxidations *By H. Mimoun*	120
Heterogeneous Catalytic Oxidation: A Review of Principles and Practice *By W.J. Thomas*	132
Ethylene and Ether from Ethanol *By P.L. Yue, O. Olaofe, and R.H. Birk*	166
Novel Pathways to Enhance Selectivity in Fischer-Tropsch Chemistry *By P.A. Jacobs*	181
In situ Electron Microscopy Studies of Catalysed Gasification of Carbon *By R.T.K. Baker*	200
Ethylene Glycol from Synthesis Gas *By J.B. Saunby*	220

Chemicals from Coal Gasification 238
By P. McBride and H.W. Patton

Biomass as a Chemical Raw Material 247
By K.J. Parker

Gas Separations by Manganese(II) Phosphine Complexes 263
By C.A. McAuliffe

Engines and Future Liquid Fuels 275
By C.C.J. French

Concluding Remarks 302
By C.F. Cullis

Historical Sessions

Priestley Lecture: The World's First Chemical Explosive - in China and the West
By Joseph Needham, FRS, FBA 305

Priestley and the Dissenting Academies 342
By J.W. Ashley Smith

Priestley in Caricature 347
By M. Fitzpatrick

Priestley in America: 1794-1804 370
By D.A. Davenport

'Fresh Warmth to our Friendship': Priestley and His Circle 384
By D.M. Knight

From Chaos to Gas: Pneumatic Chemistry in the Eighteenth Century 392
By R.G.W. Anderson

The Professional Life of an Amateur Chemist: Joseph Priestley 410
By R.E. Schofield

'A Sower Went Forth': Joseph Priestley and the Ministry of Reform 432
By J.H. Brooke

Priestley and the Manipulation of Gases 461
By W.A. Campbell

Introductory Lecture: The Elegant Use of Oyxgen

By R. Malpas[1]*, T. B. Anderson[1], and J. M. Mitchell[2]
[1] THE BRITISH PETROLEUM COMPANY PLC, B.P. RESEARCH CENTRE, CHERTSEY ROAD, SUNBURY, MIDDLESEX TW16 7LN, U.K.
[2] THE BOC GROUP PLC, GREAT WEST HOUSE, PO BOX 39, GREAT WEST ROAD, BRENTFORD, MIDDLESEX TW8 9DQ, U.K.

1. INTRODUCTION

It is indeed an honour to open this third BOC Priestley conference. It is also somewhat daunting and has involved much work; so at the outset I want to thank my two collaborators - Tom Anderson of BP and Martin Mitchell of BOC who, pressganged into helping me, did so well and with good grace. I also wish to acknowledge the help of Dr. J.V.Porcelli and John L. Erhler of the Halcon SD Group. We hope that the result is worthy and fitting to set the scene for the important papers which follow.

The scope of the conference is broad, so when confronted with it, the first task was to narrow the focus somewhat. This we have done by developing the theme of technological excellence related to the use of oxygen in converting future feedstocks.

We have also shortened the timescale to make it all more manageable. I, for example, look forward to a world when nuclear energy is abundantly cheap allowing highly efficient electro-chemical processes to predominate for the manufacture of most chemicals with perhaps hydrogen electrolysed from water as the main transport fuel. But that, if it is the future, and my guess is almost bound to be wrong, is still too far away. More immediate needs are pressing for solutions rather better than those available to-day. But they are not so pressing that they need an approach like President Carter's 88 billion dollar synfuels programme. Thank goodness the need for that has gone, at least for the time being. It would have produced some vulgar technological monuments which would quickly have become embarrassments to us all.

The time scale of this paper therefore extends only over the next twenty years or so when the world's main fuel and feedstocks for energy, transportation, and hydrocarbon chemicals will be oil, gas and coal.

2. THE CONCEPT OF PROCESS ELEGANCE

What do I mean by process elegance? First let me try and highlight the essential features by a demonstration - not of elegance - but of inelegance.

Most of you will be familiar with the Fischer-Tropsch process and I do not want to tell this fairly well-known tale yet again - but the process does illustrate remarkably well the point I wish to make. As implemented at Sasol this process has two principal stages - first, gasification using oxygen and steam to break up the coal structure and produce syngas; and second, the reassembly of that syngas back to liquid hydrocarbon products using Fischer-Tropsch chemistry. The major products range from ethylene, through LPG to gasoline and diesel fuels - none of which contain oxygen. Methane is recycled back through a Lurgi auto-thermal reforming unit to produce more syngas. As stated, this sounds relatively straightforward, but both the gasification and the Fischer-Tropsch (F-T) stages involve many processing steps. For example, this second stage includes not only the F-T reactors but a virtual refinery to then produce the liquid fuels.

Figure 1 - Sasol II and III Fischer-Tropsch Process.

Figure 1 shows the main flow of coal, steam and oxygen through the Fischer-Tropsch process, focussing on the individual flows of carbon, hydrogen and oxygen. Of the original carbon fed to the gasifiers, only 46% ends up in syngas while only 12.8% of the total oxygen fed (as pure oxygen, steam and oxygen bound up in the coal) appears in the syngas and virtually none in the final product.

Figure 2 - Sasol Oxygen Flows

Virtually all the oxygen fed to the process is rejected as carbon dioxide and water. The oxygen is being used as a key to unlock the coal structure - or, rather - as a blunt chisel to prise open the molecules. Both the gasifier operation and the Synthol reactor operation are inelegant. The essential elements of this inelegance are:

- Oxygen and carbon are converted to carbon dioxide to release the energy needed to break up the coal structure.

- Large quantities of steam are needed to moderate the temperature caused by this heat release.

- Large quantities of useful raw materials are consequently rejected as carbon dioxide and water.

- The gasifier produces a wide range of undesirable by-products requiring substantial clean-up.

- The F-T chemistry is highly non-selective and produces a range of hydrocarbons and some oxygenates.

- The methane produced in both the gasifiers and Synthol reactors is recycled for additional syngas production.

- The very wide product spectrum leads to a highly complex separation scheme.

- The liquid hydrocarbons require considerable upgrading in order to produce liquid transport fuels.

Because of these factors, large quantities of valuable feedstock material - carbon and oxygen - are rejected from the process. And the oxygen is valuable - the air separation plant would cost about US $700 M - about 8-10% of the total capital investment.

Technically the Fischer-Tropsch process is highly inelegant. The consequent process economics are adverse - probably among the worst of all coal conversion processes for liquid fuel production. But politically and strategically for South Africa, the project is a great success. Cheap coal - and hence cheap energy - and freely available oxygen are used to reduce dependence on imported oil. Further support for the Sasol projects comes from the rest of the South African economy through tax waivers on the finished products. So elegance is in the eye of the beholder.

Introductory Lecture: The Elegant Use of Oxygen

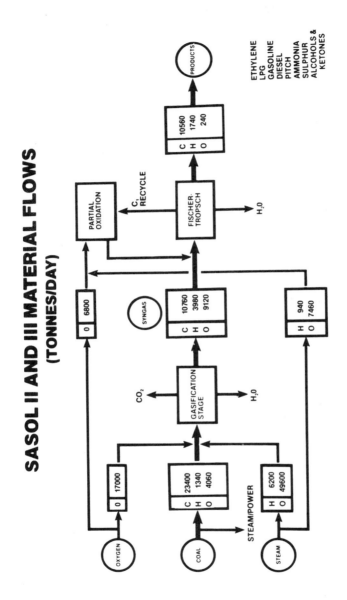

Figure 1

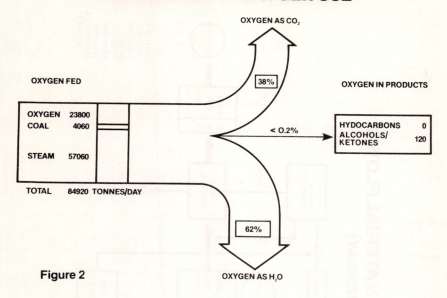

Figure 2

However, in discussing the concept I refer to technical elegance. Essentially it is a qualitative concept which attempts to identify in a simple, relatively obvious way judgements which are often bound-up in complex questions of feedstocks, process chemistry and engineering. The thesis of this paper is that processes which are recognisably 'elegant' are also those processes most likely to have favourable economics.

<u>Figure 3</u> - Process Elegance

I would say that elegance results from the marriage of good chemistry, process design and engineering and leads to an efficient and technically satisfying process. Consider these factors:

- Chemistry — we would all recognise the desirable objectives:
  Simplicity of the reaction scheme, high selectivity, high conversion, few steps, mild conditions, minimum recycle.

- Design — again the objectives are clear:
  The need for imaginative and innovative use of design concepts to meet the challenges of the process chemistry and give an optimal process design.

- Engineering — the process design has then to be engineered into an operable and reliable plant capable of meeting the cost targets.

PROCESS ELEGANCE

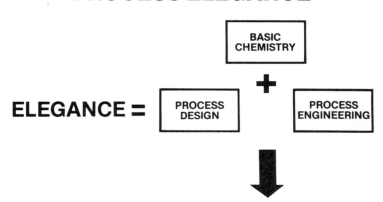

Figure 3

Elegant chemistry will not necessarily give an elegant process, but it is an essential prerequisite. The combination with elegant design and engineering should result in a process making the most effective and lowest cost use of all raw materials and energy, at the lowest capital cost.

Most processes use a number of raw materials. The extent to which one is sacrificed to optimise the total use of the others depends on their availability and relative values. In this regard oxygen is perhaps used somewhat too freely because of its ready availability and relative cheapness. I would suggest that a more elegant use of oxygen will not only result in oxygen savings but in better use of all other feedstocks and energy supplies as well.

Raw material availability is, of course, influenced by political factors and sometimes there is a price to pay, for example, subsidies for Sasol in South Africa or for ethanol in Brazil. Note, however, that air, the feed for oxygen, is free of politics.

Let me now summarise those factors which lead to process elegance:

- Small number of process steps
- High selectivity to products
- High degree of conversion
- Recycle streams limited in size and number
- Mild operating conditions
- Simplicity of the separation requirements
- Process efficiency in material and energy use
- Optimal process design
- Operability and reliability

Clearly these are highly inter-related and the most elegant process will often represent a compromise between conflicting factors, for example selectivity *versus* conversion. Further, I would suggest that technical elegance although qualitative is a reasonably consistent and universally applicable concept. You will know it when you see it.

I started this section by giving an example of a highly inelegant process. Let me close by two further examples which I think crystallise the concept.

First, the development of processes for the production of acetic acid.

Figure 4 - Acetic Acid Process Development

Initially the acid was produced from coal using the sequence of steps shown. The alternative route involved sugar and starch based feedstocks for the production of ethanol. Both of these routes are complex and fell into disfavour with the growth of a petrochemical industry based on cheap oil. Initially ethylene from steam cracking was hydrated using concentrated sulphuric acid to give ethanol, thence to acetaldehyde which can be oxidised to give acetic acid. The Wacker process - a remarkably neat piece of chemistry - eliminated the need for a sulphuric acid step by directly oxidising ethylene to acetaldehyde.

In the early sixties, this was overtaken by BP's acetic acid process - a one-step air oxidation of light naphtha to give a mixture of formic, acetic and propionic acids. The chemistry was simple (although non-specific in product formation) and the process engineering optimised by-product formation to meet the market requirements for other acids and solvents. However, this technical elegance was overtaken by economics. When the first plant using this process was installed in the early sixties, the price of gas was over three times the price of naphtha. When BP authorised its last acetic acid plant in the mid-seventies, the price of gas was one-third the price of naphtha.

BP did not build another naphtha oxidation plant. Instead we installed a Monsanto process based on carbonylation chemistry using gas as the primary feedstock. Although the acid synthesis is highly specific and involves some very subtle rhodium-based catalysis, in many ways the process is less elegant. The overall reaction scheme is more complex and involves a greater number of processing steps. The overall stoichiometry produces an excess of costly hydrogen, so the degree of elegance depends on how the process is integrated into other production facilities.

In fact the existing naphtha oxidation plants continue to run economically to meet the growing markets for propionic and formic acid by-products.

Finally, before leaving the general question of process elegance, let me me illustrate it by reference to a conversion process which is about the most elegant I know and which is not likely to pass into obsolescence for some time - the biological life cycle.

Figure 5 - The Biological Cycle

Oxygen plays a central role in the life cycle. Mammals and many other organisms use oxygen to derive energy from their food in the respiratory process. Essentially carbohydrates and other food components are burnt using the oxygen we breathe to release energy to drive all other cell processes from growth to movement. The resulting carbon dioxide is released into the atmosphere. Plants take up this carbon dioxide and convert it back to carbohydrates through photosynthesis - releasing the oxygen back into the atmosphere. (Plants also carry out a parallel respiratory process analogous to mammals). The carbon dioxide essentially acts as a basic plant food and the resulting plant biomass is the primary input to all food chains.

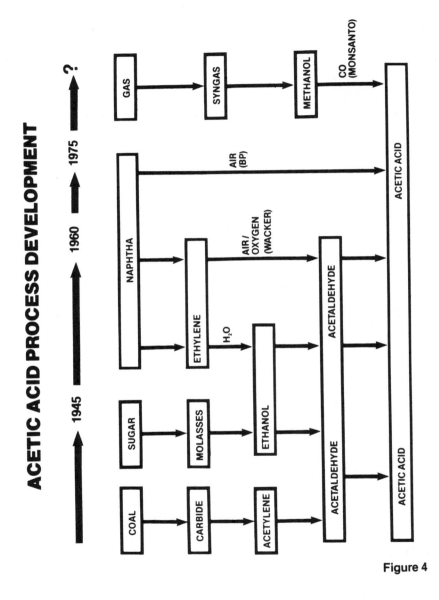

Figure 4

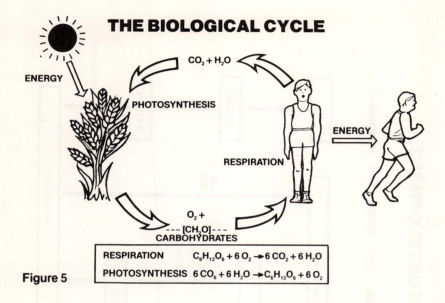

Figure 5

This cycle operates in more or less dynamic equilibrium with little change to the oxygen and carbon dioxide contents of the atmosphere – although the growing use of fuel is tending to increase the concentration of carbon dioxide and raise the temperature of the atmosphere – the well-known and controversial "greenhouse" effect.

The use of oxygen in biological systems can, therefore, be thought of as enabling the conversion of the essential feedstock – the energy in sunlight – into a form which can be used by all living organisms.

The process itself is very elegant, and the biochemical pathways of photosynthesis and respiration reflect this. Although complex, with numerous individual steps and intermediates, the reactions occur under mild physiological conditions, they are self-controlling, clean and rapid due to the specificity of the enzyme catalyst. The last Priestley conference dealt with "Oxygen and Life" and amply illustrated the elegance of the biochemical steps involved.

The biological cycle must be a model of process elegance. If we could emulate it in our industrial processes by recovery and use of the carbon in the carbon dioxide from combustion, our feedstock problems would disappear. Would cheap fusion power give us the energy source to do this? Truly this would be an elegant use of oxygen. Regrettably I do not think it will happen in my lifetime.

3. <u>THE USE OF OXYGEN AS A PROCESS FEEDSTOCK</u>

The biological cycle conveniently leads me back to oxygen.

Let us put its use as a process feedstock into the context of total oxygen consumption.

Figure 6 - World Oxygen Consumption

By far the biggest consumption of oxygen - some 87% - occurs in the biological cycle and is more-or-less balanced by photosynthesis. Combustion of fuels accounts for a further 12% or more. Industrial consumption of oxygen (as both air and oxygen) is estimated to account for only about 0.2% of the total; this amounts to some one million tonnes per day. And of this, only about 270,000 te/day is supplied as pure oxygen.

However this is still a large quantity of oxygen, and in the USA oxygen is the fourth largest tonnage chemical.

Figure 7 - US Tonnage Chemicals

Between the twenties and the early seventies, US oxygen demand increased on average by 12%pa. Growth dropped through the seventies to about 4%pa and the current recession has temporarily halted growth.

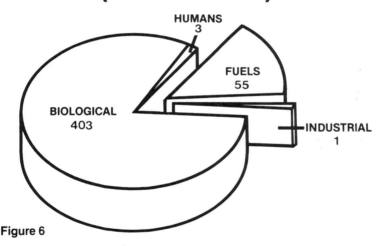

WORLD CONSUMPTION OF OXYGEN (10^6 TONNES/DAY)

Figure 6

Figure 8 - US Oxygen Demand and Average Selling Price

At the same time, the real price of oxygen has fallen steadily in response to the increased demand, advances in production technology (bigger plants, better designs, improved efficiency) and, possibly even more important, the growth of markets for the argon and nitrogen by-products from air separation units.

Figure 9 - Prices of Commodity Chemicals and Feedstocks

This compares bulk UK oxygen prices with naphtha, ethylene and chlorine prices. Unlike naphtha and ethylene, the oil price rises of the seventies had relatively little effect on chlorine and oxygen prices

1982 MAJOR US CHEMICALS

Rank	Chemicals	10⁹ lbs/annum	Rank	Chemicals	10⁹ lbs/annum
1	Sulphuric Acid	64.74	26	Ethylene oxide	4.87
2	Nitrogen	35.00	27	Formaldehyde	4.68
3	Ammonia	30.99	28	Ethylene glycol	3.99
4	Oxygen	29.22	29	Ammonium Sulphate	3.57
5	Lime	28.40	30	p-Xylene	3.20
6	Ethylene	24.66	31	Acetic acid	2.75
7	Sodium hydroxide	18.45	32	Cumene	2.69
8	Chlorine	18.27	33	Aluminium Sulphate	2.38
9	Phosphoric acid	17.10	34	Carbon black	2.29
10	Sodium carbonate	15.79	35	Phenol	2.13
11	Toluene	15.25	36	Acrylonitrile	2.02
12	Nitric acid	15.24	37	Vinyl acetate	1.88
13	Ammonium nitrate	14.67	38	Butadiene	1.83
14	Propylene	12.30	39	Sodium sulphate	1.79
15	Urea	11.84	40	Acetone	1.76
16	Ethylene dichloride	9.99	41	Calcium chloride	1.75
17	Benzene	7.87	42	Propylene oxide	1.48
18	Carbon dioxide	7.38	43	Isopropyl alcohol	1.30
19	Methanol	7.25	44	Sodium silicate	1.30
20	Ethylbenzene	6.61	45	Titanium dioxide	1.28
21	Vinyl chloride	6.50	46	Cyclohexane	1.27
22	Styrene	5.87	47	Sodium tripolyphosphate	1.27
23	Xylene	5.29	48	Adipic acid	1.20
24	Terephthalic acid	4.97	49	Acetic anhydride	1.06
25	Hydrochloric acid	4.97	50	Ethanol	1.02

Figure 7

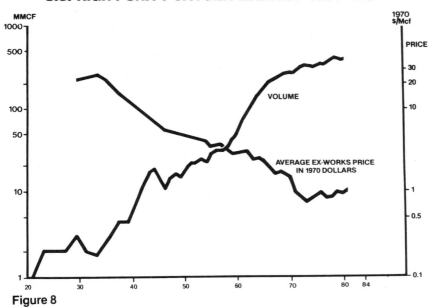

Figure 8

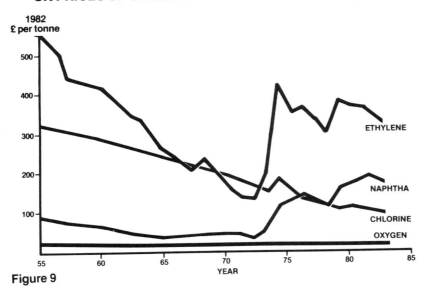

Figure 9

where electricity is the main energy input. The price of oxygen is low
compared to the others (about £22/te) and this reflects the relative
simplicity of its production process and the zero-cost feed material.
Nevertheless the contribution of oxygen production costs to final product
cost can still be substantial. We have already seen that oxygen
contributes about 8-10% to the capital requirement for Sasol and in a
coal to methanol plant oxygen production facilities could account for
some 20% of the final product cost. Clearly, despite its relatively low
cost, the more 'elegant' the use of oxygen, the better.

How do we use oxygen?

Figure 10 - Pure Oxygen Use in the USA

In the USA, total chemical and other process uses probably account for
only 3500-4000 te/day ie about 6-7% of the total pure oxygen use. In
contrast Sasol II and III require about 24,000 te/day of oxygen and the
Great Plains gasification project for production of synthetic natural gas
from coal will use about 4,000 te/day oxygen. Current use of pure oxygen
in the process industries is relatively small, but could grow very
dramatically if even one or two synfuels projects ever got off the
ground.

Of course 3,500 te/day pure oxygen used in the production of chemicals in
the USA does not represent anything like the amount of oxygen actually
incorporated into or used in the production of organic chemicals.

Of the top 50 volume chemical products in the USA (Figure 7) 29 are
organics produced directly from oil and gas feedstocks. Of these 29, 15
contain oxygen in their molecules while several more use oxygen directly
in their production.

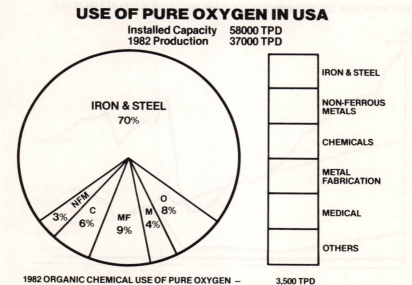

Figure 10

Figure 11 - Oxygen Content of Feedstocks and Products

This compares the weight proportions of carbon, hydrogen and oxygen in typical oil, gas and coal raw materials with the proportions in the 'average' organic chemical product and some typical high tonnage products.

The addition of oxygen to hydrocarbon feedstocks is a fundamental operation of the chemicals industry and the amount of oxygen in major chemical products is more than the nitrogen, sulphur and chlorine etc. added together. The total oxygen "traffic" in the production and use of these high tonnage organic chemicals amounts to some 24,000 te/day of which only some 3000 te/day or less comes as pure oxygen.

The balance does not even all come from air. It comes from a number of sources and a wide range of different types of reaction are used to incorporate the oxygen into the product molecules.

Figure 12 - Mechanism of Oxygen Addition

Oxygen can be added directly either as air or oxygen or it can be added indirectly using oxygen-containing compounds. Oxygen can also be used in the reaction without ever appearing in the principal product. Ammoxidation of propylene to give acrylonitrile or oxychlorination of ethylene to give ethylene dichloride are two important industrial processes where the oxygen ends up as water rather than as product.

Direct oxygen addition routes range from the brutal conditions of combustion through to the subtle chemistry of liquid phase oxidation using homogeneous catalysts under mild conditions. Co-oxidation in this context means the oxidation of a suitable molecule which can then act as an oxygen donor eg. ARCO's co-oxidation technology for the epoxidation of propylene via t-butyl hydroperoxide derived from isobutane.
The severe conditions of combustion and partial oxidation are directed towards breaking up the molecular structure of the feedstock while at the other extreme the liquid-phase reactions are generally used to incorporate oxygen directly into the structure. All of the mechanisms shown find extensive industrial use.

Of the indirect oxygen addition routes, steam reforming and carbonylation are the most relevant to our current interest. The carbon monoxide used in carbonylation processes is in turn derived from steam reforming of gas or naphtha or from partial oxidation of gas, hydrocarbon liquids or coal.

The ultimate source of the oxygen taking part in these reactions is either air or steam. Often the direct oxidation reactions can use either air or oxygen. Reforming and partial oxidation are alternative routes for the production of syngas. These alternatives are central to the problem of the elegant use of oxygen and I will return to them shortly.

4. THE CONVERSION OF FUTURE FEEDSTOCKS

Let us now consider briefly future feedstocks.

In 1982, oil accounted for some 47% (220 million tonnes) of the total non-communist world's energy use. As a liquid hydrocarbon with a high energy density, its light and middle distillates provide the best transport fuels; the hydrogen to carbon ratio of about 2 : 1 makes them ideal feedstocks for chemicals production and the residues can be used as heating fuels or upgraded to lighter products. In the non-communist

OXYGEN CONTENT OF FEEDSTOCKS AND PRODUCTS

FEEDS			
RELATIVE WEIGHTS	GAS	OIL	COAL
CARBON	100	100	100
HYDROGEN	33	14	6
OXYGEN	0	0	10

↓ CONVERSIONS TO CHEMICALS ↑

PRODUCTS				
RELATIVE WEIGHTS	AVERAGE	FORMAL-DEHYDE	VINYL-CHLORIDE	BENZENE
CARBON	100	100	100	100
HYDROGEN	14	17	12	8
OXYGEN	18	134	0	0
OTHERS	16	0	148	0

AVERAGE USA OXYGEN TRAFFIC IN TOP 50 US CHEMICALS = 8.8 million t.p.a.

Figure 11

MECHANISM OF OXYGEN ADDITION

TYPE OF USE	MECHANISM	TEMP (°C)	TYPICAL PROCESSES	
			FEED	PRODUCTS
DIRECT USE OF OXYGEN OR AIR	COMBUSTION	1500-2000	FUELS	CO_2
	COMBUSTION/PARTIAL OXIDATION	1300-1400	FUEL OIL	CARBON BLACK
	PARTIAL OXIDATION	700-1500	COAL, GAS	H_2, CO, CO_2
	VAPOUR PHASE OXIDATION	200-600	ETHYLENE	ETHYLENE OXIDE
	LIQUID PHASE OXIDATION	100-200	CUMENE	PHENOL, ACETONE
	LIQUID PHASE CO-OXIDATION	100-200	PROPYLENE/ISOBUTANE	PROPYLENE OXIDE/T-BUTYL ALCOHOL
ADDITION USING OTHER OXYGEN CONTAINING COMPOUNDS	STEAM REFORMING	700-1000	GAS	H_2, CO, CO_2
	HYDRATION	200-400	ETHYLENE	ETHANOL
	CARBONYLATION	100-300	METHANOL	ACETIC ACID
	OTHER INDIRECT ADDITIONS (HYPOCHLOROUS ACID)	100-200	PROPYLENE	PROPYLENE OXIDE
OTHER USES	OXYCHLORINATION AMMOXIDATION	200-400	ETHYLENE/CHLORINE	ETHYLENE DICHLORIDE

Figure 12

world, oil contributes 98% of the energy used in transportation and 85% of the chemical industry feedstocks. Nearly half of the oil consumed goes into these two premium uses - about 40% to transport and 7% to chemicals.

Gas and coal, currently used mainly as heating fuels, could make a substantial contribution to meeting the transport fuel and chemical feedstock demand. At present gas makes up most of the other 15% of chemical industry feedstocks.

Resources of oil and gas are sufficient to last well into the next century while reserves of coal are truly vast. However much of the oil reserves will remain within OPEC and most of the gas reserves are in remote areas a long way from markets.

Remote gas is expensive to transport, either by pipeline or as liquefied natural gas, and the best prospects at present for its exploitation may be through conversion to methanol. The methanol can then be sold either as transport fuel - assuming it gains acceptance as such - or as a petrochemical feedstock. Current economics make the transport fuel use a medium-term prospect and the chemical use is starting to happen now.

Coal on the other hand is much more difficult, and hence more expensive, to convert either by liquefaction to a syncrude or by gasification to syngas for production of methanol or fuel gas. The adverse economics suggest coal will continue to be used mainly as a heating fuel with only limited conversion use according to specific local circumstances - unless of course, there is some radical improvement in conversion technology. Growing environmental pressures could work against this view however. Coal is a dirty fuel and gasification to fuel gas allows easy removal of sulphur. This trend could favour the introduction of combined cycle power generation schemes.

Biomass will also see some limited use in specific local circumstances, as in Brazil, but its contribution in global terms will be strictly limited by resource availability and very high processing costs.

The rate at which any of these developments will come about depends not only on production costs in relation to changing oil prices but also on political, strategic and more general economic considerations. Although in the short term oil is in plentiful supply, it is still the most finite of todays's energy resources and nations rich in alternative resources like gas and coal will naturally wish to exploit this.

Therefore although I see oil continuing to dominate our energy supplies for at least the next twenty years - particularly for premium uses - the contribution of gas, coal and nuclear energy to the energy mix is expected to increase. Residue conversion and the use of heavier and unconventional oils (shale oil, bitumens) is likely to grow. In particular, I see the very real possibility emerging of a substantial process industry based on natural gas aimed at transport fuels and chemical feedstocks which will challenge the current dominance of oil.

For some time to come this alternative will be based on syngas, which will allow coal or other raw materials to contribute in specific, favourable circumstances.

Figure 13 - Conversion of Primary Energy Resources

This diagram summarises the routes for conversion of oil, gas or coal to fuel or chemical products using current technology. It is a simplified

Introductory Lecture: The Elegant Use of Oxygen

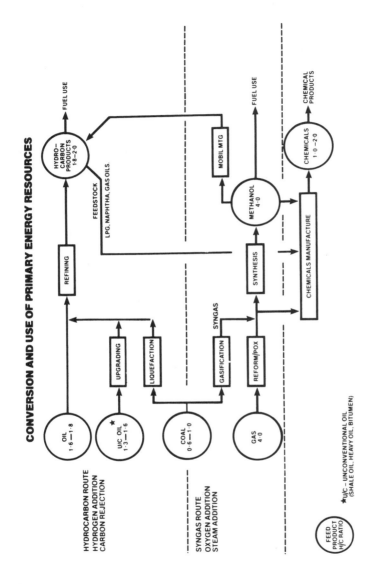

Figure 13

representation of a very complex pattern, for example; the use of ethane and LPG's as steam cracker feedstock is not shown and the conversion of coal to fuel gas is not included.

The diagram identifies the two main routes to transport fuels, feedstocks and chemicals: the hydrocarbon route from conventional or unconventional oils and coal syncrude and the syngas route from gas or coal.

The processes used in these routes are very different. The hydrocarbon routes use carbon rejection, hydrogen addition refinery processes to alter the hydrogen to carbon ratio of the raw materials to suit liquid fuel requirements or to give chemical feedstock. Syngas routes use oxygen or steam to break down the molecular structure of the feedstock prior to synthesis back to the desired fuel or chemical product. Methanol plays a central role in the syngas route to chemicals and could itself become a major transport fuel.

Feedstocks can therefore be classified on the basis of their suitability for the hydrocarbon or syngas routes.

<u>Figure 14</u> - Future Feedstocks

This lists the most important feedstocks for the hydrocarbon or syngas routes. It is interesting to note that the syngas feedstocks tend to lie at the extremes of the range of H/C ratio - from natural gas at 4:1 to residues at 1:1 with coal, lignite and coke even lower; these, of course, are the difficult ones.

FUTURE FEEDSTOCKS

Route	Feedstock
Hydrocarbon	NGL's (C_2 to C_5) LPG (C_3 and C_4) Naphtha Gas Oils
Syngas	Methane Residues Petroleum Coke Coal

Figure 14

5. PRODUCTS FROM THE HYDROCARBON AND SYNGAS ROUTES

The major products of both the hydrocarbon routes and the syngas routes are transport fuels and chemicals and in a sense these routes are rivals. Chemical products which contain substantial quantities of oxygen can start from many feedstocks – coal, gas or oil. On the other hand, transport fuels start from the reality that mixtures of liquid hydrocarbons (paraffins, naphthenes and aromatics predominantly) are very convenient fuels with a very high energy density which can be readily produced from conventional oil.

This could limit the potential for syngas chemistry in the production of transport fuels which at present hinges on the role of methanol.

Methanol can be converted to gasoline *via* the Mobil MTG process. This route is highly inelegant and involves converting a hydrocarbon to an oxygenate and back to a hydrocarbon mixture. It is a three-step process – gas to syngas, syngas to methanol, methanol to gasoline. Even with the cheapest gas, the costs are very high and economic prospects for the process poor. Direct conversion of syngas to gasoline is an obvious development target but it has not been achieved yet.

Methanol would have to see widespread acceptance as a transport fuel in its own right before the syngas route could become significant for transport fuels. Methanol has a much lower energy density than gasoline and hence gives lower miles per gallon, and there is the water miscibility problem. To overcome these disadvantages would require a substantial degree of political and institutional willingness and cooperation at a national and international level.

What seems clear however is that there is an important role for methanol or other oxygenated compounds as gasoline additives. These would function both as gasoline extenders and, more importantly, as replacements for lead to boost octane. Many oxygenated compounds have very high octane numbers and hence are valuable materials for blending into the gasoline pool. The pressure to reduce or eliminate lead is growing rapidly.

The major oxygenates available at present are:

- Methanol
- Ethanol
- Methyl tertiary-butyl ether (MTBE)
- Tertiary-butyl alcohol (TBA)
- Mixed alcohols (C_1 to C_4)

As gasoline oxygenate demand increases, it would not be surprising if better oxygenated products from more elegant processes appear on the market.

It is not unreasonable to expect that refiners will eventually be adding 5% or more oxygenates to the gasoline pool. In the USA this could amount to 10000 te/day oxygen added into gasoline – depending on blending levels and oxygenate used. If this were supplied solely as pure oxygen into the oxygenate manufacturing process, the demand for pure oxygen would increase by at least 30% over current levels.

However oxygenates in this role should properly be seen as chemical products sold back into the fuel market rather than transport fuels as such. And it is in the production of chemicals where the underlying competition between syngas-based routes and hydrocarbon-based routes is most evident. Certainly syngas chemistry has clearly established itself for the production of major oxygen-containing products like acetic acid, acetic anhydride and, of course, methanol itself and this use is likely to grow.

Indeed one extreme point of view suggests that syngas chemistry could replace steam cracking within ten years for production of our bulk primary chemical products. I doubt this because of the technial inelegance of using carbon monoxide to make end-products not containing oxygen. And, in fact syngas chemistry has a long way to go to overtake established hydrocarbon routes.

Let me explain.

Figure 15 - Chemicals from Syngas and Hydrocarbons

The chemical products shown here are divided into three groups - those made solely from syngas, those which are currently made mainly from hydrocarbons but for which syngas processes are available or are being developed and those for which only hydrocarbon routes exist. When talking about syngas-based routes, incidentally, I also include those processes using methanol as a feedstock, eg formaldehyde.

The list is not exhaustive but contains most of the high-tonnage possibilities. Not shown are those products which require both hydrocarbons and syngas, for example, OXO alcohols, MTBE.

I want to focus briefly on those processes for which both the syngas and hydrocarbon routes exist, in particular, those at or close to commercialisation or which are the subjects of papers at this conference.

Over a very large range of important products, the overlap between the two routes is very clear. However, comparison with the third group - products from hydrocarbons only - shows just how far syngas chemistry has to go to supplant the hydrocarbon routes. For most of the major high-tonnage organic products, syngas chemistry does not yet provide any real hint of alternative routes. However there is considerable effort directed towards the production of olefins and aromatics from syngas, essentially extensions of Fischer-Tropsch or modifications of the Mobil MTG Zeolite technology, which if successful would change the picture dramatically.

Hydrocarbon chemistry is going to be with us for some time to come. What I suspect will happen is that both routes will happily co-exist with the syngas routes gradually increasing in importance as the oil/gas supply situation changes and technology develops. And ultimately I can imagine more direct gas conversion routes emerging to challenge both the hydrocarbon and syngas routes.

·6. THE ROLE OF OXYGEN IN HYDROCARBON ROUTES

Oxygen plays only a very limited, but growing, part in the refining processes used for production of transport fuels and hydrocarbon feedstocks for the chemical industry. However, oxidation processes are the major route for conversion of these feedstocks to bulk organic chemical intermediates.

In refinery operations there are two major uses of oxygen developing:

- The use of oxygen-enriched air for debottlenecking cat crackers.
- Partial oxidation of residues.

CHEMICALS FROM SYNGAS AND HYDROCARBONS

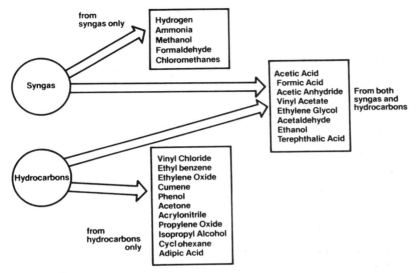

Figure 15

The longer-term trend towards heavier cat cracker feeds and a lighter product slate means more coke lay-down on the catalyst and hence a bigger load on the catalyst regeneration system. By reducing the flow of inert nitrogen, oxygen enrichment can increase the capacity of the regeneration system to burn coke off the catalyst.

Partial oxidation of residues neatly eliminates the problems of refining heavy residual material with high levels of sulphur, metals and asphaltenes. The product is syngas - which would probably be used for hydrogen production in a refinery context, at least initially. Refinery hydrogen is conventionally produced by steam reforming of refinery gases and this is more logical because of the extra hydrogen available from the steam. Nevertheless the cost gap between steam reforming and partial oxidation of residues seems to be narrowing. Texaco are installing a residue gasifier at their Convent Refinery as part of a residue conversion scheme. I suspect there will be more of these to come.

Turning now to hydrocarbon oxidation processes for production of chemical intermediates - either air or oxygen can be used as the oxidant with air supplying most of the oxygen requirement.

Figure 16 - Major Organic Intermediates Needing Oxygen

This shows the 15 major organic products which use air or oxygen directly in their current production processes. Of these fifteen, pure oxygen might be used for only seven products. And of the seven, acetic acid *via* acetaldehyde is being superseded by the methanol carbonylation route while acetic anhydride from acetaldehyde is also now under threat from a syngas-based route. Inelegant processes are always vulnerable!

Major Organic Intermediates Needing Oxygen

PRODUCT	AIR OR OXYGEN
Ethylene Dichloride	New plant — [oxygen]
Terephthalic Acid	Air
Ethylene Oxide	New plant — [oxygen]
Formaldehyde	Air
Acetic Acid	{ Acetaldehyde route — air or [oxygen] Hydrocarbon route — air
Phenol	Air
Acrylonitrile	Air
Vinyl Acetate	New plant — [oxygen]
Acetone	Air
Propylene Oxide	{ Isobutane route — [oxygen] Ethylbenzene route — air
Adipic Acid	Air
Acetic Anhydride	Air or [oxygen]
Acetaldehyde	Air or [oxygen]
Methyl Methacrylate	Air
Phthalic Anhydride	Air

Figure 16

Of the fifteen, only five products can be considered as major oxygen (as distinct from air) consuming products:

- Ethylene Dichloride
- Ethylene Oxide
- Vinyl Acetate
- Propylene Oxide
- Acetaldehyde

Air still predominates as the source of oxygen for most chemical processes. Is there not a greater scope for the more elegant use of oxygen than this implies? The chemical industry makes surprisingly little use of oxygen as such. Why is this?

Figure 17 - Comparison of Air and Oxygen Processes

The general factors favouring use of oxygen and the factors favouring use of air are contrasted in Figure 17. It is by no means an exhaustive list, but it includes the major factors to be considered in all cases.

First - oxygen. Use of oxygen rather than air generally results in much lower gas flows through most of the process and hence smaller equipment for new plants, or increased capacity for existing plants converted from air to oxygen. Increased oxygen concentrations, using pure oxygen, lead to increased reaction rates, hence, reduced catalyst volumes and reactor sizes for the same production rate. This increase in reaction rates can also be used to increase selectivity by reducing the conversion per pass. The use of pure oxygen will, therefore, tend to give more flexibility in arriving at optimal reactor design and operating conditions, a side effect of which will very often be increased catalyst life.

Introductory Lecture: The Elegant Use of Oxygen

Comparison of Air and Oxygen Processes

FACTORS FAVOURING OXYGEN
- Smaller Equipment
- Higher Yields
- Longer Catalyst Life
- Lower Purge Gas Losses
- Reduced Environmental Impact
- Greater Energy Efficiency

FACTORS FAVOURING AIR
- Safety
- Operability
- Heat Removal
- Cost of Oxygen

Figure 17

Because the volumes of gas traffic are so much lower, gas purges are lower and hence losses of reactants and products to the atmosphere are generally lower. Raw material efficiency is improved and the need for expensive knock-out or incineration equipment is much reduced. The environmental impact will be lower because of the reduced level of emissions. For example, in the production of vinyl chloride monomer, the vent gas clean-up problem is reduced 100-fold by the use of oxygen rather than air. This is one of the single most important factors dictating the use of oxygen - despite the inelegance that none of the oxygen ends up in the product but is downgraded to water.

A major benefit of using oxygen rather than air is, of course, that by eliminating the need to heat up and compress large volumes of 'free-ride' nitrogen, overall process energy consumption will be lower.

All these benefits will be realised on a new plant, but retrofitting an existing air-based plant to use oxygen will also obtain many of them.

These are very powerful arguments in favour of the use of oxygen, but against them must be set the very real problems of safety.

There are hazards in operating with pure oxygen and the need to avoid explosive mixtures is paramount both in design and operation. However, the handling of oxygen is well understood and safety can be designed into oxygen-using plants.

A major problem with many oxidation processes is the need to remove the large quantities of heat released. The presence of nitrogen in air provides a mechanism for removal of this heat both because it will absorb some of the heat itself and also because of the extra heat needed to vaporise the increased quantity of reactants and products carried out in the nitrogen stream.

Finally, of course, oxygen costs rather more than air.

Each of these factors for or against the use of oxygen will affect different processes in different ways and to different degrees. How the trade-off will work depends on individual processes in specific contexts, for example, oxygen availability and costs, emission regulations, the skill of the operating team. However, I feel there must be greater scope for the use of oxygen than at present. I find it difficult to accept that the costs of the inelegance of allowing such large quantities of nitrogen a completely free ride through the system are in so many cases lower than the costs of using oxygen.

Of course, oxygen enrichment could be the proper solution for several processes and this is now being actively pursued.

As a Director of BOC I have an interest in the greater consumption of oxygen; however I do offer these views objectively. To demonstrate this objectivity, I would urge researchers and process designers to try to obtain the benefits of oxygen over air in more processes than at present and at the same time seek to eliminate the use of oxygen where it does not appear in the final product but is rejected as carbon dioxide or water. Reduction of production costs through a more elegant use of oxygen offers a real challenge to process developers.

7 THE ROLE OF OXYGEN IN SYNGAS ROUTES

With the exception of relatively minor oxidative carbonylation reactions and the direct air oxidation of methanol to formaldehyde, the scope for use of oxygen in syngas-based routes is limited to the production of the syngas itself. Natural gas is likely to remain the major feedstock and steam reforming is currently the dominant process for production of the syngas.

The future use of oxygen in syngas routes will thus depend on:

- The use of partial oxidation rather than steam reforming for conversion of significant quantities of natural gas to syngas.

- A much greater broadening of the feedstock base to include coal, coke and residues, which require a partial oxidation process for gasification to syngas.

I shall return to these two points as part of the theme on the elegant use of oxygen, but first let me focus briefly on four of the products which can be made *via* either the syngas route or the hydrocarbon route. The syngas processes are either commercialised or nearly commercialised except for one still in the development stage. They are:

- Acetic Acid
- Acetic Anhydride
- Vinyl Acetate
- Ethylene Glycol

Two of these are the subjects of papers in this conference. All of them involve methanol synthesis and carbonylation stages.

Figure 18 - Acetic Acid and Acetic Anhydride

The figures also compare the hydrocarbon-based route to the same product. The diagrams understate the complexity of the hydrocarbon route because steam cracking is treated as a single step and given the same apparent weight as the much simpler gasification or reforming step.

First, acetic acid.

I have already discussed this in some detail. The syngas route is already commercial and is rapidly overtaking hydrocarbon routes. The Monsanto carbonylation step uses a homogenous rhodium/iodine catalyst and is highly specific. The number of reaction steps is small. Syngas is separated to give carbon monoxide for the methanol carbonylation step. Also shown is the overall stoichiometry of the reaction steps. This indicates the most appropriate syngas composition to match the process requirement, *ie* a hydrogen:carbon monoxide molar ratio of 1:1 in the syngas.

Introductory Lecture: The Elegant Use of Oxygen

ACETIC ACID

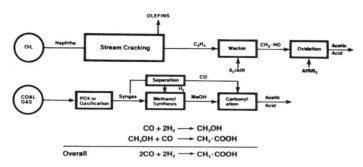

$$CO + 2H_2 \longrightarrow CH_3OH$$
$$CH_3OH + CO \longrightarrow CH_3 \cdot COOH$$

Overall $\quad 2CO + 2H_2 \longrightarrow CH_3 \cdot COOH$

ACETIC ANHYDRIDE

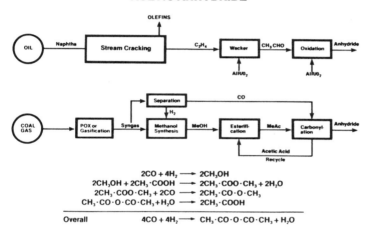

$$2CO + 4H_2 \longrightarrow 2CH_3OH$$
$$2CH_3OH + 2CH_3 \cdot COOH \longrightarrow 2CH_3 \cdot COO \cdot CH_3 + 2H_2O$$
$$2CH_3 \cdot COO \cdot CH_3 + 2CO \longrightarrow 2CH_3 \cdot CO \cdot O \cdot CH_3$$
$$CH_3 \cdot CO \cdot O \cdot CO \cdot CH_3 + H_2O \longrightarrow 2CH_3 \cdot COOH$$

Overall $\quad 4CO + 4H_2 \longrightarrow CH_3 \cdot CO \cdot O \cdot CO \cdot CH_3 + H_2O$

Figure 18

The acetic anhydride route is also shown on this figure. There is a paper later in the conference from Tennessee - Eastman who are close to commissioning a coal-based complex for production of acetic anhydride and I do not want to steal Mr Patton's thunder. This is a very significant investment and represents the first truly <u>commercial</u> application (that is, no government assistance) of an advanced coal gasification technology. The process is elegant and fits very well their requirements.

<u>Figure 19</u> - Vinyl Acetate and Ethylene Glycol

Vinyl acetate from syngas bears a marked resemblance to the acetic anhydride route but involves reductive carbonylation of the methyl acetate to give ethylidene diacetate instead of acetic anhydride. This is then cracked to give vinyl acetate plus acetic acid recycle. This route is developed and awaits commercialisation.

VINYL ACETATE MONOMER

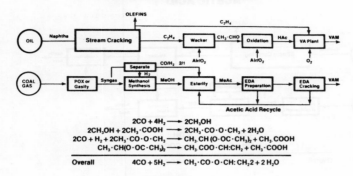

$$2CO + 4H_2 \longrightarrow 2CH_3OH$$
$$2CH_3OH + 2CH_3 \cdot COOH \longrightarrow 2CH_3 \cdot CO \cdot O \cdot CH_3 + 2H_2O$$
$$2CO + H_2 + 2CH_3 \cdot CO \cdot O \cdot CH_3 \longrightarrow CH_3 \cdot CH(O \cdot OC \cdot CH_3)_2 + CH_3 \cdot COOH$$
$$CH_3 \cdot CH(O \cdot OC \cdot CH_3)_2 \longrightarrow CH_3 \cdot COO \cdot CH:CH_2 + CH_3 \cdot COOH$$

Overall $4CO + 5H_2 \longrightarrow CH_3 \cdot CO \cdot O \cdot CH:CH_2 + 2H_2O$

ETHYLENE GLYCOL

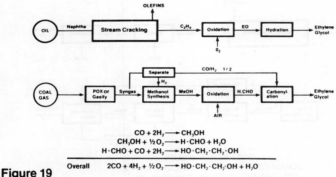

$$CO + 2H_2 \longrightarrow CH_3OH$$
$$CH_3OH + \tfrac{1}{2}O_2 \longrightarrow H \cdot CHO + H_2O$$
$$H \cdot CHO + CO + 2H_2 \longrightarrow HO \cdot CH_2 \cdot CH_2 \cdot OH$$

Overall $2CO + 4H_2 + \tfrac{1}{2}O_2 \longrightarrow HO \cdot CH_2 \cdot CH_2 \cdot OH + H_2O$

Figure 19

The stoichiometry suggests a hydrogen:carbon monoxide molar ratio in the syngas of 1.25:1. The inelegance is in the large recycle of acetic acid and the rejection of 50% of the oxygen in the syngas as water.

The ethylene glycol route is still very much under development. I have selected the route using the reductive carbonylation of formaldehyde formed by air oxidation of methanol (Exxon, Monsanto, Chevron and Arco all seem to have development interests here). I leave Dr Saunby to give a much more expert view on Union Carbide's interest later in the conference.

So what are the general conclusions from this very brief scan of emerging syngas routes?

- Carbonylation chemistry is remarkably flexible in the way it can be used to build-up carbon chains in molecules.

- The syngas processes are simpler and more specific in the way feedstocks are converted to products. Steam crackers are complex plants producing a range of primary products and tend to have a whole array of downstream processes operating in tandem with them.

- The syngas processes tend to have overall more reaction stages.
- The use of oxygen in syngas processes (apart from syngas generation) is limited compared to the alternative hydrocarbon routes.

8. THE PRODUCTION OF SYNGAS

Let us now turn to the subject of syngas itself.

Products from syngas range from virtually pure carbon monoxide (with no hydrogen) through to hydrogen (with no carbon). Thus the range of ideal hydrogen:carbon monoxide molar ratios required in the syngas is almost infinite, though the majority of products require the ratio to lie between 1 and 2.

Syngas can be obtained from a wide variety of feedstocks by a range of partial oxidation technologies (including autothermal reforming) and this results in wide variations in the hydrogen:carbon monoxide molar ratio of the raw gas. Steam reforming is the most common process today for production of syngas from natural gas.

Figure 20 - Syngas and Product H_2/CO Ratios

This shows the hydrogen:carbon ratio in the feedstock and the hydrogen: carbon monoxide molar ratio in the raw syngas product for the range of available feedstocks and processes. Below these is shown the stoichiometric hydrogen:carbon monoxide molar ratios required for a range of products.

The raw syngas will very rarely match-up with that required for the product. To achieve the stoichiometric match, carbon monoxide has to be shifted to increase the hydrogen content or hydrogen removed (either before or after the synthesis sections) to increase the carbon monoxide content. For example, in the production of methanol from natural gas by steam reforming, nearly 30% of the syngas hydrogen content is rejected in the purge gas and burnt as fuel. This cannot be an elegant use of hydrogen. With hydrogen at about 8 $/m Btu and fuel gas at 4 $/m Btu in Western Europe, separation of hydrogen would seem very worthwhile - given a suitable outlet, for example as feed to an ammonia plant as in the ICI Metham project of a few years ago. Similarly, the need for extensive carbon monoxide shift represents a waste of valuable carbon and oxygen as carbon dioxide and increases capital, operating and feedstock costs.

Clearly, the most elegant use of raw materials is to use a syngas production process which gives a hydrogen:carbon monoxide ratio closest to that required by the downstream synthesis.

The pressure for the elegant use of raw material will become greater and greater as raw material prices rise. This clearly points to partial oxidation of gas as the most appropriate syngas production route. However, at present reforming is the more developed and established route and is thus much further along the learning curve than partial oxidation.

If gas becomes the most likely alternative feedstock to liquid hydrocarbons, and if partial oxidation becomes a more widely accepted process, the increase in demand for oxygen will be enormous - particularly if the fuel use of methanol were to develop.

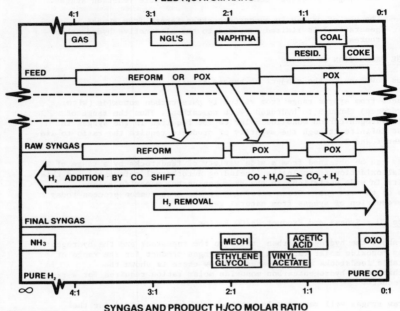

Figure 20

9. CONCLUSIONS

The elegant use of oxygen implies a parallel elegance in the use of feedstock carbon and hydrogen. Process elegance will result from the appropriate combination of good chemistry, process design and plant engineering. Its pursuit should result in a cost - effective plant making the most efficient use of all the feedstocks, including oxygen.

Over the next 20 years or so, these feedstocks will continue to come mainly from oil, gas and coal. Premium uses for these resources are the production of transport fuels and chemicals. Oil will continue as the dominant resource but the contribution of gas is likely to increase substantially - particularly from the exploitation of remote gas reserves. Compared to oil and gas, coal conversion to transport fuels and chemicals is inelegant and consequently suffers fairly severe cost penalties, so it is likely to see only limited use. However, gasification of coal to a clean fuel gas is a potential application which could lead to increased demand for oxygen.

The trend towards increasing use of gas could gradually bring about a shift away from the mainly liquid hydrocarbon feedstocks used by the chemical industry to syngas/methanol feedstocks. Syngas is the route to methanol and ammonia and syngas-based conversion processes using carbonylation chemistry are already in competition with current hydrocarbon oxidation processes for a wide range of products. However syngas chemistry is unlikely to supplant hydrocarbon chemistry for chemicals production other than for oxygenated products.

Spectacular growth in the use of syngas will occur if methanol is
accepted as a transport fuel, and quite substantial growth should result
from the move to phase-out lead in gasoline and the consequent use of
oxygenates as gasoline blending components to boost octane.

The hydrocarbon and syngas routes imply very different uses of oxygen.
The present hydrocarbon-based industry uses relatively small amounts of
pure oxygen in its oxidation processes and there is probably scope for a
much greater and more elegant use of oxygen. But even this would not
dramatically increase the overall oxygen demand. A growing syngas-based
industry, however, could lead to either a very large increase in oxygen
demand or alternatively, a possible contraction.

The outcome depends on the method of generating syngas. Currently steam
reforming is the main route to syngas. If this were to continue, then
the growth in alternative syngas/carbonylation processes would mean that
the main source of oxygen in the petrochemical industry products would
swing from air to water and could ultimately mean a drop in oxygen
demand. Partial oxidation however produces a syngas better matched to
the requirements of the downstream processes. It does seem a more
elegant way of producing the syngas/methanol feeds for carbonylation
processes. The increased use of oxygen hinges on the adoption of
partial oxidation rather than steam reforming for the generation of
syngas.

Of course, if coal gasification becomes more economically attractive,
oxygen demand would grow rapidly.

How far, how fast and in what direction these possibilities will
ultimately resolve themselves is not at all clear. The outcome could
have quite substantial commercial implications for the oxygen supply
business.

For the next twenty years or so, the elegant use of oxygen is constrained
by two major limitations:

- Transport fuels are liquid hydrocarbons with a hydrogen:carbon ratio
  around 2 : 1. Coal and gas are at the opposite extremes of this
  ratio.

- Steam reforming and partial oxidation are the only industrial scale
  technologies currently available to convert gas and coal to transport
  fuels and chemicals. Both involve introducing oxygen (either from
  water or air) but only a small proportion of the final products we
  want, some chemicals, actually require this oxygen.

If we believe that the supply of crude oil is finite and that more
advanced technologies involving fuel cells or hydrogen are a long way
off, then the conclusion is obvious. Either transport fuels will
eventually move away from hydrocarbon liquids to methanol or other
oxygenates, or better processes will emerge to convert gas and coal to
liquid hydrocarbon. I believe both will happen, but only slowly.

My ideal of a better process – if we could find it – would be:

It is simple in concept, but very difficult in practice. It is only likely to come about well beyond my limited time horizon today.

Meanwhile, within the compass of today's technology, there are clear challenges to process developers both to continue the development of cheaper and more elegant hydrocarbon oxidation processes and to develop more elegant syngas production and utilisation technologies.

The Catalysis of Synthesis Gas Production

By D. A. Dowden
12 DUNOTTAR AVENUE, EAGLESCLIFFE, STOCKTON-ON-TEES, CLEVELAND TS16 0AB, U.K. (FORMERLY OF ICI)

Introduction

During the mid-fifties it became evident that coal had become uneconomic as a raw material for the production of hydrogen and carbon oxides for the synthesis of ammonia and methanol. However, the catalytic steam-reforming of methane:

$$CH_4 + H_2O \rightleftharpoons CO + 2H_2$$

had been known since 1912[1] and the process with natural gas:

$$C_nH_{2n+2} + nH_2O \rightleftharpoons nCO + (2n+1)H_2$$

since ca. 1930[2], so that routes from the petroleum-derived alkanes, especially from the lighter naphtha in the absence of native natural gas, were adjudged to be both technically feasible and cost effective. The problem was to expedite the introduction of such processes with durable hardware and long-lived catalysts suited to the high temperatures indicated by thermodynamics, and the medium pressures demanded by the engineering costs of the several associated processes. There are three stages[3,4], primary steam reforming (30 atm, $T \leqslant 1123$ K) to a gas mixture containing CH_4, CO, CO_2 and hydrogen, secondary reforming (~30 atm, $T \leqslant 1573$ K) to reduce further the equilibrium concentration of methane and to introduce nitrogen (from added air) for ammonia syngas, and water-gas shift (CO + $H_2O \rightleftharpoons$ CO_2 + H_2, 30 atm, T ~500 K and ~700 K) to lower the carbon monoxide concentration or to adjust CO/H_2 ratios.

The development of adequate catalysts followed, for the main part, the usual course favoured by industry in a hurry: catalysts of well-established types were modified by rational trial and error to give higher activities resistant to the pneumo-thermal atmospheres within the reactors. But parallel, fundamental studies were also done (1958-59) and these form the basis of this paper which describes the chemistry of steam reforming, outlines a system of catalyst design and demonstrates the relevance of academic research (if its results are noted and applied).

The System of Catalyst Design

The stoicheiometries and the thermodynamics of the desired reaction(s) are written down together with the typical alterations of the reactants, intermediates and products in a series of steps which can be linked in chains to form possible 'virtual' molecular mechanisms[5,6]. The nature of each necessary step identifies a catalyst function (dehydrogenation, hydration, etc.) essential to that mechanism, so that most catalysts are multifunctional to accord with the requirements of the various links of the reaction chain. The gas compositions in the steady state suggest those classes of solid (metals, oxides) which will be stable, in the bulk in situ. The known patterns of catalyst characteristics indicate the phases with selectivity appropriate to individual steps and also provide a rough ranking of the corresponding areal activities; effective catalysts result from the juxtaposition of the chosen solid phases in proper ratios and intimate admixtures. As the activity of unit mass of the solid phase(s) depends upon specific area as well as upon areal activity, the active phases are dispersed and stabilised against sintering by extension upon a refractory compound, itself of high specific area. In as much as the selectivity of a solid phase is seldom perfect and admixture introduces secondary effects, the preliminary composites may speed parasitic side reactions leading to unwanted products; these may now be minimised by the inclusion of promoters or inhibitors, again selected after consideration of the mechanism of byproduct formation and the patterns of activity.

Finally, the molecular mechanisms are translated into mechanisms in the chemisorbed state on the chosen catalysts, a stage which can be investigated at as deep a level as the problem may require or the researcher desire, and which leads to further insights into the probable performance of the catalyst and its improvement.

It will be seen that the design procedure is based upon a morphological analysis of the relevant chemistry and upon recorded empirical data collated and interpreted with the aid of the fundamental chemistry.

Derivation of the Virtual Mechanisms

In a limited space it is neater to outline the procedure for butane rather than for heptane which would better model naphtha.

The Target Transformation. The chemical objective is written down alongside the relevant thermodynamics:

$$C_4H_{10} + 4H_2O \rightleftharpoons 4CO + 9H_2$$

whence it is noted that the reaction requires a H_2O/C ratio not less than unity. The knowledge of methane reforming, and calculation, show that the reaction is highly endothermal and will require a temperature of 1000-1200 K with elevated pressures to achieve adequate equilibria and rates. Already it is seen that the combination of high temperatures and high steam pressures must accelerate sintering of the catalyst[8].

Characteristic Chemistry. In this development of the likely mechanisms, only molecules are included in equations of molecularity not exceeding two; each reaction is awarded as many class names as possible and its thermodynamics derived at some median reaction temperature (but not included here). Thermodynamically improbable reactions are not excluded as it has been shown that even very small partial pressures of intermediates are adequate to sustain a mechanism on a multifunctional catalyst[7].

Primitive Processes. Herein are all the possible reactions of single molecules of reactants:

$nC_4H_{10} \rightleftharpoons iso\text{-}C_4H_{10}$ (isomerisation) 3.2.1.1
$C_4H_{10} \rightleftharpoons C_4H_8 + H_2$ (dehydrogenation) 3.2.1.2
$C_4H_{10} \rightleftharpoons C_3H_6 + CH_4$ (demethanation, cracking) 3.2.1.3
$C_4H_{10} \rightleftharpoons C_2H_4 + C_2H_6$ (de-ethanation, cracking) 3.2.1.4

As the alkanes are relatively non-reactive these derivatives should be the primary products of the reaction.

Self-Interactions.

$2C_4H_{10} \rightleftharpoons C_8H_{18} + H_2$ (dehydrogenation) 3.2.2.1
$\rightleftharpoons C_7H_{16} + CH_4$ (demethanation) 3.2.2.2
$\rightleftharpoons C_6H_{14} + C_2H_6$ (de-ethanation) 3.2.2.3

Cross-Combinations.

$$C_4H_{10} + H_2O \rightleftharpoons C_4H_9OH + H_2 \qquad \text{(dehydrogenation, 3.2.3.1} \\ \text{hydration,} \\ \text{hydroxylation)}$$

Because of the relative lack of reactivity of alkanes and water this reaction is probably slow.

Derived Primitive Processes. A set of dehydrogenations, dealkylations, crackings and dehydrations of the products of 3.2.1 to 3.2.3 yielding highly unsaturated molecules and small molecules (eg CH_4) capable of no further molecular primitive processes. Thus:

$$\begin{aligned}
C_4H_8 &\rightleftharpoons CH_2=CH-CH=CH_2 & \text{(dehydrogenation)} & \quad 3.2.4.1 \\
&\rightleftharpoons 2C_2H_4 & \text{(cracking)} & \quad 3.2.4.4 \\
C_3H_6 &\rightleftharpoons CH_2=C=CH_2 & \text{(dehydrogenation)} & \quad 3.2.4.5 \\
&\rightleftharpoons C_2H_2 + CH_4 & \text{(demethanation)} & \quad 3.2.4.6 \\
C_nH_{2n+2} &\rightleftharpoons C_nH_{2n} + H_2 & \text{(dehydrogenation)} & \\
C_4H_9OH &\rightleftharpoons C_3H_7CHO + H_2 & \text{(dehydrogenation)} & \quad 3.2.4.10 \\
&\rightleftharpoons C_4H_8 + H_2O & \text{(dehydration)} & \quad 3.2.4.11 \\
CH_4 &\rightleftharpoons C(s) + 2H_2 & \text{(dehydrogenation)} & \quad 3.2.4.12
\end{aligned}$$

The unstable polymers and oxygenated compounds are not expected to be detectable in the gas phase, but may exist at least transiently as the corresponding chemisorbates.

Derived Self-Interactions. Eg:

$$2C_nH_{2n} \rightleftharpoons C_{2n}H_{4n} \qquad \text{(dimerisation)} \qquad 3.2.5.1$$

Derived Cross-Combinations. These steps are mainly reactions of water with the reactive unsaturated molecules formed in 3.2.1 to 3.2.5, eg:

$C_nH_{2n} + H_2O \rightleftharpoons C_nH_{2n+1}OH$ (hydration) 3.2.6.1

$C_nH_{2n-2} + H_2O \rightleftharpoons C_nH_{2n-1}OH$ (hydration) 3.2.6.1

$C_2H_2 + H_2O \rightleftharpoons CH_3CHO$ (hydration) 3.2.6.2

$CH_4 + H_2O \rightleftharpoons CO + 3H_2$ (hydration, dehydrogenation decarbonylation) 3.2.6.3

<u>Sequential Primitive Processes</u>. Subsequent reactions of the same kind lead to still smaller molecules and greater unsaturation but can also give rise to parasitic species:

$C_2H_2 \rightleftharpoons C_2 + H_2$ (dehydrogenation) 3.2.7.1

$C_6H_8 \rightleftharpoons C_6H_6 + H_2$ (dehydrogenation) 3.2.7.2
         benzene

$C_nH_{2n+1}CHO \rightleftharpoons C_nH_{2n+2} + CO$ (decarbonylation) 3.2.7.3

$C_nH_{2n+1}OH \rightleftharpoons C_nH_{2n} + H_2O$ (dehydration) 3.2.7.4

$C_6H_{12} \rightleftharpoons$ cyclohexane (ring closure) 3.2.7.5

$\rightleftharpoons C_6H_6 + 3H_2$ (dehydrogenation) 3.2.7.6

<u>Sequential Self-Interactions</u>. Unsaturated and oxygenated molecules may undergo further polymerisation, condensation and dehydration but most of the products will not desorb, <u>eq</u> depolymerisation is more likely to lead to gaseous products than is polymerisation. The free energy of chemisorption of large unsaturated molecules suggested that such processes should not be ignored:

$2CH_3CHO \rightleftharpoons CH_3CH(OH)CH_2CHO$ (condensation) 3.2.8.1

$2C_nH_{2n} \rightleftharpoons C_{2n}H_{4n}$ (polymerisation) 3.2.8.2

$\rightleftharpoons$ cycloalkanes (ring closure) 3.2.8.3

$2CO \rightleftharpoons C(s) + CO_2$ (Boudouard reaction) 3.2.8.4

$2C_2 \rightleftharpoons C_4$ (polymerisation) 3.2.8.5

Sequential Cross-Combinations.

ROH + CO	$\rightleftharpoons$ RCOOH	(carbonylation)	3.2.9.1
H_2O + CO	$\rightleftharpoons$ HCOOH	(carbonylation, hydration)	3.2.9.2
C_2 + H_2O	$\rightleftharpoons$ CH_2 = CO	(hydration)	3.2.9.3
C_6H_6 + C_nH_{2n}	$\rightleftharpoons$ alkylbenzenes	(alkylation)	3.2.9.4
C_2 + H_2	$\rightleftharpoons$ C_2H_2	(hydrogenation)	3.2.9.5
C(s) + H_2O	$\rightleftharpoons$ HCHO	(hydration)	3.2.9.6
CO + H_2	$\rightleftharpoons$ C(s) + H_2O	(hydrogenolysis)	3.2.9.7

Terminal Primitive Processes.

C_6H_{12}	$\rightleftharpoons$ C_6H_6 + $3H_2$	(dehydrogenation)	3.2.10.1
RCOOH	$\rightleftharpoons$ R'CH = CO + H_2O	(dehydration)	3.2.10.2
HCHO	$\rightleftharpoons$ CO + H_2	(decarbonylation, dehydrogenation)	3.2.10.3
alkylbenzenes	$\rightleftharpoons$ naphthalenes + H_2	(dehydrogenation, aromatisation)	3.2.10.4

($\longrightarrow$ polynuclear aromatics
$\longrightarrow$ 'graphitic carbon')

Tabulation of Reaction Types

The grouping of the types of reaction of the characteristic chemistry is necessary because this then indicates the catalyst functions which are appropriate to each, and points to the classes of solid possessing such activities. Metals have markedly different specificities from insulators whereas oxide semi-conductors have intermediate properties[9]. The tabulation then summarises the desirable and undesirable reactions and the corresponding catalyst functions; where both are catalysed by the same class of solid a more precise control of specificity will be required. The tabulation is here omitted.

Virtual Mechanisms. All reaction chains linking reactants with products must be considered as possible virtual mechanisms. If the cross-combinations of water with alkanes are slow as compared with the rate of formation of alkanes, then Figure 1 shows the only molecular reactions yielding primary products, as in the uncatalysed pyrolysis of butane.

Figure 1 : Primary Products

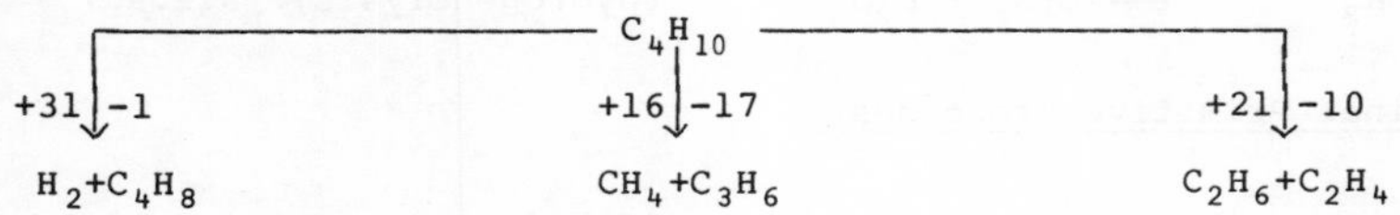

```
          ┌──────────── C₄H₁₀ ────────────┐
   +31│-1              +16│-17              +21│-10
   H₂+C₄H₈             CH₄+C₃H₆             C₂H₆+C₂H₄
```

Stoicheiometric coefficients and indications of reversibility are omitted from these diagrams but the numbers to the left and the right of the arrows are rounded values (kcal mol^{-1} at 1000 K) of enthalpy and free energy changes respectively. The primary reactions are evidently feasible and endothermal and the activation energies cannot be less than the enthalpies.

The secondary reactions (Figure 2) of the alkanes should be relatively fast but the decomposition of methane and the onset of hydrogenolysis are delayed until partial pressures (CH_4, H_2) have risen. Processes effecting increase of molecular weight (broken arrows) are counterproductive.

Figure 2 : Secondary Reactions

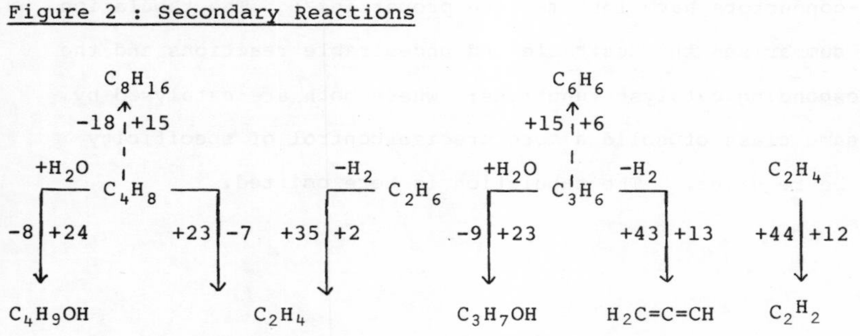

```
         C₈H₁₆                                    C₆H₆
          ↑                                        ↑
        -18│+15                                  +15│+6
           │                                        │
    +H₂O   │                      -H₂      +H₂O     │      -H₂
   ┌──── C₄H₈ ────┐              ┌──── C₂H₆ ────┐ C₃H₆ ────┐           C₂H₄
   -8│+24      +23│-7          +35│+2         -9│+23    +43│+13     +44│+12
     ↓            ↓                ↓             ↓          ↓           ↓
   C₄H₉OH       C₂H₄            C₃H₇OH                  H₂C=C=CH      C₂H₂
```

The Catalysis of Synthesis Gas Production

Only ethene among the virtual intermediates might be detectable in the gas phase; the reaction remains endothermal.

The subsequent changes of these reactive transients will be fast (Figure 3).

Figure 3 : Tertiary Reactions

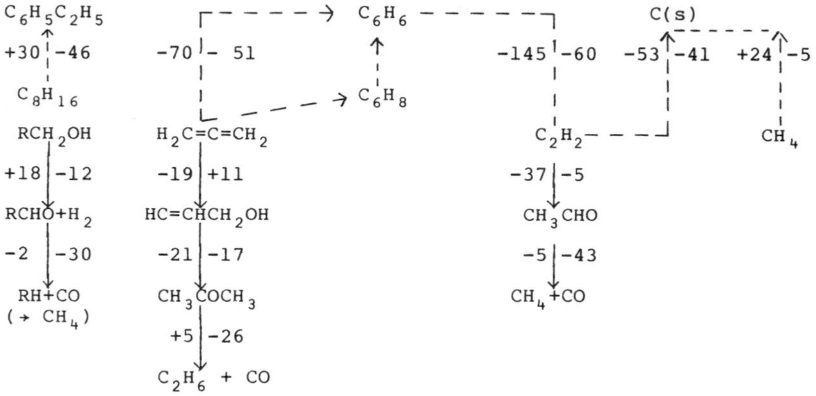

```
C6H5C2H5         ┌ ---- → C6H6 ----- ┐        C(s)
  ↑                        ↑          ↑        ↑
 +30│-46       -70│-51              -145│-60  -53│-41   +24│-5
  │              │                    │         │        │
 C8H16           │     __ → C6H8      │         │        │
                 │    ↗                │         │        │
 RCH2OH        H2C=C=CH2              C2H2 ---- ┘        CH4
 +18│-12        -19│+11               -37│-5
  ↓              ↓                     ↓
 RCHO+H2       HC=CHCH2OH             CH3CHO
 -2 │-30       -21│-17                -5│-43
  ↓              ↓                     ↓
 RH+CO         CH3COCH3               CH4+CO
 (→ CH4)       +5│-26
                 ↓
               C2H6 + CO
```

The main lines of the reaction erode the organic molecules one carbon atom at a time, but the tendency to carbon laydown is notable. The steps giving rise to gaseous products continue to be endothermal. Finally the reactions of the residual entities are given in Figure 4.

Figure 4 : Terminal Reactions

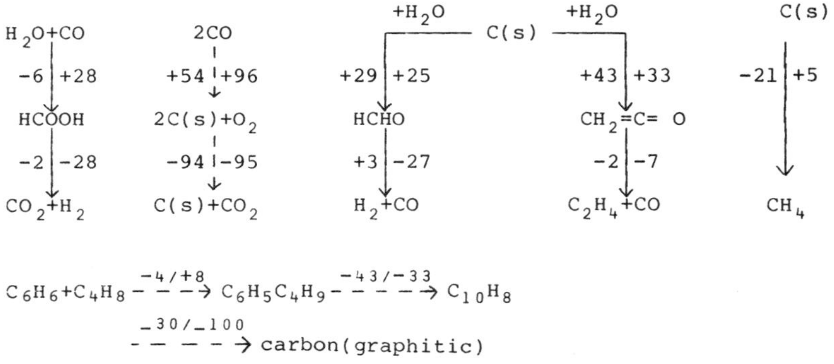

```
                              +H2O           +H2O            C(s)
 H2O+CO        2CO         ┌─── C(s) ───┐
   │            │
  -6│+28      +54│+96      +29│+25      +43│+33         -21│+5
   ↓            ↓            ↓            ↓               ↓
 HCOOH       2C(s)+O2      HCHO         CH2=C=O
  -2│-28      -94│-95       +3│-27       -2│-7
   ↓            ↓            ↓            ↓
 CO2+H2      C(s)+CO2      H2+CO        C2H4+CO          CH4
```

$$C_6H_6 + C_4H_8 \xrightarrow{-4/+8} C_6H_5C_4H_9 \xrightarrow{-43/-33} C_{10}H_8$$

$$\xrightarrow{-30/-100} \text{carbon(graphitic)}$$

$$n(C, C_2, C_2H_2, \underline{etc.}) \longrightarrow C_xH_y \text{ (carbon)}$$

'Black box' descriptions are avoided where possible so to offer further clues to essential catalyst functions. Thus, hydrogenolysis of an alkane is written as a cracking (dealkylation) followed by hydrogenation of the alkane fragment; the water-gas shift is shown as proceeding <u>via</u> formic acid, methane combining with steam by way of deposited carbon and carbon with steam through formaldehyde and ketene. As radicals and atoms are excluded from the molecular mechanism, reactions including solid carbon are not well represented.

The steps of the virtual mechanism, after the formation of the primary products, accord with two main paths: (a) a sequence of alternate dehydrogenations (to olefines), hydration (to alcohols) dehydrogenations (to carbonyl compounds) and decarbonylations (to lower alkanes) leading finally to methane and thence to carbon species which undergo the water-gas and water-gas shift reactions or (b) a fast stripping of the hydrogen from the hydrocarbons to form carbon which then reacts as in (a). The second is the older mechanism but if the steps in (a) linking alkane with the water-gas reaction are fast it is difficult to distinguish between them <u>a priori</u>.

Not surprisingly the thermodynamically most disfavoured reactions in (a) are the carbon, carbon monoxide and alkane hydrations followed by the dehydrogenations to highly unsaturated hydrocarbons and the dimerisations and alkylations. Methane is the most stable of the hydrocarbon intermediates and should be found in the gas phase. The preferred routes retain the unsaturated intermediates down to the smallest hydrocarbons, as in the chains:

The Catalysis of Synthesis Gas Production

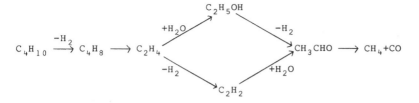

thereby maintaining the reactivity toward water, but the methane-steam and carbon-steam reactions cannot be examined further without some consideration of chemisorbed states. The scheme implies that the required catalyst functions are dehydrogenation, alkane cracking, hydration and decarbonylation, but that initial dealkanation, hydrogenolysis and the alternating formation of alkanes should be inhibited. Mechanism (b) in its simplest form requires only dehydrogenation: ($C_4H_{10} \longrightarrow 4C + 5H_2$) and hydration ($C + H_2O \longrightarrow CO + H_2$), but the deposition of carbon (as also in (a)), taken together with know-how concerning carbon lay-down, already suggests that the intrusion of these processes will impair catalyst activity by chemical and physical (eg encapsulation) means.

Catalyst Selection (Mark I)

The objective of this stage of the programme is to devise catalysts which will induce the reaction to follow the proposed routes at the maximum rates.

Known patterns of activity[5,9,10] are inspected to seek solids possessing the catalyst functions entailed by the virtual mechanisms; if the patterns are lacking then they must be assembled from the available data for related reactions, taking care to select information obtained under conditions as close as possible to those of the target transformation.

Metals. High areal activities for hydrogenation and dehydrogenation are found over electronic conductors. Metals are the most active but although all the metals of the periodic table should be considered only those which are stable under the reaction conditions are admissable. The thermodynamics can be used to find such metals if it is recognised that metal-gas reactions are fast and can be considered to reach a quasi-equilibrium with the local ambients in a time which is very short compared with the life of the catalyst. Table 1 lists the metals whose bulks will remain unoxidised in situ, together with their melting points (T_m/K):

Table 1 : Metals

		GROUPS		
7A		8		1B
	Fe	Co	Ni	Cu
	1808	1768	1726	1356
Tc	Ru	Rh	Pd	Ag
	2583	2239	1825	1235
Re	Os	Ir	Pt	Au
3456	3318	2683	2045	1337

The areal activities of congeners may differ by a factor of as much as 10^3 whereas the specific area of an individual metal, in a well-made catalyst, can rarely be adjusted or sintered to such extents. The more critical selection for areal activity is therefore made first. The metals having unfilled d-orbitals ('holes in the d-band') are always more active than those with full d-orbitals[9]; hence the metals of Group 1B, and indeed all B-group elements, could be ignored except as inhibitors or poisons. There tend also to be relative minima in Group 7A for the reactions of hydrocarbons so that a maximum, which varies somewhat with the catalysed reaction, occurs in Group 8 in each long period. Table 2 summarises the activity series for each function as recorded or inferred from published reports; activity decreases from left to right. Little was known about osmium but the volatility and toxicity of its oxide are always impediments to its use.

Table 2 : Metals, Activity Series

Dehydrogenation :	Rh	Pt	Pd	Ni	Co ⟫ Fe			
Dealkylation (Cracking) :	Rh			Ni	Co	Fe		
Hydrogenolysis :	(Rh			Ni	Co)	Fe	Pd	Pt
	Ru	Rh	Ir ~ Ni Re	Co		Fe	Pt	Pd
Hydration :	Pt	Rh	Pd ＞ Ni			Fe		
Decarbonylation :	Ni	Rh			Co	Fe		
Aromatisation :	Pt ＞ the rest							
CO - H_2O :	Cu ＞ the rest							
C - H_2O :				(Ni	Co	Fe)		

(Hydration-dehydration is not a characteristic activity of metals; the series is based on the exchange of deuterium with water.)

The precious metals, especially rhodium, were placed high in most series, with nickel outstanding among the base metals. Besides the oxygen from water, the only other elements present which might react with metals to form less active solid phases are carbon and hydrogen yielding solid solutions, carbides, hydrides, carbohydrides, <u>etc</u>. At 1,000 K hydride formation, even with palladium, could be neglected, but carbide (not adatoms of carbon), although confined to surface layers was expected, from electronic theories[9] of catalysis, to diminsh the activity of these metals: such an inhibition should be less for the more noble precious metals of highest activity. The resistance of the metals to loss of area was estimated with the use of Tammann's rough rule[11] (marked bulk sintering of particles occurs at temperatures near $0.5\ T_m$) which divides the series into a more stable set ($0.5\ T_m > 1,000$ K), Ru, Rh, Re, Os, Ir and Pt and a less stable set ($0.5\ T_m < 1,000$ K), Fe, CO, Ni, and Pd. Loss of area due to crystal growth by interparticle transport of metal vapour must be very slow, although less for the precious than the base metals, but the possibility of transfer as a volatile compound had to be considered for these extreme conditions. Under dioxygen at high temperatures (>1173 K) gaseous oxides of the precious metals exist and are known to mediate the transfer of platinum and rhodium, but such conditions are not pertinent to steam reforming. Yet the possibility that small vapour pressures of hydrous metal oxides might arise[12] could be minimised by again choosing the more noble metals; oxide stability increases on moving to the left in the horizontal triads of Group 8.

On all counts therefore, (except cost) the precious metals, especially rhodium, were preferred to nickel, but in catalyst design even the most expensive components should be considered in preliminary studies.

<u>Semiconductors</u>. The semiconducting elements, oxides and sulphides have some of the relevant functions to a degree. The elements are in the B-subgroups and could be eliminated for the reasons already given; the sulphides were set aside because of their relative inactivity and instability, as well as their deleterious effects (due to attendant hydrogen sulphide) on downstream catalysts and product purity. Only the oxides stable <u>in situ</u> remained (Table 3, with absolute melting points);

all are more or less non-stoicheiometric, but in the absence of thermodynamic data approximate compositions in situ could only be guessed. However, the patterns of activity[9,10] reveal that the oxides of Group 6A, expecially dichromium trioxide, are the most active in dehydrogenation and that they also have

Table 3 : Oxide Semiconductors (T_m/ K)

		GROUPS			
3A	4A	5A	6A	7A	2B
	TiO_2	$\sim V_2O_5$	Cr_2O_3	MnO	ZnO
	≮2103	963-2240	2539	1923	2248
		Nb_2O_5	MoO_{2-3}		
		≮1758	≮1068		
$\sim CeO_2$		Ta_2O_5	WO_{2-3}		
≮1965		≮2145	≮1746		
			UO_{2-3}		
			≮1573		

hydration-dehydration activity[13], as do many of the oxides of high valency and consequent greater acidity. The other necessary functions were known to be present, but with dealkylation at a minimum in Group 6 where dehydrogenation is a maximum; unwanted cracking due to acidity could always be limited by neutralisation with an alkali. Resistance to loss of area by bulk diffusion is satisfactory only for TiO_2, Cr_2O_3, ZnO, Ta_2O_5 and, perhaps, $\sim CeO_2$ as judged from the melting points. The use of the higher oxides of molybdenum and tungsten would have been ill-advised as they have an appreciable volatility, especially in steam[12], but UO_x was just possible as a substitute if x ~ 2-2.7. Unless zinc oxide could be kept stoicheiometric, n-type ZnO would be expected to lose zinc atoms to chemisorbed states on the surface of an admixed metal and to inhibit the very active d-metals. The set of usable oxides is thus reduced to dichromium trioxide alone which, despite its inferiority to metals, might have been introduced to promote hydration reactions or to act as a support.

The Insulators. The solid oxides which are electronic insulators at moderate temperatures because of the stable valencies of their leptons are necessarily poor catalysts for redox reactions. They sustain active sites which range from basic to acidic, depending upon the low or high formal ionic potentials, respectively, of their lattices; like their homogeneous counterparts, they have the capacity to catalyse reactions mediated by carbanions and carbonium ions[13,14]. A given oxide of high specific area carries centres which are for the most part basic or acidic in accord with the above subdivision, but it nevertheless exposes centres of varying base or acid strength distributed somewhat about the mean. The more refractory oxides are obtainable in high area and are commonly employed as supports and stabilisers. Table 4 contains the typical oxides with their melting points.

In atmospheres containing hydrogen and carbon monoxide (but also water and carbon dioxide) at high temperatures some of these compounds, especially those of Group 1A, may be somewhat reduced[15,16] especially in the presence of active metals due to hydrogen spillover[17].

Table 4 : Acidic and Basic Oxides (T_m/ K)

GROUPS

1A		2A		3B	4B	5B
(Li_2CO_3)	(LiOH)	BeO				
996	723	2803				
(Na_2CO_3)	(NaOH)	MgO		Al_2O_3	SiO_2	P_2O_5
1124		3125		2288	1883	842
(K_2CO_3)	(KOH)	CaO	($CaCO_3$)			
1164	633	2887	~1610			
(Cs_2CO_3)	(CsOH)	SrO	($SrCO_3$)			
>883	545	2693	~1770			
(Rb_2CO_3)	(RbOH)	BaO	($BaCO_3$)			
1110	574	2191	~2013			

Acids are outstanding catalysts for, <u>inter alia</u>, hydration-dehydration, dealkylation (cracking) and decarbonylation reactions; they also have a small, atypical activity of the semiconductor kind which one expects to be inhibited by water and carbon dioxide. Silica and diphosphorus pentoxide have not only low melting points, but high volatilities (with or without steam) and could not be admitted in the free state to catalysts. An additional serious objection to the use of strongly acidic oxides (including SiO_2-Al_2O_3, P_2O_5-Al_2O_3, P_2O_5-SiO_2, <u>etc.</u>) arose from the widespread empirical observation that they cause very rapid lay-down of carbonaceous deposits at high temperatures while possessing small activity for the reaction of carbon with steam. Good resistance to bulk and vapour phase sintering, plus multiple bond hydration-activity, resides only in alumina.

The electropositive elements will exist in situ mainly as carbonates; they catalyse some carbonylation reactions, have a small activity in the carbon monoxide-steam reaction[15,18,19] at elevated temperatures and some dehydrogenation activity[13], but have high activity for the reactions of carbon with oxygen, steam and carbon dioxide[20] and for decarboxylation[13]. At 1000 K, in the presence of reducing agents, the Group 1B oxides would probably undergo some reduction to a condition M_2O_{1-x} with gradual loss of metal atoms from F-centres into the gas stream. The hydroxides were also known to be volatile in steam and a potential cause of instability in view of the reaction:

$$M_2CO_3(s) + H_2O(g) \rightleftharpoons 2MOH(g) + CO_2(g)$$

and vaporisation of the small equilibrium concentrations of free hydroxide.

Supports. This discussion could then be adapted to the procedure for support selection but it is evident that many oxides used as supports cannot be inert.

Some relatively low-melting aluminates and silicates are included in Table 5 as these are among the constituents of the cements used in bonding catalyst aggregates; but for a given melting point it seems that the more complex the structure the less rapid the sintering[5]. Despite their high melting points, the alkaline earth oxides do not generally make good supports for operation in steam; eg, because magnesia hydrates at low temperatures, lattice hydration and dehydration on multiple

Table 5 : Oxide Supports, (T_m/ K)

Bases:	BaO, 2191; SrO, 2693; CaO, 2887; MgO, 3125
Amphoters:	CeO_x, <1965; Al_2O_3, 2288; Cr_2O_3, 2539; ZrO_2, 2988; ThO_2, 3493
Ternary Oxides: (neutral)	$Ca_3Al_2O_6$, >1800; $CaSiO_3$, 1813; $CaAl_2O_4$, 1873; Mg_2SiO_4, 2183; $CaCr_2O_4$, 2323; Ca_2SiO_4, 2403; $MgAl_2O_4$, 2408; ($ZnAl_2O_4$, $MgCr_2O_4$, $ZnCr_2O_4$, >2000) $CaZrO_3$, 2823

passage through this temperature region in steam tends to weaken the catalyst aggregate. The ideal support would be a high-melting, stoicheiometric, non-defective, complex oxide[5], but these are not easily made with large specific areas; also they must not break down in steam to form strong-acid overlayers, eg $MgSiO_3$ giving an $MgO-SiO_2$ mixed oxide[14]. Residual acidity in a support can always be neutralised by the addition of alkali and such alkali will be bound if the acid centres are strong.

Catalysts; Mark I.

Active Phases:	Rh > Pt > Ni > Ni-Cu alloys
Supports:	$MgAl_2O_4$, $MgCr_2O_4$, $MgAl_{2-x}Cr_xO_4$
Promoters:	$\gamma-Al_2O_3$, Cr_2O_3, (CeO_x, $UO_{2.7}$) alkaline oxides (hydroxides, etc.)

Later experimentation gave results in reasonable accord with this ordering of the activities of the metals[21,22], and catalysts containing spinels and uranium oxide were claimed[2,23]. Recent work on the steam dealkylation of alkylbenzenes gave a very similar activity series, Rh > Ru > Pt > Pd > Ir > Ni > Os[24].

Application of the quasi-equilibrium principle to the formation of carbon by reactions 3.2.4.12, 3.2.8.4 and 3.2.9.7

yielded a minimum steam ratio ($H_2O/C \sim 1.2$) the maintainance of which was advisable as an aid to limit the production of so-called 'thermodynamic' carbon.

The Chemisorbed Virtual Mechanism. Further progress needs an interpretation of the molecular mechanisms in chemisorbed states on the surfaces of the Mark I catalysts.

The dehydrogenation function is a prime requirement: it is a general rule that all reactions of saturated hydrocarbons are preceded by the breaking of at least one carbon-hydrogen bond, although not necessarily with evolution of dihydrogen. The primary reactions of Figure 1 were then rewritten as in Figure 5; asterisks denote leptons of the solid surface and the symbol ⇌ a pi-bond. Adsorbates can be shown with equal justification binding two or more surface leptons. The reactive intermediates were seen to be sigma-bonded alkyl groups and pi-bonded or di-σ-bonded alkyl groups and pi-bonded or di-σ-bonded methane, alkanes, alkynes, carbonyl, etc.

Figure 5 : Primary Reactions - Chemisorbed

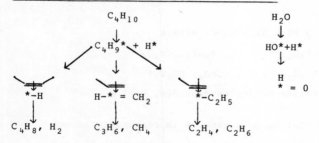

Bond-breaking was taken to be homolytic on metals and heterolytic on oxides; the complete dissociation of water was presumed to be more extensive on the base metals than on the noble metals and oxides. Subsequent experiment showed that these primary products could be detected at 873 K[25]. That they were formed catalytically (not pyrolytically as was later objected[26]) was demonstrated in blank runs and by the observation that a nickel catalyst gave mainly demethanation whereas a rhodium catalyst mainly dehydrogenation[27]. The unsaturated chemisorbates react next with the fragments of water (Figures 6 and 7):

Figure 6 : Secondary Reactions - Chemisorbed

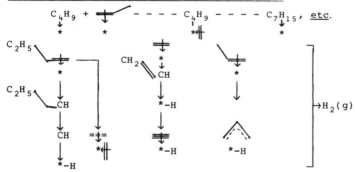

The growth of high molecular weight adsorbates is shown in Figure 6 as an alkene insertion reaction and the dehydrogenation continues as far as alkylidene and alkyne species. Strongly acidic oxides having been excluded, carbonium ion cracking reactions do not appear in Figures 5 and 6, but such proton transfers and switches must be considered for the hydration reactions on the insulator oxides.

In Figure 7, major differences of mechanism arise. The addition of the elements of water and further reactions may occur either by (a) homolytic fission and forming of bonds within adsorbates confined to the metal, (b) the same, but assisted by spillover[17] of dissociatively chemisorbed water from the oxide, or (c) via the series of steps: desorption of alkene from the metal, readsorption on the oxide with hydration to

Figure 7 : Tertiary Reactions - Chemisorbed

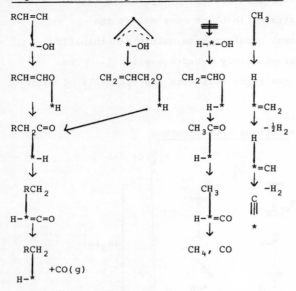

alkanol and gas phase transfer of the latter back to the metal for dehydrogenation and decarbonolyation. The stages (b) and (c) could be restricted to the metal/oxide interface, but in view of the mobility of adsorbates at high temperatures, this seemed unduly limiting. The whole process could take place on the surface of a sufficiently active metal but promotion by oxide must involve either spillover (b) or springover (c).

The slow atypical reactions of the same kind wholly on the solid insulator oxides cannot compete, but the sequences leading to polymeric adsorbates will occur. Aromatisation of the polymers is not shown but can be adequately represented as a progress through alkanes to bound cyclohexyl radicals, with subsequent stepwise dehydrogenation, or from methene and alkenes to polymers with final dehydrogenative ring closure to aromatic rings pi-bonded to the surface.

In Figure 8, cross-conbinations with the chemisorbed

hydroxyl radical could be written with an adatom of oxygen.

Figure 8 : Terminal Reactions - Chemisorbed

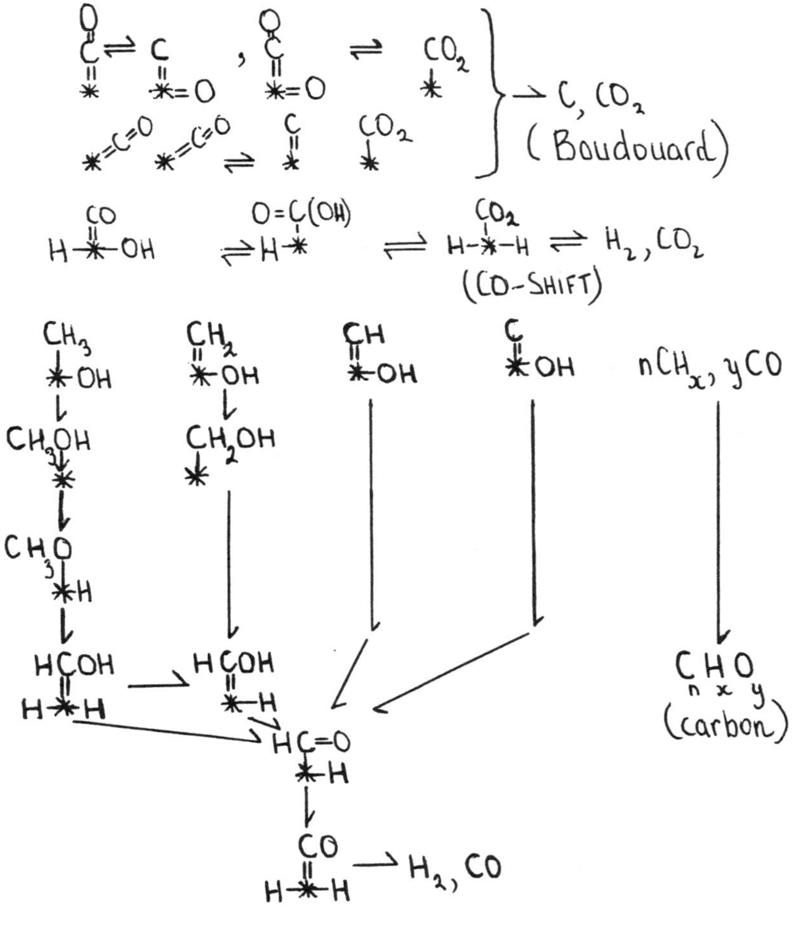

$$\begin{matrix} \overset{O}{\underset{*}{C}}=C \\ \overset{\shortparallel}{*}=O \end{matrix} \rightleftharpoons \begin{matrix} \overset{O}{\underset{*=O}{C}} \end{matrix}, \begin{matrix} \overset{O}{\underset{*=O}{C}} \end{matrix} \rightleftharpoons \begin{matrix} CO_2 \\ * \end{matrix} \Biggr\} \rightarrow C, CO_2 \quad (Boudouard)$$

$$*=C=O \quad *=C=O \rightleftharpoons \begin{matrix} C \\ \shortparallel \\ * \end{matrix} \quad \begin{matrix} CO_2 \\ \downarrow \\ * \end{matrix}$$

$$\underset{H-\overset{\shortparallel}{*}-OH}{CO} \rightleftharpoons \underset{H-*}{O=C(OH)} \rightleftharpoons H-\overset{CO_2}{*}-H \rightleftharpoons H_2, CO_2 \quad (CO\text{-SHIFT})$$

$$\begin{matrix} CH_3 \\ *OH \\ \downarrow \\ CH_2OH \\ * \\ \downarrow \\ CH_3 \\ *H \\ \downarrow \\ HCOH \\ \shortparallel \\ H*H \end{matrix} \quad \begin{matrix} CH_2 \\ \shortparallel \\ *OH \\ \downarrow \\ CH_2OH \\ * \\ \downarrow \\ \\ \\ \\ HCOH \\ \shortparallel \\ *-H \end{matrix} \quad \begin{matrix} CH \\ \shortparallel \\ *OH \\ \downarrow \\ \\ \end{matrix} \quad \begin{matrix} C \\ \shortparallel \\ *OH \\ \downarrow \\ \end{matrix} \quad nCH_x, yCO$$

$$\rightarrow \underset{*H}{HC=O} \downarrow \quad \downarrow \quad \underset{(carbon)}{C_n H_x O_y}$$

$$\underset{H*H}{\overset{\shortparallel}{CO}} \rightarrow H_2, CO$$

The scheme is based entirely upon the dissociation of single atoms <u>seriatim</u>, in accord with the principle of minimum motion and upon the interception of the radicals so formed by other fragments to yield metastable intermediates (Ostwald's rule).

The reactions of carbon which could not be adequately visualised in the molecular mechanism are thereby largely evaded and the undesirable demethanation steps are replaced by splitting out methene groups which react more quickly with water.

The Boudouard reaction, which is fast only on metals, probably occurs by the dissociative rather than the associative mechanism because of competition for active sites by water and, in any case, because of the small surface coverages at these high temperatures. The carbon monoxide shift reaction could follow the redox or the associative (formate) chains on the metals and semiconductors, but only the latter chain on the insulators (but see below); the amphoteric and acidic oxides have negligible activity whereas the bases have some activity at high temperature which could be associated with the formate route when the oxides are irreducible.

The activity of the less 'irreducible' alkali metal compounds might then be due to their base strength or to some small reduction, in situ, sufficient to initiate a redox mechanism.

Inspection of the hydroxylation of the C_1 radicals gave no hint that strong bases could be catalytic unless the acidic character of formaldehyde were implicated and the alkalies bound to the metal surface. The old literature[16] refers to the reaction:

$$2CH_2O + 2NaOH + CuO \rightarrow Cu + 2NaOOCH + H_2 + H_2O$$

but if this is rewritten as formaldehyde decomposing to hydrogen and carbon monoxide followed by:

$$2NaOH + 2CO \rightleftharpoons 2NaOOCH$$
$$CuO + H_2 \rightleftharpoons Cu + H_2O$$

wherein the metal (oxide) catalyses the decomposition of the aldehyde, it gives no advantage. It was concluded that any positive catalytic function of alkali must be confined to speeding the reaction of the 'carbon' with steam. Most of the early definitive work on this catalysis had used essentially pure carbons (chars, graphite) so the composition of 'carbon' had to be guessed, and it was assumed to be $C_nH_xO_y$ (x=small and y=very small) some of an amorphous fluffy (filamentary) structure, some graphitic and slow growing on the catalyst surface. Radicals and unsaturates formed by pyrolysis in the gas phase and the C-O and C-H bonds of catalytic carbon will be attacked by water according to the mechanisms already outlined provided that these species are in contact with active surface.

Carbon which does not contact with active surface (metal) must react by other mechanisms and catalyses. Alkali metal hydroxides, being volatile in steam and mobile in the adsorbed state (in general surface mobility parallels volatility), were potential catalysts for the removal of the superficial carbon and the limitation of its accumulation. The extant theories about the efficacy of alkali in the reactions of carbon involved the liberation of alkali metal which either modified the electronic structure of the carbon[28], reacted transiently with water to yield hydrogen[15], or decarboxylated carboxylic acids in $C_nH_xO_y$[29] (for carbon oxidation). It had been recorded[16] that fused sodium hydroxide at 893 K dissolved carbon and liberated hydrogen and water and reactions of the type

$$4NaOH + 2C \rightleftharpoons Na_2CO_3 + 2Na + CO + 2H_2$$

were claimed which looked more feasible when broken down as follows:

$$4NaOH \rightleftharpoons 2Na_2O + 2H_2O$$
$$2Na_2O + 2C \rightleftharpoons 4Na + 2CO$$
$$2Na + 2H_2O \rightleftharpoons 2NaOH + H_2$$
$$2NaOH + 2CO \rightleftharpoons 2NaOOCH$$
$$2NaOOCH \longrightarrow Na_2CO_3 + CO + H_2$$
$$Na_2CO_3 \rightleftharpoons Na_2O + CO_2 \text{ etc.}$$

Generation of alkali metal adatoms was thus possible, but, despite the high reforming temperatures, the excess of steam must keep the concentration low. The active species is one of M_2CO_3, MOH, M_2O and M at least, preferably taken from the last three because of their greater mobility.

The carbon must have a surface which is saturated ($\equiv CX$, $=CX_2$, $-CX_3$, X=H, OH) or variously unsaturated ($>C=C<$, $-C\equiv C-$, $>C:$, $>C=O$, etc.) and was considered to erode atom by atom according to the general scheme. It was supposed that carbon growing from the catalyst would be covered with a proportion of all these groups, whereas that deposited from gas phase might initially retain more unsaturation. As carbon proper possesses some hydrogenation-dehydrogenation activity it was also supposed that 'carbon' growth, after nucleation, was to some extent autocatalytic. Bases have modest hydrogenation-dehydrogenation and hydration-dehydration activities at moderate temperatures so this was taken to be their role at the high temperatures of steam reforming, eg:

$$\begin{matrix} \diagdown \\ / \end{matrix} C = C + H_2O \rightleftharpoons -CH - C(OH) \rightleftharpoons = C - C = O + H_2$$

$$\rightleftharpoons \equiv C + CO$$

$$\begin{matrix} \diagdown \\ / \end{matrix} C: + H_2O \rightleftharpoons C\begin{matrix} H \\ OH \end{matrix} \rightleftharpoons = C = O$$

The alkaline compounds have the same catalytic function here as the metals and transitional oxides, but to a much smaller degree. In this model, the advantage resides in their mobility which enables them to reach parts which other catalysts cannot reach. But it also implies loss from the catalyst, especially at high temperatures, and some inhibition of the active metals, as a fraction of their surface must be covered by migrating, less active, alkali species.

Catalysts; Mark II. Components. Leaving aside all other considerations but activity and stability, and assuming that the most active metal possesses all the required functions in sufficient degree, the selected metal was rhodium, and the supports one of $MgAl_2O_4$, $MgCr_2O_4$ and $MgAl_{2-x}Cr_xO_4$. Analogy with existing precious metal catalysts (eg Platforming Catalysts) suggested that the amount of metal must be <1% wt. The spinels have little hydration activity[30].

If a hydration promoter is included then a transitional alumina (eg gamma-alumina) was preferred. The catalyst is then rhodium dispersed on two or three monolayers of gamma-alumina themselves supported on magnesium aluminium spinel of high area. Experience taught that carbon laydown must be expected (if only due to plant difficulties) and it could be controlled by addition of a small amount (<5% wt) of an alkali metal carbonate or hydroxide. Base strength increases as Group 1A is ascended so that Cs_2CO_3 would have been chosen if activity alone had been the criterion; the addition of base inhibits the hydration function of alumina, but also carbon growth on residual acidic centres.

The intrusion of the cost factor, the unknown problems associated with the large-scale production of spinels of high area and the existing know-how concerning nickel catalysts resulted in little attempt to develop entirely new kinds of catalyst. The standard steam reforming catalyst comprises nickel supported upon gamma-alumina alkalised with potassium compounds and bound with cement. Potassium compounds have the highest base strength at a tolerable cost. Although the design scheme can give an approximate catalyst composition and point to possible difficulties, these can only be optimised and quantified, respectively, by experiment.

It is claimed that similar catalysts can be made of such high activity that alkalisation is unnecesary[4], but the mechanism suggests that adventitious carbon can then be removed with steam slowly or not at all.

<u>Poisons</u>. The most dangerous poisons are compounds of sulphur, halogens and Group B elements (<u>eg</u> arsenic) which may elude or intrude from the gas purification stages. The parallelism between melting point, volatility and sintering has already been noted and can be extended to include the properties of chemisorbed layers whose mobility also varies, roughly in the same way[8]. The nickel-sulphur phase diagram contains some eutectics of low melting point and the chlorides are relatively volatile so that the mentioned compounds poison not only by blocking active surface, but also by accelerating sintering. Attempts were made to estimate maximum allowable concentrations of poisons in the reactants, but these were no more than guesses. Precursors of the poisons had also to be excluded from the catalyst preparation.

It is interesting to discover that small amounts of polycyclic aromatics appear in the products when catalyst activity declines[3,4].

<u>Hazards</u>. The loss of free silica and alkali from some catalysts and 'inert' packings was subsequently observed. Free silica was eliminated from the catalysts, as suggested by the design, and the loss of alkali by adjustment of the alkali content to suit the higher or lower temperatures within the reactor. The silica content of catalysts operating at higher temperature (1123-1573 K) or with natural gas and methane (<1123 K) is essentially zero[1].

Loss of catalyst activity can cause the appearance of 'hot bands' some 3 m below the reactor inlet due to the local decrease of the endothermicity of the reaction; the situation can be rectified by placing more active catalysts in this vicinity.

Conclusions. Experience teaches that thorough application of a design procedure is a necessary preparation for and adjunct of a catalyst research and development programme. It discovers the relevant information systematically, exposes desiderata, proposes the catalysts worth testing and the problems likely to be encountered, as illustrated, in summary, for the steam-reforming of naphtha. Consistent recourse to its ammassed recorded and current results and concepts convinces that academic research is invaluable if applied[31].

References

1 A. Mittasch and C. Schneider, German Patent 296866, 1912
2 P. J. Bryne Jr., R. J. Gohr and R. T. Haslam, Ind. Eng. Chem., 1932, 23, 1129
3 "Catalyst Handbook", Wolfe Scientific Books, London, 1970
4 J. R. Rostrup-Nielson, "Steam Reforming Catalysts, Teknisk-Forlag, Copenhagen, 1975
5 D. A. Dowden, Chem. Eng. Progr. Symp. Ser., 1967, 63, 90; Chimica e Industria, 1973, 55, 639
6 D. L. Trimm, "Design of Industrial Catalysts", Elsevier, Amsterdam, 1980
7 P. B. Weisz, "Advances in Catalysis", Academic Press, New York, 1962, Vol. 13, Chapter 3, p. 137
8 D. A. Dowden, I. Chem. Eng. Symp. Ser., 1968, No. 27, 18; "Progress in Catalyst Deactivation", ed. J. L. Figueiredo, Martinus Nijhoff, The Hague, 1982, p. 281

9   D. A. Dowden, J. Chem. Soc (London), 1950, 242; "Catalysis Revs", Marcel Dekker, New York, 1972, Vol. 5, Chapter 1, p. 1

10 D. A. Dowden, ll. Mackenzie and B. M. W. Trapnell, Proc. Roy. Soc. (London), 1956, A237, 245

11 G. Tammann and Q. A. Mansuri, Z. anorg. allg. Chem., 1923, 126, 119

12 O. Glemser and H. G. Wendlandt, "Advances in Inorg. and Radiochem.", Academic Press, New York, 1963, p. 215.

13 P. Sabatier, "Catalysis in Organic Chemistry", trans. E. Emmet Reid, Library Press, London, 1923

14 K. Tanaka, "Solid Acids and Bases", Academic Press, New York, 1974

15 D. A. Fox and A. H. White, Ind. Eng. Chem., 1931, 23, 259; ibid., 1934, 26, 95

16 "Gmel:ns Handbuch d. Anorg. Chem.", Verlag Chemie, Berlin, 1928, System Nummer 21, p. 209

17 D. A. Dowden, "Chemisorption", ed. W. E. Garner, Butterworths, London, 1957, p.p. 3, 55.

18 G. Natta and R. Rigamonti, Chimica e Industria, 1936, 18, 623

19 F. Schussl, Gas u. Wasserfach, 1939, 82, 359

20 H. H. Lowry, "Chemistry of Coal Utilisation", John Wiley, New York, 1945, Vol. 2

21 P. Davies, D. A. Dowden and C. M. Stone, British Patent 1 029 235, 1966

22 Ref. 4, p. 101

23 T. Nicklin and R. J. Whittaker, Inst. Gas Eng. J., 1968, 8, 15

24 D. C. Grenoble, J. Catalysis, 1978, 51, 203

25 C. R. Schnell, J. Chem. Soc., B, 1970, 158

26 Ref. 4, p. 120

27 C. R. Schnell, Unpublished Results, 1965

28 F. J. Long and K. W. Sykes, Proc. Roy. Soc. (London) 1948, A193, 377

29 H. Harker, Proc. 4th Conf. Carbon, Buffalo 1959, 1960, 125

30 B. C. Alsop and D. A. Dowden, <u>J. Chim. Phys.</u>, 1954, <u>51</u>, 678; S. G. Szabo, B. Jover and J. Jahasz, <u>Z. Phys. Chem.</u>, 1977, <u>108</u>, 73

31 "Catalysis", Science Research Council Working Party Report, Science Research Council, London, 1975

Development of the Shell Coal Gasification Process

By M. J. van der Burgt* and J. E. Naber
SHELL INTERNATIONAL PETROLEUM MAATSCHAPPIJ B.V., CAREL VAN BYLANDTLAAN 30, THE HAGUE, THE NETHERLANDS

1. REQUIREMENTS FOR MODERN COAL GASIFICATION TECHNOLOGY

To determine the economic viability of coal gasification both in comparison with other energy sources and in comparisons between different technologies, the following general requirements are considered important:

(1) High process efficiency

(2) Low technical complexity

(3) High coal flexibility

(4) Good product gas suitability for various applications

(5) A high degree of environmental acceptability

(6) Cost effectiveness

The above requirements will now be discussed in some detail.

1.1 <u>High Process Efficiency</u>

Coal gasification is both a fuel-conversion and an upgrading process: a solid fuel is converted into a gaseous fuel or fuel base material which is free of solids and from which sulphur can be removed to the extent required by the intended use. The major part of the heat of combustion of the coal fed to the plant is recovered as heat of combustion of the product gas, whereas the remainder is used for the operation of the gasifier. The thermal efficiency of the gasifier is often defined as the percentage of the heat of combustion recovered in the product gas. For a good comparison between processes the total energy balance of the whole complex should be considered, relating the thermal efficiency to the net production of gas and the total coal consumed, thus including the energy required for raising steam, oxygen and electricity. For a high process efficiency it is mandatory that gasifiers have a capacity of 50-100 tonnes/h or more in order to reduce reactor heat losses. Such large capacities can be attained only at high temperatures and pressures. Although every application of the process is likely to have its specific optimum set of process conditions, these tend to move to pressure levels of 20-40 bar and temperatures of at least 1350°C, <u>i.e.</u> above the melting range of the coal ash ('slagging operation'). Apart from the beneficial effect of the elevated pressure on reactor capacity, there are spin-offs in increased rates of heat transfer in the waste-heat boiler, cheaper gas treating and a significant reduction in gas compression costs.

Last but not least, a high thermal efficiency dictates an efficient heat integration of the various process units.

1.2 Low Technical Complexity

Coal gasification is basically a simple process and, although in real life it becomes more complex owing to the presence of ash, sulphur and nitrogen, the aim should remain to keep it simple.

1.3 High Coal Flexibility

Coals are found in a very wide range of ranks and qualities (Table 1). Depending on the degree of coalification, calorific values range from 30 MJ/kg for a high volatile bituminous coal to 10 MJ/kg for a brown coal. Apart from coal rank and caking properties, there are differences in ash content and ash composition, resulting in different melting characteristics and limiting the temperature range for gasification.

Some processes like moving-bed and fluid-bed technologies have the drawback of requiring non-caking coals in order to maintain a proper gas/solid contact. In all cases where coal has to be shipped overseas, there is a strong preference for a process capable of converting any type of coal. Preferably, the process should also be capable of treating the total run-of-mine output, of which about 30% wt is smaller than 10 mm. When the coal fines cannot be processed in the coal gasifier, they may be considered to be a large by-product stream for which an outlet has to be found. Further, it is an advantage when coals with a very high ash content up to 40% can be processed, especially for plants at the mine site.

1.4 Good Product Gas Suitability for Various Applications

Counter-current moving-bed type coal gasifiers yield large amounts of by-products such as coke, tar, phenols and cresols. For modern large-scale processes, such large by-product streams are unacceptable because they increase the production cost per unit of syngas and have the unpleasant feature of process economics based on different markets. Even when formed in small quantities, they add to process costs if environmental standards necessitate their removal. Ash, whose production cannot be avoided, should preferably take the form of an inert molten slag with a glass-like appearance.

This objective can best be realized in a high-temperature gasification process where coal volatiles are completely converted into small molecules, methane being the only hydrocarbon component remaining. If methane is an undesirable constituent in the production of synthesis gas, its formation can be avoided by further increasing the gasification temperatures above 1200°C. It is characteristic of today's gasification developments that there is a general tendency towards elevated gasification temperatures.

1.5 A High Degree of Environmental Acceptability

As coal will be gasified by tomorrow's technology, the process has to meet tomorrow's high standards for each type of environment where it can conceivably be required. These places may range from an installation at the mine site in Australia or North America to such densely populated areas as Japan and Western Europe. An important factor in solving the environmental problems is the minimization of both the cooling water and process water consumption. Coal gasification is a net water-consuming process; with a

Table 1 Expected SCGP Performance on a Commercial Scale

	Acland	Texas Lignite	Rheinbraun Brown Coal	Illinois No. 6	W. Virginia (Pittsburgh) seam)	Auguste Victoria	Goettelborn	Fluid Coke (tar sands bitumen)
Proximate Analysis (as received, % wt)								
Ash	25	9	2	9	12	5	8	5
Moisture	4	33	60	16	6	6	4	0
Volatile matter	36	31	20	34	36	31	34	4
Fixed carbon	35	27	18	41	46	58	54	91
Ultimate Analysis (moisture- and ash-free, % wt)								
C	80	72	71	77	83	84	83	88
H	6	6	5	6	6	6	6	2
O	12	19	23	11	6	7	9	2
N	1.2	1.2	0.6	1.4	1.6	1.6	1.5	1.6
S	0.6	1.0	0.3	4.3	3.4	1.1	0.9	6.9
LHV, M.A.F. coal, MJ/kg	32.1	28.6	26.7	31.0	33.4	34.3	32.4	32.3
Gasifier Intake (mt/MM Nm³ H_2+CO)								
M.A.F. Coal	477	523	577	496	456	439	465	492
A.R. Coal[*]	672	900	1518	661	556	493	528	518
Oxygen (99% pure)	482	445	480	444	428	431	433	474
Steam	-	-	-	25	81	105	62	175
Gasifier Output								
Cold gas efficiency[**], %	79	81	78	81	81	80	81	80
HP superheated steam[***]	815	693	816	710	716	726	714	781
Dry gas composition, % vol.:								
H_2	30.2	32.4	28.6	30.5	31.7	32.6	31.2	22.3
CO	67.0	61.8	64.5	65.1	64.3	63.2	65.6	70.9
CO_2	1.8	4.6	6.1	2.0	2.0	3.0	2.0	3.7
CH_4	-	-	-	-	-	-	-	-
H_2S + COS	0.2	0.3	0.1	1.5	1.1	0.3	0.3	2.2
N_2 + A	0.8	0.9	0.7	0.9	1.0	0.9	0.9	0.9
Overall process eff.[****], %	76	77	72	78	78	78	78	77

[*] Intake of the complex, corresponding to gasifier intake. [**] Defined as LHV of the product gas/LHV of the M.A.F. coal to the gasifier. [***] Metric tons/MM Nm³ H_2 + CO; 100 at, 520°C. [****] Defined as LHV of the product gas/LHV of total as-received coal.

proper water economy, bleeds of process water can be largely avoided.

Coal gasification gases may contain a range of unwanted components depending on the type of technology and the gasification temperature. They always contain hydrogen sulphide, carbonyl sulphide, hydrogen cyanide and ammonia. Low-temperature gasification involves formation of phenols, tar and aromatic hydrocarbons, whilst high-temperature gasification in particular produces traces of hydrogen cyanide. In general, the product gas from high-temperature gasification appears to be much easier and cheaper to clean rigorously than that from low-temperature gasification. Removal of H_2S, NH_3, COS and HCN is readily achieved by commercially available processes; for special applications new routes towards the removal of trace components are under development. The ideal process produces elemental sulphur and the unavoidable ash from the feed coal as the only by-products.

1.6 Cost Effectiveness

The most important aspect of a process is its cost effectiveness in the given situation. Once a decision has been taken, a choice has to be made among competing technologies. This choice will depend not only on the coal feedstock and the product required, but also on whether the plant will be built at the mine site where the coal is relatively cheap or whether the plant is to be built for expensive imported or deep-mined domestic coal.

2. THE SHELL COAL GASIFICATION PROCESS

An example of a new coal gasification process meeting the above requirements is the Shell Coal Gasification Process (SCGP), a flow scheme of which is given in Figure 1. This development is based on the experience gained from the design and operation of Koppers-Totzek atmospheric coal gasifiers and the long-standing expertise of Shell in the high-pressure gasification of oil. The programme has advanced to the stage where a 6 tonne/day pilot plant at Shell Laboratories in Amsterdam and a 150 tonne/day experimental plant at Harburg Refinery are in continuous operation (Figure 2).

The Shell Coal Gasification Process employs the entrained-feed, high-temperature gasification technology. Coal is ground to a size of less than 100 microns (90% wt $<90 \mu m$); when required, the feed is also dried in the mill to a water content of typically 2-8% wt. After pressurization, coal is fed into a reaction chamber through diametrically opposed burners, where it reacts with the blast, consisting of oxygen and steam or air, in a flamelike reaction.

An oxygen demand of 0.9-1.0 tonne/tonne of moisture-free and ash-free coal is fairly typical of hard coals, whereas for low-rank coals a figure of 0.7 tonne/tonne coal is more representative. Steam consumption is generally less than 0.1 tonne/tonne hard coal. Lignites generally do not require steam addition. Reactor outlet temperatures will not normally exceed 1400 to 1500°C. The reactor is an empty vessel providing a residence time of the order of a few seconds at pressures between 20 and 40 bar. Currently, the wall temperature is controlled by means of an intricate tube-wall construction in which high-pressure steam is raised. The extreme conditions in the reactor cause most of the ash from the coal to melt; it is subsequently collected in the water-filled bottom compartment of the reactor as non-leachable glass-like

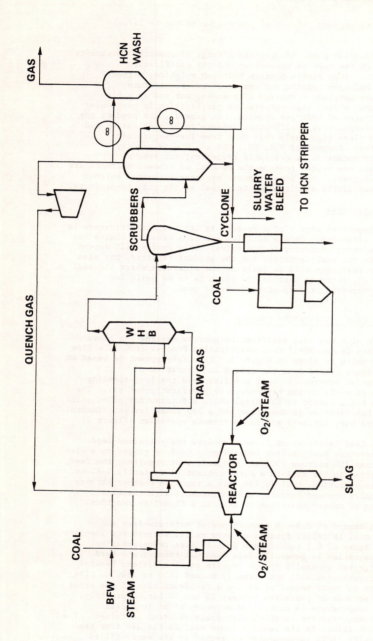

Figure 1 Shell coal gasification process: typical flow scheme

Figure 2   150 tonne/day pilot plant at Deutsche Shell's Harburg refinery

granules. Part of the ash, however, will be entrained in the
product gas. It is a strict requirement to solidify these
entrained ash droplets before they enter the waste-heat boiler.
Cooling is achieved by recycling of cold cleaned product gas to the
reactor outlet.

From a thermodynamic point of view it seems that the
entrained-feed, co-current high-temperature technology has obvious
disadvantages in terms of reactor energy balance relative to the
counter-current and/or low-temperature processes. However, with
oxygen/steam gasification, the SCGP product gas contains about 80%
of the heat of combustion of the coal, with the remainder mainly
converted into latent heat, the major part of which is used for
raising high-quality superheated steam in a waste-heat boiler
(Figure 3). When the complete, integrated installation is
considered, including energy and heat for coal drying and grinding,
for running equipment and for gas heating, it appears that the SCGP
is fully competitive in energy efficiency. This is, <u>inter alia</u>,
because of the very low process steam requirement and the simple
energy-saving gas cleaning that can be applied as a result of the
low CO_2 content of the gas (2-4% vol.). The amount of steam
produced in the waste-heat boiler is roughly sufficient to drive
the compressors of the air-splitting plant for the production of

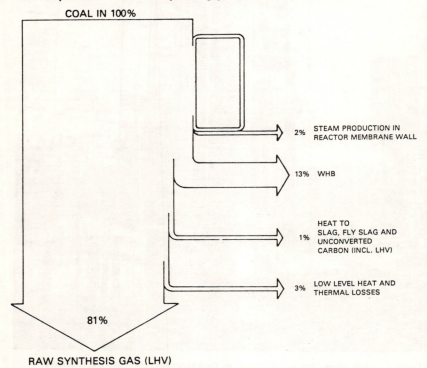

Figure 3   Typical heat balance of gasifier proper. Basis: lower heating value and sensible heat

Development of the Shell Coal Gasification Process 69

oxygen. The efficiency of the gasifier proper for a 10% wt. ash, 10% wt. moisture coal is about 81%. This efficiency is scarcely affected by coal rank (Table 1).

An integral part of the process is the removal of particulate matter from the raw gas. The proprietary system used for this purpose consists of a cyclone and scrubbers. This system, which is relatively cheap, removes solids to a level of less than 2 mg/Nm^3.

About 92-97% of the raw product gas consists of CO and hydrogen in a molar ratio of 2.0-2.4 when the reactor is operated at the preferred minimum steam dosage rate. Other components present in the gas are CO_2 (2-4% vol.), some unreacted steam (1-3% vol.), hydrogen sulphide (0.1-0.3% vol.), COS (200 ppm vol.), ammonia (500 ppm vol.) and hydrogen cyanide (150 ppm vol.) (see Table 1). The raw gas production is about 2000 Nm^3/tonne for a good-quality bituminous coal; the calorific heat content of the gas for O_2/steam gasification is of the order of 11 MJ/Nm^3 (2700 kcal/Nm^3). The solids-free raw gas has to be stripped of sulphur and other components. The nature of such treatment will very much depend on the ultimate use of the gas produced. Some situations require very deep sulphur removal to less than one ppm; others call for less deep or selective removal of sulphur, leaving CO_2 in the product gas. It is therefore of importance that the gasification process can be tied in with the treating process most suitable for the intended use. Commercially available treating processes can be applied under most conditions, but improved processes are under development for some specific uses.

3. THE SCGP DRY FEED SYSTEM

One of the characteristic features of the SCGP process is its dry feed system. The coal is ground and dried to a moisture content which depends on the quality of the coal. It is pressurized in a lock-hopper system and pneumatically fed to the reactor. The lock-hopper system is cyclic, and requires the frequent opening and closing of valves in a dust environment.

An alternative would be a "wet" system. The ground coal can be suspended in water to a pumpable slurry and be compressed and transported direct into the reactor. Mainly on the basis of process efficiency considerations, Shell have given preference to the dry feed system, as will be illustrated below.

In a wet feed system, the water-to-coal ratio of the mixture fed to the gasifier is determined by the pumpability and spraying characteristics of the slurry. This will always lead to an excess of water in the reactor as compared with the stoichiometric and kinetic requirements for the gasification. As a consequence, an excess of mass has to be carried through the high-temperature reactor, resulting in generation of unnecessary large amounts of sensible heat. In addition, the water has to be evaporated inside the reactor, which is another heat sink reducing the cold gas efficiency and increasing the oxygen requirement of the gasifier.

It is evident that from a thermodynamic point of view a wet feed system should be operated at minimum water addition and maximum slurry concentration. High slurry concentrations, however, are more difficult to disperse and put constraints on the particle size distribution of the coal to such an extent that it influences gasification kinetics adversely. At similar reactor outlet temperatures, wet feed systems will therefore produce lower coal conversions per pass than dry feed systems. The unconverted carbon can of course be recycled to the reactor, albeit with additional water for feeding.

With decreasing coal rank (*i.e.* increasing inherent moisture of the coal) the water content of an appropriate coal/water slurry will increase and the efficiency advantage of a dry over a wet feed system will increase. In the extreme case of lignites, wet feeding is considered to be very unattractive, if at all feasible.

Table 2 shows a calculation example illustrating the effects of the choice of feed system as discussed above. In this example, two identical reactors are fed by the two different feed systems. In the wet feed system it has been assumed (*cf.* Ref.) that 65.5% solids concentration can be applied in the feed slurry, resulting in 95% coal conversion per pass and a reactor outlet temperature of 1350°C. Recycle of unconverted carbon is applied to an overall

Table 2   Dry versus wet feed systems

		Dry feed	Wet feed
Water (moisture) to gasifier	KG/KG MAF coal	0.10	0.63
Oxygen to gasifier	"	0.89	1.02
Steam to gasifier		0.09	0
Cold gas efficiency *	% LHV	81.5	73.6
Latent heat of water evaporation	% LHV	0.8	4.8
Sensible heat of products	% LHV	15.5	20.3
H_2 + CO produced	Nm^3/kg MAF coal	2.10	1.91
Raw gas composition	% vol		
H_2O		4.5	20.4
H_2		32.4	28.3
CO		57.5	37.7
CO_2		3.5	11.7
H_2S		1.2	1.0
N_2+A		0.9	0.9

* LHV of raw product gas, including H_2S, divided by the LHV of the MAF coal

carbon conversion of 99.4%. In the dry feed case the coal is dried down to 10% wt. moisture, and 99% coal conversion can be obtained at the same temperature level of 1350°C. Illinois No. 6 (Ref.) is used as the feed coal.

In this calculation example, the difference in cold gas efficiency amounts to some 8 percentage points. In order to judge the overall process performance, some other factors have to be taken into account as well: the wet feed system involves a considerably higher oxygen requirement, but no process steam requirement, no drying

energy, less power consumption in coal pressurization and
potentially more steam production from the sensible heat in the
product gas. The resultant difference in overall process efficiency
between dry and wet feeding is at present estimated to be some 5
percentage points (6.5% relative).
A second aspect of a "dry" versus a "wet" system is the product
gas composition (Table 2). For "wet" systems the CO_2 content
rises sharply. Even if the CO_2 itself does not have to be
removed (as in most fuel gas applications), it is a disadvantage,
as it significantly increases the costs of H_2S removal by
requiring an extra enrichment stage before the sulphur recovery
unit.

For synthesis gas applications requiring CO shift, a somewhat
higher H_2/CO is beneficial. In addition, in the "wet" feed case
the raw gas already contains part of the steam required for the
shift. If a sulphur-resistant shift is applied, this has a
favourable effect on the shift economics. On the other hand, the
high CO_2 content remains a disadvantage and, more generally,
introducing water into the gasifier is an expensive way of raising
steam for the shift reaction.

Taking the above aspects together it is evident that a "dry" feed
system offers remarkable advantages. The dry feed system with
lock-hoppers has performed satisfactorily in both the Amsterdam and
Harburg plants.

4. CURRENT SCGP DEVELOPMENT STATUS

The 6 t/d capacity pilot plant (built in Phase 1 of the programme)
has been in operation at Shell Laboratories in Amsterdam since
December 1976. To date, the plant has successfully operated for
over 10,000 hours on a number of feedstocks, including various
German hard coals, Illinois No. 5 and No. 6 coal, Pittsburgh seam
coal, Acland coal, German Rheinbraun brown coal, Texas lignite, and
coke from Athabasca tar sands.

The present function of the pilot plant is to widen the range of
feedstocks tested, to continue collecting fundamental process data
and to test new equipment.

Besides the operation of the pilot plant, extensive research
activities directly related to the gasification project continue at
Shell Laboratories in Amsterdam. These include:

- the operation of component test facilities, *e.g.* for burners, feeder
  systems and valves;
- the development of advanced measurement and control techniques;
- reactor model studies (both mathematical and physical).

The 150 t/d capacity pilot plant (built in Phase 2 of the
programme) started operation at Harburg Refinery before the end of
1978. The plant has now completed 5500 running hours, including an
uninterrupted run of over 1000 hours on Goettelborn coal and one of
700 hours on Illinois No 5 coal.
A summary of the operational experience is given in Table 3.

The most important equipment components which were successfully
tested in the 150 t/d pilot plant were:

- the fully automated lock-hopper system,
- the coal feeding and dosing system,
- various types of burners,
- the reactor, namely the reactor wall and slag tap,
- the waste heat boiler,
- the gas quench,
- the production of superheated steam (500°C, 50 bar) in the waste heat boiler,
- the solids removal system,
- the HCN slurry stripper.

Table 3   Operational Experience,   SCGP Pilot Plant Harburg

General Information

Accumulated run hours (1/4/83):   5500 hours
Amount of coal processed   : 25,000 metric tons
Coals tested: Griesborn Duhamel
                Goettelborn
                West Virginia (Pittsburgh Seam)
                Illinois No. 5

Typical Results

Coal: Goettelborn (see Table 1)   kg/h

In : Coal + carrier gas   4847
      Oxygen   4553
      Steam   700

Out : Gas   9685
      Solid residues   415

Carbon conversion (without ash recirculation)   99%
Slagging efficiency (slag carbon content 0.1% wt.)   50-70%
Cold gas efficiency (LHV basis)   79%
Thermal efficiency (incl. HP steam)   94%
Dry gas composition: H_2 29.4% vol.
                     CO 61.0% vol.
                     CO_2 3.6% vol.
                     CH_4 0.1% vol.
                     H_2S 0.2% vol.
                     N_2 5.8% vol.

Development of the Shell Coal Gasification Process 73

During the first two years of operation the main problems encountered related to waste heat boiler fouling and process control.

In the period March-June 1982 the unit was overhauled, after which the availability improved dramatically. In the period July 1982 - March 1983 a total of 3000 h was accumulated during which period no waste heat boiler cleaning was required and a good gas quality was maintained thanks to improvements in process control.

Surveying the total operating experience in the two plants in Amsterdam and Harburg, it may be concluded that the basic concept of the Shell gasification process has been demonstrated. Important in this respect are the following:

- a conversion of 99% is obtained;
- the CO_2 content in the product gas can be maintained between 2 and 4%;
- the reactor conditions (temperature) can be controlled accurately.

5. WASTE WATER TREATMENT

The potential sources of water in an SCGP unit are:
1. steam in the product gas leaving the reactor;
2. steam condensed in the HCN stripper (see Figure 1);
3. storm run-off;
4. incidental waters.

Other water streams like the slag bath water, water from the wet solids removal system and water from the HCN wash are recycled within the gasification and treating stages.

Deep solids removal before biotreatment is achieved in a flocculation/sedimentation unit. The sediments are filtered and sent to solids work-up. The last treatment steps before the water discharge into the environment are biotreatment under extended aeration followed by after-aeration. With some coals there may be levels of inorganic pollutants remaining that require control technology beyond biotreatment. Generally, the ultimate water bleed will be determined by the chlorine content of the coal and the allowable salt concentration of the water discharge.

6. PROCESS APPLICATIONS

There are many future applications for the SCGP. The gas produced (93-98% vol. hydrogen and carbon monoxide) is suitable for the manufacture of hydrogen or reducing gas and, with further processing, substitute natural gas (SNG), ammonia, methanol and liquid hydrocarbons. The combination of SCGP and methanol or Fischer-Tropsch synthesis has the potential of producing gasoline, aviation fuel and diesel fuel from coal with efficiencies of 45-55%.

Another application of the process is as a fuel gas supplier to a combined-cycle power station, which will allow of electricity generation at more than 40% efficiency for a wide range of coals.

In Figure 4 simplified block schemes are shown for the production of methanol from syngas as produced in the Shell Coal Gasification Process and methanol synthesis followed by conversion into gasoline.

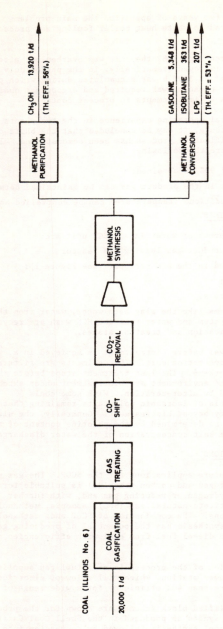

Figure 4 Coal conversion schemes for the production of methanol and liquid hydrocarbons

7. ECONOMICS

The economics of coal gasification will be determined by investment costs, operating costs, coal cost and last but not least the cost of oil and natural gas. As the price differential between coal and oil is the most important factor, it is obvious that the current uncertainty in this price differential is not exactly advantageous for the introduction of new technologies for coal conversion. This is especially true in Europe, where the combination of high coal cost and a relative abundance of natural gas does not seem to favour the introduction of coal gasification on a significant scale for a considerable time to come.

Nevertheless there are locations in the world which, apart from strategic/political reasons, offer much more favourable conditions and it can be expected that some larger-scale commerical applications will develop in the near future.

8. REFERENCE

EPRI Report AP-2488, July 1982.

Development of the Fixed Bed Slagging Gasifier

By C. T. Brooks
BRITISH GAS CORPORATION, WESTFIELD DEVELOPMENT CENTRE, CARDENDEN, FIFE, SCOTLAND KY5 0HP, U.K.

Introduction

The world reserves of coal are vast, and greater by an order of magnitude than those of oil and natural gas. Against this background there is likely to be increased use of coal and nuclear energy into the 21st century. The application of energy conservation measures will also be an important part of any energy strategy.

Support for the coal industry in Europe is considerable. In Britain, there is a long-term commitment of investment expenditure, in addition to an operating subsidy together with a government scheme of grants for converting factories to coal. British Gas Corporation share in this national interest in the future exploitation of coal. They are committed[1] to a vigorous substitute natural gas (SNG) development programme, with great emphasis on coal gasification, and see SNG becoming important to the nation when supplies of natural gas are depleted, well into the 21st century; it should be realised, however, that the present day cost of SNG from coal is high, and it will not be a competitor to alternative fuels for some considerable time. This is just as well, as coal gasification is a complicated and expensive technology, and the lead time required to put down working plants of considerable size (say, to gasify 1 million tonnes/year of coal) will be considerable, and it is desirable that further development is carried out.

Coal gasification to SNG is, of course, the main reason for the existence of the current British Gas Research, Development and Demonstration Programme on coal gasification. There are, however, many other potential uses for gas from coal, major ones being :

* combined cycle power generation
* iron and steel manufacture
* synthesis gas manufacture

Interest in these other avenues has been shown both in the UK and abroad, and it is British Gas Policy to co-operate in the exploitation of its technology with other companies to the mutual benefit of both parties.

The particular benefits of coal gasification, with special regard to the production of SNG, have been reviewed in a previous lecture[2] to this society, by Dr. J. A. Gray, OBE. In a world increasingly more conscious of the environment and of energy conservation, the advantages of gasification of coal are even more relevant six years later.

The simplest way to make gas from coal is to heat it in the absence of air[3]. In this pyrolysis/carbonisation process the volatile matter in the coal decomposes and leads to the production of 'coal gas', tar and aqueous liquor, leaving the fixed carbon behind as coke. This process has been available to the Gas Industry for many years, being developed commercially using a single horizontal retort in 1808 by William Murdoch and further refined in the early 1900s as a vertical retort process. The carbonisation process is basically thermally efficient but suffers from very large yields of byproducts (only 25% of the thermal value of the coal appeared in the make gas). Other problems were limited throughputs of units and the need for a coal having suitable coking properties, in order to produce a suitable marketable byproduct.

The need for complete gasification of coal, with particular reference to the fixed carbon, requires the involvement of a gasification agent. This is normally oxygen and/or steam, which react in the following ways :

$$C + \tfrac{1}{2} O_2 = CO + \text{Heat}$$
$$C + H_2O(g) = CO + H_2 - \text{Heat}$$

These gasification principles were applied from the 1890s onwards, with a cyclic process in which the bed of coke was heated to incandescence ($1500^\circ C$) by air blowing, followed by a gas making period with steam as the reactant. This make gas CV in this second stage could be increased by spraying oil into the reactor. This water gas process was inefficient, and produced a gas of high CO content unsuitable for direct use as a town gas, but nevertheless acted as a useful (20% of supply) supplement to the total.

The cyclic nature of the water gas process was undesirable and led to the development of reactors capable of producing complete gasification of coke in a single stage (producer gas). Examples of this type of reactor are

Tees, Mond, Marischka and Kerpely[4]. These gasifiers worked on air and steam at low pressure and produced a nitrogen-laden gas of low Btu which was uneconomical to distribute. The reactors were later designed to accept some low-grade coals but the process still remained inadequate for many needs.

The above problems with regard to the poor quality of the gas produced could be elegantly alleviated by the use of oxygen in place of air as was pointed out as early as 1920 by Hodsman & Cobb[5]. Unfortunately, at that time oxygen could not be produced in the quantities required and was prohibitively expensive. A decade later, this was no longer the case, and this led to the development of a new generation of gasifiers which worked on oxygen and gave a continuous output of intermediate Btu gas from coal.

There are three basic methods of gasifying coal depending on how the coal is contacted with the reactants, and these methods have been described in more detail elsewhere[6]. The fixed bed reactor used lump coal as its basic feed and is exemplified by the Lurgi reactor which was developed commercially in the 1930s. The fluidised bed systems use granular coal, typically $\frac{1}{8}$" particle size, and entrained systems use pulverised fuel. The Winkler fluidised bed system was developed commercially as early as 1926 and the Koppers-Totzek entrained gasifier was operating commercially by 1952. These three basic processes still remain as technically viable and despite more elegant processes being designed (the so called second generation processes) still remain as the most likely future commercial coal gasification processes, with the Texaco coal gasifier as a front runner in the entrained field.

Of the processes mentioned above, the Lurgi process has dominated the commercial scene. It was, as first designed, similar to the producer gas reactor with the important exceptions that it worked at high pressure (20 bar) and used oxygen. Thus reactor throughput was increased and several cost and process advantages accrued. The Lurgi gasification process today constitutes commercial technology based on more than forty years of continuous development work and the experience gained from the operation of 16 commercial plants with a total of 140 gasifiers.

British Gas have used the Lurgi gasification to produce town gas at their Westfield[7] and Coleshill plants, operated during the sixties, and in parallel with this operation of the commercial Lurgi gasifier, had been developing a slagging version at the Midlands Research Station[8]. This slagging version of the fixed bed gasifier was shown to have numerous advantages over the ashing Lurgi version, and has been further developed to a commercial status at the British Gas Westfield Development Centre[9].

The slagging gasifier represents the first of the new generation coal
gasifiers which may be operated by British Gas to make SNG in the future
and the present paper therefore discusses in some detail how the fixed bed
gasifier works and makes particular reference to British Gas Corporation's
Development and Demonstration programme currently being carried out at the
Westfield Development Centre.

The Fixed Bed Slagging Gasifier

The British Gas Industry has had experience of fixed bed gasification for
over 150 years[3]. Early experience was gained from the horizontal retort
which was developed at the beginning of the 19th century and was used for
over a hundred years, to be followed later by the vertical retort. Modern
experience of advanced processes was gained with the operation of Lurgi
gasifiers at Westfield[7] and Coleshill, which were used in the sixties to
produce town gas from British coals.

This fixed bed Lurgi gasifier, as has been mentioned in the previous section,
is the dominant commercial gasifier today. It operates at pressures
typically between 25 and 32 atm, and is shown schematically in Figure 1.
The system and its mode of operation has been described fully elsewhere[9].

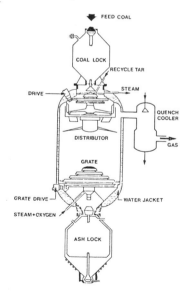

Figure 1 : The Lurgi Gasifier

The main disadvantage of the Lurgi system stems from the necessity to supply with the gasification oxygen a large quantity of steam far in excess of that required to gasify the carbon. This excess is required to maintain the temperature in the combustion zone below the level at which ash can clinker, allowing the ash to then be extracted via the mechanical grate. This maximum temperature is affected by :

* the steam/oxygen ratio
* the reactivity of fuel to steam
* the rate of gasification

Thus, high temperatures will be promoted by low steam/oxygen ratios, low char reactivities and high gasification rates. This means that for certain coals, such as unreactive coals of relatively low ash fusion point, the Lurgi process will suffer from high steam usage (with related high byproduct liquor yields from the reactor) and low reactor throughputs.

In Britain, the commonly available bituminous coals fall into the above category and there were thus considerable incentives for the British Gas Industry to want to improve upon the Lurgi gasification process. The British Gas Corporation have put a large effort, since the 1950s, into developing a fixed bed gasifier which will overcome the constraints that the ashing Lurgi gasifier is subject to. This has been done by eliminating the excess steam and only supplying the steam required for gasification. Under these conditions the ash temperatures rise above melting point and the ash has to be dealt with as a molten slag at the gasifier bottom.

This mode of gasification creates radical differences as compared to the standard Lurgi and the development is referred to as the British Gas/Lurgi fixed bed slagging gasifier. The gasifier is shown schematically in Figure 2. The fixed bed gasifier works normally on sized coal in the range 6 to 50 mm and is fed on to the bed top via a coal locking system and a distributor fitted with chutes. The distributor acts to buffer out the intermittent nature of coal feeding from the lock hopper, and is essentially a rotating box which is cooled to prevent the coal caking and swelling until it is laid down on the bed top. The countercurrent nature of the reactor means that the lump nature of the coal must be preserved for its passage down the shaft, so caking coals have to be broken up by a stirrer arrangement at the bed top.

Development of the Fixed Bed Slagging Gasifier

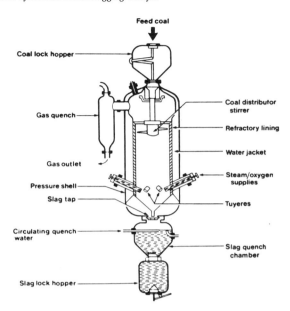

Figure 2 : The Slagging Gasifier

The zones down the bed of the gasifier, and their approximate temperatures for operation on a typical bituminous coal, are shown in Figure 3. Although these zones can overlap considerably the following discrete zones are recognisable.

- Drying and devolatilisation and pyrolysis zone
- Gasification zone
- Combustion zone
- Slag removal zone

The drying and devolatilisation zone is the principal heat exchange zone for the hot gases which rise up the reactor. This sensible heat, generated essentially by oxygen combustion at the gasifier bottom, is gainfully employed in driving reactions ranging from simple drying to complex physical and chemical reactions of the coal structure. It is this feature which is a major contributor to the attractive efficiencies of the fixed bed process.

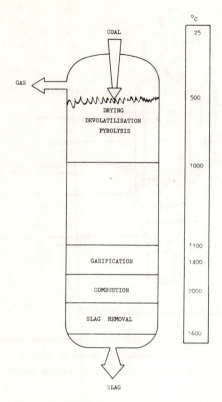

Figure 3 : Zones in the Slagging Gasifier

The temperature of the crude gas leaving the bed top is particularly influenced by the moisture and ash content of the coal[9]. It is theoretically possible, for high-moisture coals (lignites can have well over 35% moisture) with moderate ash contents, for the dew-point of liquid hydrocarbon products to be reached, and for there thus to be some restriction to the gas flow at the bed top, although this phenomenon has never been reported as having been observed in practice, and in any case does not occur for bituminous coals, which typically operate at bed top temperatures of 500 °C. At about these temperatures complex pyrolysis reactions occur, coupled with caking and swelling of the coal, and these will start to occur close to the bed top, depending upon various factors (moisture content, size, gas/solid contact) which will determine when the coal reaches gas temperature during passage down the bed at typically 10 cm per minute.

Development of the Fixed Bed Slagging Gasifier

The mechanism of caking and swelling in this zone involves the release of substances which can cause the coal mass to become plastic and hence the need for a stirrer to preserve the countercurrent flow. The extent of this effect depends upon a number of factors, in particular the rank of the coal, see Table 1, which lists some of the coals tested by British Gas in their Slagging Gasifier.

Table 1 : Coals Run through the Westfield Slagging Gasifier

Coal	Manton	Rossington	Gelding	Ohio 9	Ohio 9
Source	England	England	England	USA	USA
Size	$\frac{1}{4}$" - $1\frac{1}{2}$" (with up to 35% under $\frac{1}{4}$")	$\frac{1}{4}$" - $1\frac{1}{2}$" (with up to 35% under $\frac{1}{4}$")	$\frac{1}{4}$" - 1"	$\frac{1}{4}$" - $1\frac{1}{4}$" washed	$\frac{1}{4}$" - $1\frac{1}{4}$" unwashed
Moisture %	3.0	6.9	10.2	1.4	4.7
Ash %	7.3	4.3	5.8	12.0	20.8
VM %	31.9	33.3	31.5	39.7	32.3
FC %	57.8	55.5	52.5	46.9	42.2
Swelling	$6\frac{1}{2}$	$1\frac{1}{2}$	1	6	$3\frac{1}{2}$
Caking	G6	E	C	G4	G

Coal	Pittsburgh 8	Hucknall	Comrie	Killoch
Source	USA	England	Scotland	Scotland
Size	$\frac{1}{4}$" - $1\frac{1}{4}$" (with up to 25% under $\frac{1}{4}$")	$\frac{1}{4}$" - 1"	$\frac{1}{4}$" - 1"	$\frac{1}{4}$" - 1"
Moisture %	2.1	5.0	2.7	7.7
Ash %	11.5	5.1	6.3	4.5
VM %	36.1	35.4	32.1	33.8
FC %	50.3	54.5	58.9	54.0
Swelling	7	$3\frac{1}{2}$	$2\frac{1}{2}$	$3\frac{1}{2}$
Caking	G6	G	F	E

Pyrolysis reactions in this region will generate a range of hydrocarbons, varying from CH_4, C_2H_6, C_2H_4 at the low boiling end, through benzene and its derivatives to the very high molecular weight constituents of tar. Pyrolysis products containing the N, S and O in the coal will be represented at the low end by CO, CO_2, NH_3, H_2S, COS, CS_2 through thiophenes, phenols to again high molecular weight substituted hydrocarbons.

For the particular application of the production of SNG, and also with respect to reducing the specific oxygen consumption per therm of gas produced, the formation of methane in the pyrolysis zone is of particular importance. The mechanism of methane formation, and the role that hydrogen formed in the reaction zone below plays in this mechanism, is not completely understood. There is a correlation between rank, and hence oxygen content of the coal as Figure 4 shows. This is in inverse correlation to reactivity of the char produced in the devolatilisation zone, as the lower rank, higher oxygen containing coals invariably give chars of higher reactivity.

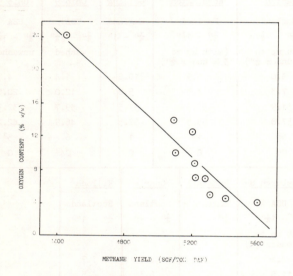

Figure 4 : Methane yields from various coals as a function of coal oxygen content

The char produced by the devolatilisation zone will continue on its journey down the bed through an idle zone, where little reaction takes place, before reaching the main gasification and combustion zones in the gasifier at and just above the tuyere level. Steam and oxygen blown into the gasifier down the tuyeres forms a void in the fuel bed and around the perimeter of this void in particular the oxygen and char react very rapidly in a strongly exothermic reaction :

$$C + O_2 = CO_2$$

$$\Delta H_{1000\ K} = -94.3\ \text{kcal/mol}$$

Temperatures in and at the raceway perimeter will thus be very high, up to around 2000 °C, and all the oxygen will be consumed.

This released heat then supports the highly endothermic gasification reactions :

$$C + H_2O = CO + H_2 \quad \cdots\cdots\cdots \quad \Delta H_{1000\ K} = +32.5\ \text{kcal/mol}$$

$$C + CO_2 = 2CO \quad \cdots\cdots\cdots \quad \Delta H_{1000\ K} = +40.8\ \text{kcal/mol}$$

These reactions rapidly cool the upflowing gases and hence the bed temperature rapidly decreases. Reactions in this area appear to be limited by equilibrium factors.

The exothermic CO-shift and carbon hydrogenation also play a minor role in this zone :

$$CO + H_2O = CO_2 + H_2 \quad \cdots\cdots\cdots \quad \Delta H_{1000\ K} = -8.31\ \text{kcal/mol}$$

$$C + 2H_2 = CH_4 \quad \cdots\cdots\cdots \quad \Delta H_{1000\ K} = -21.4\ \text{kcal/mol}$$

The extent of the first reaction is limited by the small amount of steam, whereas the left-hand side of the hydrogenation equation is much favoured at the high temperatures involved.

The physics and chemistry of the fixed bed is thus complex and multi-staged, with many processes occurring. It is thus different to entrained and fluidised systems, which seek to impose reasonably constant conditions across the reactor.

Gas and solid residence times vary widely across the various zones in the slagging gasifier, because of the different volumes of various reaction zones, and some indication of the order of residence time is shown in Table 2. The complex picture of the fixed bed given in this section still represents an over-simplification. The model assumes good gas/solid contact and plug flow of solids down the gasifier shaft. In practice there will be deviation from ideality and overlap of the gasifier zones.

Table 2

Orders for Residence Times in the Fixed Bed Slagging Gasifier

Zone	Solids	Gases
Drying/Devolatilisation	15 minutes	10 seconds
Reaction	2 minutes	1 second
Combustion	seconds	milliseconds
Slag	minutes	-

Despite these factors, the large inventory of carbon in the fixed bed, representing typically 30 minutes to 1 hour's gasification, represents a relatively constant and stable system from the operator's point of view. The gasifier will thus run stably at a wide variety of loadings (30% - 140%) and can change rapidly from one loading to another within this range. This feature is an advantage when the gasifier is considered in a power generation context[10].

The Process in Action

The development of the British Gas/Lurgi Slagging Gasifier is described in depth elsewhere[9]. Currently, British Gas, at Westfield are just completing Phase II of a programme, using a 6' shaft diameter gasifier, after which Phase III will commence with operation of a 7'6" commercial prototype gasifier. Elements of the British Gas Programme are shown in Table 3. The six foot shaft gasifier is a conversion of one of the standard 9' shaft diameter Lurgi gasifiers which operated at Westfield from 1960-73 and operates at 350 psig. Standard Lurgi technology is used to feed coal to the bed top via a lock hopper. Under standard operating conditions the throughput is 300 tonnes/day for good quality British bituminous coals, although the gasifier has performed 40% above this loading.

Table 3 : The British Gas Slagging Gasifier Programme

	Phase	Elements
Operation on 6' shaft 300 ton/day gasifier	I	Testing of various coals for slagging gasification
	II	Development of proprietary equipment
		90-Day Demonstration Run
		Fines Handling
		Hydrocarbon Handling
		Liquor Handling
Operation on 7'6" shaft 500 ton/day gasifier	III	Commercial demonstration of elements of Phase I and II
	IV	Long Demonstration Run
		Demonstration of Hicom Process as a route to SNG
	V	Testing foreign coals

The size range at the gasifier top is generally quoted as 6-50 mm, but a considerable amount of fine, -6 mm material can be tolerated at the bed top, depending upon the rank of coal fed to the top. Table 1 indicates that strongly caking coals can tolerate 30% material less than 6 mm in the top of the bed feed. Any solids carried over from the gasifier are washed/ cooled and contained in the tar product. This tar product is separated from liquor at low pressures and the tar plus coal dust recycled back to the coal as it enters the reactor. Thus, the tar can act as a vehicle to ensure that there is no net escape of carbon solid from the bed of the gasifier.

This tar that is recycled to the gasifier top is in part also gasified by pyrolysis with the residue being brought to the gasifier bottom. However, it is not possible to gasify the net hydrocarbon product in this way.

At the bottom of the 6' gasifier shaft a number of tuyeres inject the steam/oxygen gasification medium into the gasifier. Gasification has been demonstrated in the Westfield gasifier at steam/oxygen ratios ranging from 0.9 to 1.7 (v/v) over which conditions slagging gasification of a 702 Rank British Coal, Rossington, is sustainable at standard loadings. Performance data across this ratio are given in Table 4. Note the slight effect upon product gas composition. The thermal efficiency of the gasifier is nearly constant over the range, with a small increase in specific oxygen consumption across the range. The gasifier outlet temperature across the range is shown in Figure 5. This can be regarded as a simple indicator of overall efficiency; the rise towards higher steam/oxygen ratios is indicative of more and more high grade heat being carried away from the reaction zone as sensible heat in the undecomposed steam, whereas towards the lowest steam/oxygen ratios there is 100% or close to steam decomposition and therefore an excess of high grade heat at the gasifier bottom, which again is carried away as sensible heat in the rising gases.

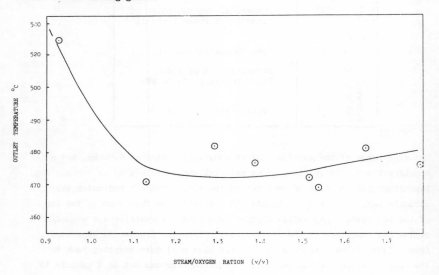

Figure 5 : Gasifier Offtake Temperature as a function of H_2O/O_2

Table 4

Performance Data for the British Gas/Lurgi Slagging Gasifier

Rossington Coal at various Steam/Oxygen Ratios

Steam/oxygen ratio (v/v)	0.93	1.14	1.29	1.38	1.52	1.66	1.78
Gasifier pressure (atm)	24	24	24	24	24	24	24
Outlet Gas Temperature (°C)	529	471	482	476	472	481	473
Crude Gas Composition (% v/v)							
CO	61.0	60.4	57.9	56.9	56.0	54.4	53.5
H_2	25.7	26.4	27.1	27.2	27.7	28.6	28.6
CH_4	6.9	6.5	6.8	6.6	6.8	6.4	7.2
CO_2	1.3	1.6	2.9	3.7	4.4	6.1	5.5
C_2H_6	0.5	0.6	0.5	0.6	0.5	0.5	0.5
C_2H_4	0.2	0.2	0.2	0.2	0.1	0.2	0.1
N_2	3.7	3.5	3.8	4.0	3.5	2.9	3.6
H_2S	0.5	0.5	0.4	0.4	0.6	0.4	0.5
Derived Data							
DAF coal gasification rate (lbs/hr/ft^2)	817	855	847	826	840	866	854
Oxygen consumption (SCF/therm)	55.2	52.0	52.1	51.2	52.7	51.9	50.0
Liquor production (lbs/lb coal)	0.16	0.18	0.21	0.19	0.22	0.26	0.27

Table 4 gives oxygen consumption as a function of steam/oxygen ratio. Decrease in oxygen consumption per therm of gas produced is attractive, but it must be noted that a decrease in specific oxygen consumption can only be gained at the expense of a less efficient use of gasification steam, which, moreover will contribute to higher liquor yields. In fact, a steam/oxygen ratio of 1.1 - 1.2 will yield the most desirable process efficiencies for most coals.

For the highest efficiencies, heat losses at the hearth bottom should be minimal. This is achieved in practice, with the only problem being the fact that no heat recovery is possible at present from the molten slag which is tapped from the reactor bottom.

The slagging gasifier working in the basic mode is an attractive process, and has fulfilled the promise of early development. A variety of coals have been tested during the Phase I of the British Gas Programme and these coals show a marked insensitivity to reactivity properties, as the oxygen consumption figures listed in Table 5 show.

Table 5

Oxygen Consumption per Therm of Gas
produced for various British Coals

Coal	Steam/Oxygen Ratio (v/v)	Oxygen Consumption (scf/therm)
Markham Main	1.30	51.6
Manvers	1.31	53.0
Rossington	1.28	52.7
Hucknall	1.30	54.4
Manton	1.39	55.9
Lynemouth	1.30	54.6

Development of the Fixed Bed Slagging Gasifier

The oxygen plays an important role in the fixed bed slagging gasifier, supplying the heat for the removal of ash as molten slag, for driving the gasification reaction and for the complexity of processes in the fixed bed, which involve a multitude of pyrolysis reactions, caking and swelling processes, devolatilisation and drying. Most of this heat is effectively used in the fixed bed, with only modest amounts escaping as sensible heat in the upcoming gases, as gasifier heat losses and as heat losses in the molten slag.

The fixed bed system thus makes effective use of the oxygen in the gasification medium and its specific oxygen consumptions are only half those of entrained systems such as Koppers-Totzek and Texaco. Nevertheless, the oxygen usage does have a significant impact on the efficiency of the system, as the simple heat balance of Table 6 shows. Despite this disadvantage, however, the fixed bed slagging gasifier represents the best option for the conversion of British Coals to SNG. The 6' diameter shaft gasifier has already demonstrated its commercial potential in a 90 day demonstration run, which gasified over 27,000 tons of various British coals. Phase II of the British Gas Programme is now nearing completion, and has addressed itself to aspects aimed at increasing the flexibility of the fixed bed gasifier. These are :

(a) Although there is 100% carbon solid gasification in the fixed bed gasifier, with tar acting as a vehicle for no net escape at the bed top, and although no char escapes with the slag, coal conversion to gas is not complete as there is a significant hydrocarbon yield. This may be acceptable if there is a premium market for these products, but this is unlikely to be the case. The results of Phase II indicate that the net hydrocarbon yield can be gasified by injection down the tuyeres.

(b) Coal mined by modern methods will be likely to contain a considerable amount of material less than $\frac{1}{4}$". Although a modest amount can perhaps be ingratiated with the coal feed to the bed top there is likely to be surplus of fine material available for gasification. This fine coal can be pulverised, entrained and injected down the tuyeres.

(c) A feature of the fixed bed gasifier is the ability of the bed to act as an efficient heat exchanger to the gases rising from the reaction zone. The coal feed is dried in this manner and the water in the coal appears as a dirty liquor downstream. Liquor yields will be dependent upon coal moisture, with other small contributions from pyrolysis of coal oxygen ultimate in the bed and from the undecomposed steam rising from the reaction zone. Phase II operation has shown that the net liquor yield for certain coals can replace gasification steam at the tuyeres.

The potential of the tuyeres for injection of material into the combustion zone formed a major theme of Phase II of the British Gas Programme on the attack on the above disadvantages, and these results will be published at a later date.

Table 6
Heat Balance for Gasifier Operating on Markham Coal

In		Out	
Input	Therms/hour	Output	Therms/hour
Coal	3412	Gas	3022
Steam[1]	137	Sensible Heat[3]	236
Oxygen[2]	298	Heat Losses	63
Tar	220	Tar/Oil	460

Notes

Operation on Markham Main coal at 1.17 steam/oxygen ratio.

1. Steam assumed to be raised from coal at 80% boiler efficiency.

2. Oxygen raised from electricity, which in turn is raised from coal at 30% efficiency. Equivalent thermal value is then :

    One SCF Oxygen $\equiv$ 192 Btu

3. Sensible heat includes the latent heat in the gases leaving the bed top.

Conclusions

The six foot slagging gasifier has been tested extensively in a number of modes since its commissioning in 1975 and the running experience is summarised in Table 7. The operating experience of gasification of 100,000 tonnes of coal in a cumulative time of a year's running have led British Gas to conclude that the process can now be regarded as commercial when operated in the basic mode.

Table 7

Summary of Operation at Westfield Development Centre

Project	Runs	Hours on Line	Coal Gasified (Long Tons)
American Sponsored	27	1508	17,361
DOE Programme	15	981	9,702
EPRI Trials	3	415	3,485
BGC Programme	37	5754	68,869
Totals :	82	8658	99,417

Fixed bed slagging gasification can be seen to be a versatile, efficient and adaptable system, and represents British Gas's major thrust for future base load SNG plants. Demonstration of an efficient route to SNG, the HiCoM process, is planned for early in Phase II, but it has been made clear in other forums that the process has a number of important potential uses which British Gas intend to exploit commercially if and when the opportunity arises.

Acknowledgements

Thanks are due to the British Gas Corporation for permission to publish this paper, and to the staff of the Westfield Development Centre.

References

1. J. McHugh : 1983 Major Achievement lecture, Institute of Mechanical Engineers.

2. J. A. Gray : 1st Priestley Oxygen Conference, Leeds 1977.

3. J. A. Gray and H. J. F. Stroud : "Gas from Coal – Full Circle" Institution of Gas Engineers Communication 1088 (1979).

4. Chemistry of Coal Utilisation : Second supplementary volume; John Wiley, New York, p. 161.

5. H. J. Hodsman and J. W. Cobb : <u>Trans Inst. Gas Engineer</u> 1919–20, pp 431–461

6. Reference 4, p. 1607

7. T. S. Ricketts : "The Westfield High Pressure Coal Gasification Plant" IGE Communication 567 ; (1960)

8. D. Hebden, J. A. Lacey, R. Edge : <u>Journal Inst. Gas Engineers</u>, $\underline{5}$ 367 (1965)

9. D. Hebden and C. T. Brooks : Westfield – The Development of Processes for the Production of SNG from Coal; IGE Communication 988 (1976)

10. R. B. Sharman and J. E. Scott : The British Gas Slagging Gasifier and its relevance to Power Generation – EPRI Conference on Synthetic Fuels, San Francisco (1980)

The Purification of Synthesis Gas

By S. P. S. Andrew
IMPERIAL CHEMICAL INDUSTRIES PLC, AGRICULTURAL DIVISION, PO BOX I, BILLINGHAM, CLEVELAND TS23 ILB, U.K.

The Need for Purification

The extent to which the composition of synthesis gas requires to be modified after it has been produced is very dependent on the production process for the synthesis gas and the use to which the synthesis gas is committed. Synthesis gas purification plant can thus vary in extent and cost from the non-existent, for instance when methanol is to be produced from gas generated by the steam reforming of naphtha, to the very extensive and expensive series of coupled processes required for the purification of synthesis gas produced by the partial oxidation of coal and required for synthesising ammonia.

Purification involves, when required, the removal of catalyst poisons, the removal of finely divided solids (dust) and the modification of the gas composition so as to enhance the concentration of valuable components and diminish the concentration of inert or valueless components. The situation with respect to the various components which may be present in the gas is shown in Table 1 for synthesis gas as devoted to either ammonia synthesis or the rather similar duties of methanol synthesis and Fischer-Tropsch synthesis for the production of paraffins and aromatics. Poisons such as sulphur (as SO_2 or H_2S or COS) and chlorine (as HCl) are invariably present when the gas has been produced by the partial oxidation of coal or heavy oil. They should never be present when the gas is generated by catalytic steam reforming of naphtha or lighter hydrocarbons, as these feedstocks must be pretreated to remove sulphur and chlorine prior to steam reforming, otherwise the nickel reforming catalyst is poisoned. Oxygen or any species containing the oxygen atom is a poison only when there is a requirement to catalytically activate the nitrogen atom - as for ammonia synthesis. A species such as HCN, produced during the partial oxidation of coal, is not as much a poison,

Table 1 : Gas Purification Requirements

		NH_3 SYNTHESIS	MeOH SYNTHESIS	FT SYNTHESIS
A T O M S	H	ALWAYS WANTED		
	Cℓ	POISON - NEVER WANTED (REMOVE)		
	S	POISON - NEVER WANTED (REMOVE)		
	C	NONE REQUIRED	MUCH REQUIRED	
	O	POISON REMOVED	MUCH REQUIRED	
	N	MUCH REQUIRED	NONE REQUIRED	
M O L E C U L E S	H_2O	POISON (REMOVE)	NONE REQUIRED	
	CO	POISON (REMOVE)	NONE REQUIRED	
	CO_2	POISON (REMOVE)	SOME REQUIRED	NONE REQUIRED
	CH_4	NONE	NONE REQUIRED	

but if allowed to pass into the reactor for synthesising methanol or Fischer-Tropsch products, would lead to the inclusion of undesirable impurities in the product. In no instance is water required passing into the synthesis reactor as its presence invariably makes the equilibrium of the synthesis reaction less favourable even if it does not poison the catalyst. Carbon is clearly not required for synthesising ammonia and, though present in great quantity in the gases produced by steam reforming or partial oxidation, is removed in great measure as carbon dioxide before the gases pass to the synthesis reactor. Excess carbon is also present in all gases made from coal and heavy oil and required for paraffin, aromatics and methanol synthesis and this excess is removed as carbon dioxide. In the case of methanol synthesis from naphtha, which roughly has the formula C_nH_{2n}, the stoichiometry is exactly correct for methanol synthesis from synthesis gas produced by steam reforming, as is evident from the simple equation:

$$C_nH_{2n} + nH_2O \rightarrow nCH_3.OH$$

Bearing in mind the requirements of the synthesis catalyst with respect to poisons and the synthesis reaction with respect to stoichiometry, it is the purpose of the process designer to assemble a suitable sequence of purification and gas composition modification steps so as to secure the desired feed gas composition to the synthesis catalyst at the minimum cost in terms of lost gas, capital and energy costs. He has a number of possibilities to choose from, these often being proprietary processes which he may incorporate into his flowsheet. In many cases these processes are treated by him as 'black boxes' into which he specifies the ingoing gas stream and the proprietors of

the processes specify the exit gas streams and the inputs of
electricity and heat and the capital cost of the equipment, its
overall physical dimensions and the requirements of such
replaceable materials as are required and the frequency of their
replacement and cost (eg catalysts, adsorbents, membranes,
solvents etc.). The overall purification plant designer is
normally only informed in a very limited manner of the design
data which are used for determining the contents of these 'black
boxes' because of commercial secrecy. For the same reason,
technical articles in the literature are usually uninformative on
the contents of these 'boxes' which are only described in a very
qualitative manner. The writer of a general paper on synthesis
gas purification therefore usually is forced merely to catalogue
a chain of 'black boxes'. In order to escape from such a dull
repetition, the present author will attempt to formulate the
general principles on which gas purification processes are based
using a number of rough, but he hopes serviceable, correlations
relating performance to simple physical properties. His hope
is that this approach, though it can have no hope in being
precise, is thought provoking both for the novice and the expert
in this field. Synthesis gas purification, insofar as it is
not concerned with the removal of particulate matter, has two
aspects - gas conversion, so as to convert unwanted components
to valuable components and to move poisonous atoms from
difficult to remove molecules to those which are easy to remove,
and molecular removal processes which abstract material from the
gas stream.

Gas Conversion

Water Gas Shift. Gas conversion processes are invariably heterogeneously catalysed. The most significant are listed in Table 2. The water gas shift reaction, $CO + H_2O \rightarrow CO_2 + H_2$, is a major step in the purification of synthesis gas for ammonia production as it enhances the quantity of hydrogen in the gas

Table 2 : Heterogeneously Catalysed Purification Reactions

WANTED	$CO + H_2O \rightarrow CO_2 + H_2$	Fe_3O_4 450°C	GOOD	EQUILIBRIUM LEVEL OF CO
UNWANTED	$CO_2 + H_2 \rightarrow CH_4$ $CO + H_2$	Cu 250°C		
WANTED	$CO + \tfrac{1}{2}O_2 \rightarrow CO_2$	Pt	ADEQUATE	SHORT CATALYST LIFE
UNWANTED	$H_2 + \tfrac{1}{2}O_2 \rightarrow H_2O$			
WANTED	$CO + 2H_2 \rightarrow CH_4$ $CO_2 + 3H_2$	Ni	–	NONE
UNWANTED	–			

stream, which is desirable, whilst diminishing the amount of the difficult to remove carbon monoxide poison for the ammonia synthesis catalyst. As with all purification processes, the selectivity of the process is most important; undesired reactions such as $CO + 3H_2 \rightarrow CH_4 + H_2O$ would, if they also occurred to a significant extent, be seriously detrimental to the economics of the process. For this reason the catalytically active species must be chosen amongst those metals which under the conditions of the reactions are surface oxidised and not surface carbided. Thus, Fe, Cu and Mo are selective catalysts whilst Ni, Co and Ru are not. The iron-containing catalyst operates as Fe_3O_4, the copper-containing catalyst has a chemisorbed layer of oxygen on its surface and the molybdenum catalyst is usually employed in sulphur-containing gases as MoS_2. The extent to which CO can be converted to CO_2 by this reaction is determined by the equilibrium of the reaction, conversion to CO_2 becoming more favourable as the temperature is reduced. Typical values for the fraction of the total CO + CO_2 unconverted to CO_2 as a function of temperature for an ammonia synthesis gas produced by steam reforming are shown in Figure 1. Such a gas has a very low sulphur content and the reaction is operated in two stages, a high-temperature stage (<u>circa</u> 450°C) using Fe_3O_4 catalyst which effects conversion down to an unconverted level of about 20%, followed by a low temperature stage (<u>circa</u> 250°C) using a Cu catalyst, where conversion is raised until only about 2% of the CO + CO_2 is unconverted to CO_2 leaving about 0.4% of CO in the gas stream.

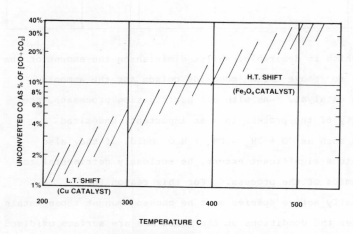

FIG. 1 **TYPICAL CONVERSIONS BY WATERGAS SHIFT IN AMMONIA PLANTS**

The Purification of Synthesis Gas

Selective Oxidation. Now the NH_3 synthesis gas catalyst is significantly reversibly poisoned in a short term by 100 ppm of O in the synthesis gas, and in the long term, if deterioration is to be reduced, O levels of greater than 1 ppm are to be avoided. Means must therefore be included in the purification system to reduce CO from 4000 ppm to 1 ppm. Two different heterogeneously catalysed processes are available to effect this. The first is the selective oxidation of CO to form CO_2, which is easier to remove than CO, the second is the methanation of CO to form methane and water by the reaction $CO + 3H_2 \rightarrow CH_4 + H_2O$. The water thus formed is more easily removed from the synthesis gas than is CO.

Supported platinum-based catalysts are normally used for promoting the selective oxidation reaction, the 'Selectoxo Process', just sufficient oxygen (as air) being added to effect the desired reaction without oxidising hydrogen to water.

Methanation. There are some doubts as to the long-term efficacy of the selective oxidation catalyst and its use is not nearly as popular as that of the methanation catalyst, a supported nickel, which is robust and where little can go wrong, so that the catalyst normally has many years of trouble-free life. It is worth mentioning here the problem which arises between the potential salesman of a novel 'black box' and the purification plant designer who is the potential purchaser. The latter is torn between the desire to obtain the professed virtues of the contents of this box and his worry of the disastrous results if they do not perform as claimed by the salesman. He therefore tries to reassure himself by enquiring how many of these 'black boxes' the salesman has already sold and for how many years have they been working. The salesman thus faces considerable resistance when the box is truly novel!

Molecular Removal Processes

Condensation. Removal processes are more varied than are the catalytic gas conversion processes. They can also be considerably greater in capital cost. Selectivity is always important and this can be secured either by utilisation of the different physical properties of the molecules in the synthesis gas, or by invoking, to a greater or lesser extent, their differing chemical reactivities. The most important physical property which is used as the basis of most physical removal processes is the condensibility of the various molecular species as indicated by their normal boiling points at 1 bar absolute pressure. These are listed in Table 3. Easily the most condensible is <u>water</u> and this molecule, quantitatively the most significant unwanted species in synthesis gas, is removed primarily by cooling the gas, condensing the water and removing it as liquid.

Clearly, with water condensation, water-soluble and readily condensible acid impurities such as HCN and SO_2, which are always in very low concentrations, will be trapped in the great quantity of water if they are present at that stage. Condensation at around ambient temperature is not, however, a satisfactory method of reducing water vapour levels to 1 ppm or less and other methods must be, in addition, employed when purifying synthesis for ammonia production. For, with a gas stream at 30 bar, cooling to 44°C gives an equilibrium water vapour content of 20 m bar or 3000 ppm. Even refrigeration to 0°C only reduces the water content to 200 ppm and below this temperature the condensed water becomes solid ice and hence difficult to remove from the condenser. The problem of ice and

The Purification of Synthesis Gas

Table 3 : Properties of Synthesis Gas Components

GAS	BOILING POINT	REMARKS
H_2	20 K	-
N_2	77	-
CO	81	-
Ar	87	-
CH_4	112	-
$(NO)_2$	122	mp 112 K
CO_2	⌐175⌐	SUBLIMES AT 195 K
HCl	188	ACID
H_2S	213	ACID
COS	223	-
NH_3	240	-
SO_2	263	ACID
HCN	299	ACID
H_2O	373	mp 273 K

of solid carbon dioxide and solid $(NO)_2$ must be overcome if low temperatures are to be employed for gas purification. It is, therefore, worthwhile to look in more detail, at this, at first glance, apparently the simplest method of removing all the higher boiling point components from a $3H_2 + N_2$ synthesis gas stream destined for the production of ammonia.

The Liquid Nitrogen Wash Process. Without doubt progressive cooling of the synthesis gas stream to around 90 K at a pressure of around 50 bar, with a progressive removal of higher-boiling component impurities produces the purest synthesis gas. It does not, however, selectively remove CO as its boiling point is too close to that of N_2. This disadvantage can, however, be bypassed if excess N_2 is introduced into the gas and the CO is condensed out of the $H_2 + N_2$ mixture along with the excess N_2. The process is known as liquid nitrogen wash and a much simplified flowsheet is shown in Figure 2. Before the unpurified synthesis gases enter the low-temperature process, water, carbon dioxide and $(NO)_2$ must all be removed otherwise, by freezing, they would in time block the plant.

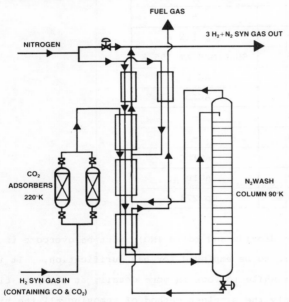

FIG. 2 **NITROGEN WASH PROCESS FOR AMMONIA SYNTHESIS GAS**
(50 BAR)

In the version of the nitrogen wash process shown, the synthesis gas contains no nitrogen prior to entering the cold part of the wash process and only mixes with pure nitrogen (coming from an air separation plant) in the cold wash column. The expansion of the added nitrogen from its injection pressure of 50 bar down to its partial pressure of 12.5 bar, due to dilution with the hydrogen, produces the Joule Thompson cooling which provides the ΔT required to drive the ingoing gas - outgoing gas heat exchange system which itself chills sufficient of the ingoing nitrogen to produce partial liquifaction at 90 K. This liquid nitrogen cascades down the wash column countercurrent to the ascending synthesis gas, condensing out methane and carbon monoxide whilst being partially vaporised so that, provided sufficient liquid flows out of the bottom of the column, all the methane and carbon monoxide are washed out with the excess nitrogen. Clearly this process can only be useful for the purification of ammonia synthesis gas as it injects a very large fraction of nitrogen into the gas stream being purified. Even for ammonia synthesis it is normally only employed when the requisite pure nitrogen stream can be readily obtained. Frequently this stream is withdrawn from the air separation plant required if the synthesis gas is produced by partial oxidation.

Gas Washing by Physical Solution of Impurities. The most common form of gas purification technique is, in effect, to perform a partial condensation of the gas stream by contacting it with a physical solvent. Such a technique is simple but only has a moderate selectivity and requires a great flow of solvent, as no good one exists.

The saturated solvent leaving the absorption tower is normally reduced in pressure, allowing much of the carbon dioxide to flash off (as it is probably saturated at about 5 bar CO_2 pressure). It is then passed to a desorption tower where it flows down countercurrent to an air flow which strips the remaining CO_2.

A rough correlation exists between the boiling point (at 1 bar) of a gas and its solubility in an ideal solvent. Solubilities of H_2, N_2, CO, CH_4, N_2O, H_2S and NH_3 in MeOH and in EtOH under 1 bar gas pressure at 25°C are shown in Figure 3, expressed as mol fraction in the solvent and plotted against the boiling point of the dissolving gas in K. It can be seen that the logarithm of the mol fraction is roughly proportional to the boiling point. Ths correlation and others derived from it will be used in many places in this paper to show the pattern of selectivity of physical processes depending essentially on condensation for separately removing different gases from the synthesis gas stream. Now it will be observed that the experimental line in Figure 3, when extrapolated to 298 K does not reach a mol fraction of unity (100% dissolved gas) but only

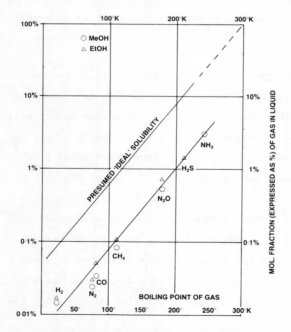

FIG. 3 **SOLUBILITY OF VARIOUS GASES UNDER 1 BAR PRESSURE AT 25°C IN MeOH & IN EtOH**

about 14%. I shall therefore presume that the 'ideal' solubility is a straight line parallel to the measured line but passing through 100% dissolving gas at 298 K. On this basis, Figure 4 can be drawn for 'ideal' solution of gases as a function of this boiling point for absorbing temperature of 250, 300 and 350 K. From this, one can immediately see that in the absence of selective chemical interaction between the dissolving gas and the solvent, the relative solubilities of different gases are solely determined by the difference in their boiling points - ie:

$$\log \left[\frac{\text{solubility of gas 1}}{\text{solubility of gas 2}} \right] \propto (T_{b1} - T_{b2})$$

where solubilities are expressed in mol fractions.

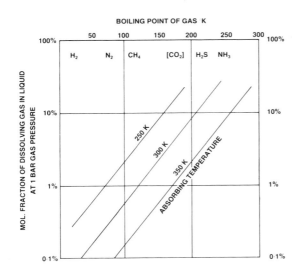

FIG. 4 **IDEAL SOLUTION OF GASES**

Along the boiling point scale of Figure 4 have been indicated the boiling points of H_2, N_2, CH_4, H_2S and NH_3. Also included is a 'boiling point' of about 175 K for CO_2. This figure must be obtained by extrapolating the behaviour of the liquid-vapour equilibrium for CO_2 to 1 bar pressure, as in reality CO_2 is solid at this pressure and sublimes at 195 K. The location of CO_2 is most important on this plot, as it is CO_2 which is the predominant species removed by 'physical solubility' dependent processes. From Figure 4 it can be seen that the selectivity for CO_2 removal by solution relative to that of H_2 is about a factor of 50 mols CO_2 per bar CO_2 relative to mols H_2 per bar H_2.

CO_2 Removal - Physical Solvents. Physical wash processes for removing CO_2 from synthesis gas are expensive and for this reason there has long been a search for better solvents. New ones emerge every few years and become the basis of licensed 'black boxes'. "Selexol", a dimethyl ether of polyethylene glycol is one of the more recent. From many points of view, water is the most convenient solvent being cheap, non-degrading, non-toxic and already present in the synthesis gas stream in considerable quantity. Unfortunately it is a very bad solvent having a dissolving capability only about 2% of the 'ideal' indicated in Figure 4. The problem is that the water molecules associate so strongly that foreign gas molecules find it difficult to gain access in the liquid phase. This very strong association is indicated in Figure 5 where the boiling points of the inert gases are related to their molecular weights, indicating that a molecule of molecular weight 18 should have a boiling point of 30°K in the absence of strong association. Water has a boiling point 13 times as great. Methanol is sometimes used instead of water as a solvent for CO_2 (the 'Rectisol' process). A

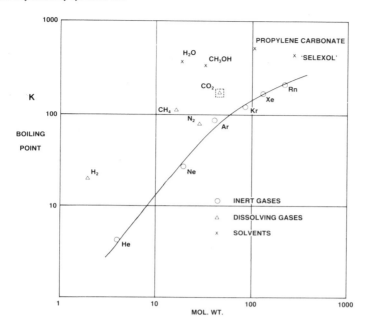

FIG. 5 **RELATION BETWEEN BOILING POINT & MOLECULAR WEIGHT**

molecule with a weight of 32 on Figure 5 would have a boiling point of 55 K if behaving like an inert gas; actually CH_3OH boils at about 6 times this figure. Methanol is thus somewhat more 'ideal' than water and reference to its measured capability for dissolving CO_2 shows it to dissolve about 12% of the 'ideal'. A more expensive solvent, propylene carbonate, sometimes employed for CO_2 removal, has a molecular weight of 102 (inert gas type boiling point 150 K), whereas its real boiling point is about 3-4 times this value. Measured CO_2 solubility is some 35% of the 'ideal'. 'Selexol' with a boiling point of 460 K, or about 1.9 times the boiling point corresponding to its molecular weight of 280, behaves virtually as an 'ideal' solvent for CO_2 comparing with the prediction of Figure 4. However, 'ideal' as seen by the physical chemist may not seem much of an improvement to the engineer who must pump the solvent through the absorber. For, as can be seen in Table 4, most of the improvement in attaining closer to 'ideal' solubilities has been offset by increasing molar volume and on the basis of CO_2 absorption per unit volume of solvent, there is little to choose between methanol, propylene carbonate and 'Selexol'.

CO_2 Removal by Adsorbents. A second form of disguised 'condensation' is the use of physical adsorbents such as active charcoal, high-area alumina or silica gel or molecular sieves. To a great extent, these materials condense liquid gases by reducing their vapour pressure as a result of the very high negative hydraulic pressures exerted by surface tension

Table 4 : Relative Capacity for CO_2 Uptake Per Bar for Various Physical Solvents

	SOLVENTS			
	H_2O	MeOH	PROPYLENE CARBONATE	'SELEXOL'
CO_2 SOLUBILITY RELATIVE TO IDEAL	2%	12%	35%	100%
SOLVENT MOLECULAR WEIGHT	18	30	102	280
RELATIVE CO_2 CAPACITY PER GRAM BAR	1	3.6	3.1	3.3
RELATIVE CO_2 CAPACITY PER cm^3 BAR	1	2.8	3.7	3.3

of the liquid in very fine capillaries. Provided the liquid gases 'wet' the adsorbent solid then the behaviour of the adsorbent with respect to a range of adsorbing gases can roughly be correlated by a diagram derived from Figure 4 and shown as

Figure 6. The vertical scale of Figure 6, though not numbered, is, in fact, the same as the vertical scale of Figure 4 and should be interpreted as the activity of the adsorbing gas reduced by adsorption at a given fractional filling of the pores in the adsorbent. Thus, with complete pore filling, the activity of the adsorbing gas is 100% of the activity of non-adsorbed liquid gas (corresponding to 100% mol fraction in Figure 4), the activity falling as fractional pore filling is reduced. With most adsorbents there is a wide range of pore sizes with a consequence that the pore filling versus gas activity curve on the right of Figure 6 is as shown for activated charcoal. With molecular sieves, which have a narrow distribution of very small pores, the curve is much sharper and moved towards the lower activities as shown.

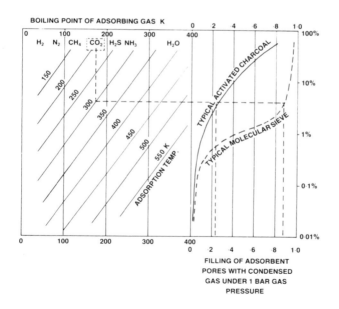

FIG. 6 **IDEALISED ADSORBENT BEHAVIOUR**

If, for instance, it was desired to remove CO_2 by adsorption at ambient temperature (300 K) then Figure 6 shows that an activated charcoal would be of little use compared with pore filling, whereas the former, little over 20%. The use of adsorption for this duty has therefore only appeared since molecular sieves became a prominent article of commerce (eg the 'Polybed PSA' process). As well as removing CO_2, molecular sieve adsorbents will simultaneously remove CO_2, CO and CH_4 from a hydrogen stream. A convenient method of operation, when the gas stream to be treated is at an elevated pressure (20-50 bar), is to employ depressurisation of the adsorbent bed to effect regeneration. By this means, adiabatic operation is secured, the adsorbent beds cycling between adsorbing at synthesis gas pressure and desorbing at atmospheric pressure.

Gas Purification by Membrane Diffusion. The similarity between condensation and purification of gases by the selective permeation through a membrane may not at first be apparent. Nevertheless, there is a considerable parallelism both with adsorption and physical solution, for easy passage of the gas through the membrane is well correlated with the product of its solubility in the membrane multiplied by the diffusion constant of the dissolved gas in the membrane. The chemical interrelation between the gas and the surface of the membrane, described by the simple physical term of 'wetting' as in adsorbents, is a crucial factor in determining the permeability of a given membrane to any individual gas. (The same is true of adsorbents where high-area small-pore adsorbents may be found which refuse to condense certain species due to failure to wet.)

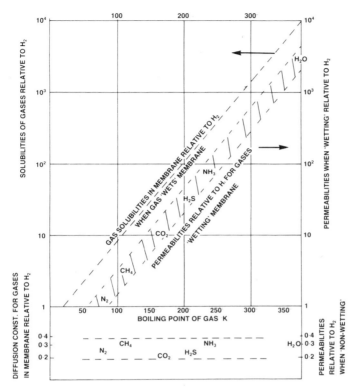

FIG. 7 **DIFFUSION CONSTANTS, SOLUBILITIES & GASEOUS PERMEABILITIES OF 'WETTING' & 'NON-WETTING' GASES THROUGH MEMBRANE TAKING H_2 AS UNITY**

Observation shows that high solubility is a result of membrane pore wetting, with the result that the relative solubility in a membrane wetted by the relevant gases generally follows the same relation of solubility versus gas boiling point shown in Figures 3 and 4. The relation has been replotted in Figure 7 showing the solubilities of various gases relative to H_2, which is taken as unity. It will be seen that CO_2 is some 50 times as soluble as hydrogen, and water some 3000 times as soluble. Assuming that the relative diffusion constants of the different gases are inversely proportional to the square root of their molecular weights, then the diffusion constant for CO_2 is 21% of that for H_2, and for H_2O is 33% of that for H_2. Multiplying relative solubility by relative diffusion constant gives relative permeability. When wetting, CO_2 is thus 10 times more permeable than H_2, and H_2O is 1000 times greater than H_2. However, there is a great range of relative permeabilities which lie below what the above theory suggests are upper bounds. When the gases do not wet the membrane, presumably all the solubilities are about equal, so that the permeabilities relative to H_2 of CO_2 and H_2O are 0.21 and 0.33 respectively. Incomplete or partial wetting presumably produces intermediate values. The above, it must be understood by my readers, is only a very rough and simple theory, designed to demonstrate the basic similarity of the phenomena involved with those of other removal processes; nevertheless, it does suggest that the selectivity for a membrane separation process based on physical solubility and wetting should be less than that for a simple physical solution process, as the selectivity is partially offset by the molecular weight effect on diffusion constant. In practice, the relatively poor selectivity of membrane separation processes has confined their use to recovery of hydrogen from purge gases ('Prism') and they have not so far found a place in synthesis gas purification.

CO_2 Removal by a Chemical Absorbent Solution. Recognising the poor absorptive capacity for physical solvents for carbon dioxide (and also for H_2S for the same reasons), chemical reagents have been added to solvents to enhance the pick up.

Choice of chemical reagent and operating conditions is dictated by the necessity both to be able to saturate the chemical reagent with CO_2 in the absorber and to be able to regenerate the reagent by desorbing the CO_2 in the regenerator. The adsorption reaction must therefore be readily reversible, preferably by merely dropping the CO_2 partial pressure. If such can be secured, then the absorber-regenerator system operates adiabatically, the heat liberated by reaction during absorption being stored in the solution as sensible heat and utilised to supply the heat of decomposition in the regenerator.

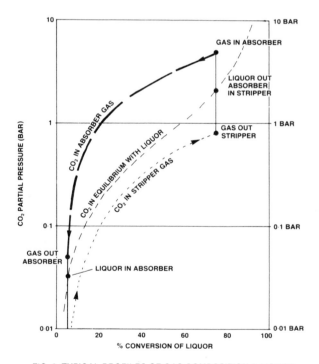

FIG. 8 **TYPICAL PROFILES OF GAS COMPOSITION & LIQUOR CONVERSION IN AN ADIABATIC ABSORBER-STRIPPER SYSTEM EMPLOYING A CHEMICAL REAGENT SOLUTION**

Diagrammatically, the partial pressure of CO_2 in the liquor would move up and down the centre line in Figure 8 as the liquor carbonated and regenerated, moving from, say, 5% conversion to 75% conversion and then back again. The CO_2 partial pressure in a typical absorber might well be about 5 bar at the inlet and 0.05 bar at the outlet, whilst stripping air in the regenerator entered with no CO_2 and left with 0.8 bar. The reagent, at 50% conversion would therefore have to exert an equilibrium CO_2 partial pressure of about 0.7 bar.

Specifying the desired equilibrium characteristics of the reagent is thus easy. Finding the reagent is more difficult. There is so far only one adiabatic process of this type, that based on the use of the alkaline carbonate to bicarbonate transition (known as the 'Benfield' process). Although adiabatic in operation, it employs a considerable amount of steam for regeneration and therefore has a large heat demand. This apparent paradox arises from the above mentioned requirement of the reagent having to exert about 0.7 bar CO_2 pressure at 50% conversion. To secure this it is necessary to operate with the <u>aqueous</u> alkaline carbonate-bicarbonate liquor at about 100°C. Of itself, this does not consume heat, but the only possible stripping gas which would not abstract a great quantity of heat from the liquor in the regenerator is steam!

The other range of reagents used for CO_2 and H_2S removal are the organic alcoholic amines, typically monoethanolamine, diethanolamine and triethanolamine. These are normally employed in aqueous solution, though one process uses an organic solvent 'Sulpholane' and di-isopropanolamine. When used in aqueous solution, these amines both react with CO_2 to form amine carbamates (in the case of primary and secondary amines) and also they form the amine H^+ ion plus HCO_3^-. The former reaction is very much faster than the latter, due to the much greater concentration of free amine than OH^-. A simple interpretation of the carbon dioxide uptake is therefore to assume that effectively for rapid absorption of CO_2, complete conversion of the amine occurs when the reaction

$$CO_2 + 2R_1R_2NH \rightarrow R_1R_2NCOO^- + R_1R_2NH_2^+$$

is complete. As with the $CO_3^= + CO_2 + H_2O \rightarrow 2HCO_3^-$ system, discussed above, the equilibrium of the absorption reaction is too far over towards the right to enable regeneration except at temperatures of above 100°C with MEA and DEA. The attempt to find reagents having both the desired equilibrium properties of 0.7 bar CO_2 partial pressure at 50% conversion and, say, 25°C as well as a desired rapid reaction rate with CO_2, is foiled by the general trend in which high kinetic constants are associated with low partial pressures for a range of similar compounds in which equilibrium and kinetic properties are altered, for instance, by changing R_1 and R_2 in R_1R_2NH. This relation is shown roughly in Figure 9. To obtain the correct CO_2 partial pressure from MEA requires 100°C, for DEA 80°C and for TEA 25°C corresponding to kinetic constants of 2×10^5 and 3×10^4 and 8×10^2 respectively.

The steam requirement for providing the stripping gas in the hot regenerator is the chief disadvantage of the chemical absorbent systems available compared with the physical absorption solvents. One advantage that they do possess, however, is markedly enhanced selectivity with respect to loss of H_2 in the CO_2 stripped as, though H_2 is lost through dissolution in the solvent, none is lost by reaction with the

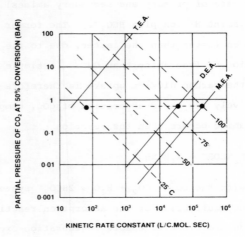

FIG. 9 **GENERAL TREND SHOWING DIMINISHING RATE CONSTANT FOR CO_2 UPTAKE WITH DIMINISHING EQUILIBRIUM CO_2 UPTAKE OVER A RANGE OF AMINES**

reagent. Selectivities some ten times greater than that for physical systems are thus attainable.

<u>The Liquid Ammonia Wash</u>. As was indicated above, simple water condensation is not capable of reducing the water content of synthesis gas for ammonia production to the desired low level. In the absence of a liquid nitrogen wash system, complete with adsorbers for residual water and CO_2 removal, adsorption by itself can be used or a liquid ammonia wash may be used. In this process the product liquid ammonia is contacted with the synthesis gas before the latter is passed to the synthesis converter. A convenient way to arrange this is to feed the synthesis gas into the circulating gases leaving the converter before these enter cooling and refrigeration. The condensation of ammonia then effectively absorbs both residual water and

residual CO_2 which are removed in the liquid ammonia product stream, the circulating gases then returning to the inlet of the converter.

Postscript

The writer is very conscious that in this brief article he has sacrificed precision and the detail which is inherent in all physicochemical phenomena as well as in all practical process flowsheets. His excuse is that if one is to stand at a sufficient vantage point to see the whole wood then it is not possible simultaneously to pick out the details of each of the trees.

Transition-metal Peroxides as Reactive Intermediates in Heterolytic and Homolytic Liquid-phase Catalytic Oxidations

By H. Minoun
LABORATOIRE D'OXYDATION, INSTITUT FRANÇAIS DU PÉTROLE,
92506-RUEIL-MALMAISON, FRANCE

1. Introduction.

Transition-metal complexes have been extensively used industrially as catalysts for the selective oxidation of hydrocarbons.[1] According to the rationalization made by Sheldon, catalytic oxidations may be divided into two types, designated as homolytic and heterolytic.[2]

a) Homolytic oxidations involve single electron processes in which free radicals are intermediates. They have been widely used in industry, e.g. for the large-scale manufacture of cyclohexanol-one mixture from cyclohexane (for adipic acid), benzoic or terephtalic acid from toluene or p-xylene, and acetic acid from butane. These processes use first-row transition metals characterized by single election change in their oxidation state, e.g. Co(II)/Co(III), Mn(II)/Mn(III), Cu(I)/Cu(II), and usually require high temperature conditions. One of the major role of these transition metals is to achieve the homolytic decomposition of the alkylhydroperoxide autoxidation product, presumably via the formation of unstable peroxidic intermediates, e.g. for cobalt (Eqs.1-3)

$$Co^{II} + ROOH \longrightarrow Co(ROOH) \longrightarrow Co^{III}OH + RO° \qquad (1)$$
$$Co^{III} + ROOH \longrightarrow Co^{II} + ROO° + H+ \qquad (2)$$
$$2\ ROOH \xrightarrow{Co^{II}/Co^{III}} RO_2° + RO° + H_2O \qquad (3)$$

It is noteworthy that homolytic oxidations are particularly well suited for obtaining products which cannot be further oxidized, such as carboxylic acids. First-generation oxygenated products, e.g. alcohols, ketones, or epoxides, are generally more oxidizable than the starting hydrocarbons and can be obtained only at very low conversion.

b) Heterolytic catalytic oxidations of hydrocarbons involve non-radical oxygen transfer pathways occurring within the coordination sphere of the metal and are particularly used for olefins. Oxygenated products generally result from the nucleophilic attack of the coordinated olefin by peroxidic or hydroxide moieties. These oxidation reactions use, as catalysts, early d° metals such as Ti^{IV}, V^V, Mo^{VI}, W^{VI} for the selective epoxidation of olefins by H_2O_2 or alkylhydroperoxides, or group VIII transition metals such as Rh,Ir,Pd,Pt for the oxidation of olefins to ketones by O_2, ROOH or H_2O_2.[3]

Industrial applications involve the Halcon Process for the epoxidation of propylene by tert-butyl or ethylbenzene hydroperoxide catalyzed by molybdenum and the Wacker oxidation of ethylene to acetaldehyde by O_2 catalyzed by $PdCl_2$- $CuCl_2$.

Heterolytic oxidations are generally more selective and stereospecific than homolytic ones and operate under mild conditions and with high conversion of the substrate.

2. Transition-Metal Peroxides.

Two categories of peroxidic complexes more frequently intervene in catalytic oxidations[4], i.e. triangular peroxo complexes **1** and alkyl or hydroperoxo complexes which may have a linear **2**, bridged **3**, or triangular **4** structure.

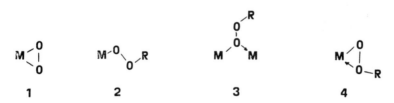

    1   **2**   **3**   **4**

Peroxo complexes can be obtained from the reaction of a reduced two-electron donor group VIII metal complex with molecular oxygen (Eq.4), or from the reaction of hydrogen peroxide with a high-valent group IV, V and VI metal-oxo compound (Eq.5).

Hydro- or alkyl-peroxo complexes result from the reaction of a metal salt or complex with H_2O_2 or ROOH (Eq.6), from the insertion of O_2 between a metal-hydrogen or a metal-carbon bond (Eq.7), or from the protonation or alkylation of a peroxo metal complex (Eq.8).

The way in which transition-metal peroxides release oxygen to hydrocarbons is therefore particularly relevant to catalytic oxidations using O_2, H_2O_2 or ROOH as the oxygen source.

$$M^n + O_2 \xrightarrow{\text{Group 8}} M^{n+2}\overset{O}{\underset{O}{|}} \quad (4)$$

$$M^{n+2}=O + H_2O_2 \longrightarrow M\underset{\underset{H}{O-H}}{\overset{O}{\diagdown}}O \xrightarrow[-H_2O]{\text{Groups 4,5,6}} \mathbf{1} \quad (5)$$

$$\underset{(R)}{MX} + HOOH \xrightarrow{-HX} \quad (6)$$

$$\underset{(R)}{MH} + O_2 \longrightarrow M\overset{O}{\diagdown}\underset{(R)}{O-H} \quad (7)$$

$$M\overset{O}{\underset{O}{|}} + \underset{(R^+)}{H^+} \quad (8)$$

3. Heterolytic Catalytic Oxidations.

Table I lists some of the most important heterolytic catalytic oxidations of olefins by O_2, H_2O_2 and ROOH using d° metal (Ti,V,Mo,W) and group VIII metal (Pd,Rh) catalysts together with the relevant stoichiometric peroxidic reagents.

Since this subject has been previously reviewed[3-6], we will only summarize the most important characteristics of these reactions:

a) The complexation of the olefin on the metal generally represents the rate-determining step of these reactions: i) The reactivity of olefins is proportional to their complexation constant on the metal. For d° metals, the olefin-metal interaction is a pure Lewis base-Lewis acid bond and the reactivity of olefins increases with their nucleophilic nature in the order: tetrasubstituted > trisubstituted > disubstituted > monosubstituted. For group VIII metals, the olefin-metal interaction obeys the Dewar-Chatt-

Table 1 Heterolytic oxidations

Catalytic oxidation		Relevant reagent
$\searrow\!\!=\!\!\swarrow + H_2O_2 \xrightarrow{(Mo,W)} \searrow\!\!\underset{O}{\diagup}\!\!\swarrow + H_2O$	(9)	$\begin{bmatrix} O=M(O)(O)(O)(L)(L') \end{bmatrix}$
$\searrow\!\!=\!\!\swarrow + ROOH \xrightarrow{(Mo,V,Ti)} \searrow\!\!\underset{O}{\diagup}\!\!\swarrow + ROH$	(10)	$d^0 \begin{bmatrix} M(O)(O)(O\text{-}R) \end{bmatrix}$
$R\!\!-\!\!CH\!\!=\!\!CH_2 + \tfrac{1}{2}O_2 \xrightarrow{(Rh)} R\!-\!\underset{\parallel\,O}{C}\!-\!Me$	(11)	$\begin{bmatrix} L_4Rh\text{-}O\text{-}O \end{bmatrix}^+$
$R\!\!-\!\!CH\!\!=\!\!CH_2 + \,{>}\!\!\!-\!OOH \xrightarrow{(Pd)} R\!-\!\underset{\parallel\,O}{C}\!-\!Me + \,{>}\!\!\!-\!OH$	(12)	$\begin{bmatrix} RCO_2\,Pd\text{-}OO\text{-}{\!\!<} \end{bmatrix}_4$
$R\!\!-\!\!CH\!\!=\!\!CH_2 + H_2O_2 \xrightarrow{(Pd)} R\!-\!\underset{\parallel\,O}{C}\!-\!Me + H_2O$	(13)	$\begin{bmatrix} L_2\,Pd\text{-}OOH \end{bmatrix}^+$

Ducanson bonding model[7] with a significant back-donation from the metal to the empty alkene $\pi*$ orbitals. The most reactive olefins are the terminal ones.[3,8] Internal olefins are either not reactive or form stable π-allylic compounds.[9] ii) These oxidation reactions are inhibited by ligands or solvents which prevent the complexation of the olefin on the metal. iii) Reactive peroxidic complexes are those having vacant or releaseable coordination sites, preferentially adjacent and coplanar to the peroxidic moiety. For example $MoO(O_2)_2HMPT$ and $[Rh(O_2)(AsPh_3)_4]^+ ClO_4^-$ are effective epoxidation and ketonization reagents, respectively, while $(dipic)Mo(O_2)$,HMPT (dipic = pyridin-2,6-dicarboxylate) and $[Rh(O_2)(Ph_2P.CH_2PPh_2)]^+ClO_4^-$ having no releaseable coordination sites are inactive.

b) These reactions are highly selective and stereospecific. Reactions 9 and 10 selectively produce cis-epoxides from cis-olefins and trans-epoxides from trans-olefins. Epoxidation of allylic or homoallylic alcohols using Reaction 10 selectively occurs syn to the hydroxyl group.[9] Using $Ti(OR)_4$-diethyltartrate as catalyst, this epoxidation reaction becomes asymmetric and produces chiral epoxy alcohols with enantiomeric excesses consistently greater than 90%.[10]

Ketonization of terminal olefins using Reactions 11,12 and 13 also occurs with a very high selectivity (up to 98%) and a high conversion of the substrate.[5] This reactivity, however, is particularly used for terminal olefins.

A common peroxymetallation mechanism depicted in Scheme 1 satisfactorily accounts for most of the data concerning Reactions 9-13.

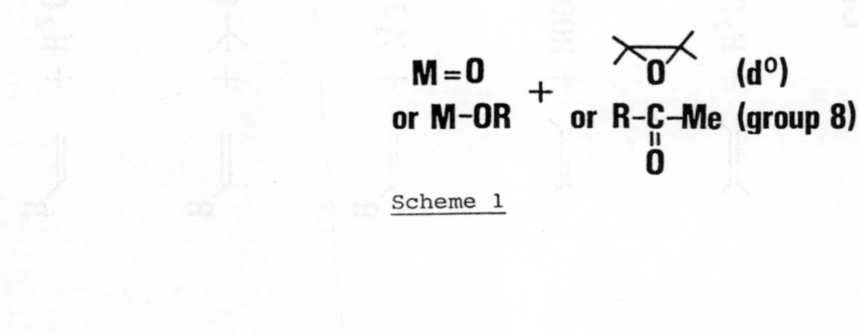

$$M=O \text{ or } M-OR + \text{epoxide} \text{ or } R-\underset{O}{\underset{\|}{C}}-Me \quad (d^0) \text{ (group 8)} \tag{14}$$

Scheme 1

The first step consists of the complexation of the metal followed by its insertion between the metal-oxygen bond (intramolecular 1,3-dipolar cycloaddition) forming a cyclic five-membered dioxametallacycle if the initial complex has a peroxo structure, or a pseudocyclic alkyldioxyethylmetal compound if the initial complex has an alkylperoxo or an hydroperoxo structure. The decomposition of the cyclic or pseudocyclic intermediate **5** occurs via a concerted 1,3-dipolar cycloreversion mechanism, resulting in the formation of an epoxide if the metal is d°, or a methyl ketone if the metal belongs to group VIII, with a β-hydrogen shift. The catalytic cycle is achieved by the regeneration of the initial peroxo compound from the reaction of the final oxo species with H_2O_2, or of the initial alkyl or hydroperoxo compound from the reaction of the final alkoxo or hydroxo species with ROOH or H_2O_2.

It is noteworthy that the peroxymetallation mechanism agrees with the general $\pi-\sigma$ rearrangement procedure occurring in most homogeneous catalytic transformations of olefins.[11]

4. Homolytic Oxidations

Table II lists some catalytic hydroxylations of hydrocarbons by H_2O_2 and ROOH in the presence of first-row transition metals (V,Cr,Co) and the relevant stoichiometric peroxidic reagent.

The use of V_2O_5 or CrO_3 as catalysts for the hydroxylation of arenes by H_2O_2 was first observed by Milas in 1937.[12] This reaction was attributed to the intermediate formation of reactive "red pervanadic" and "blue perchromic" acids, respectively.

4.1 Vanadium Catalysts

Vanadium compounds have been widely used as catalysts for the oxidation of olefins by H_2O_2 and ROOH.[2] They are generally less efficient and selective catalysts than molybdenum compounds for the epoxidation of unactivated olefins,[13] but are highly active and stereoselective for allylic and homoallylic alcohols.[9,14]

We recently synthesized stable covalent vanadium (V) peroxo complexes[15] such as $(Pic)VO(O_2),2H_2O$ **6** (Pic = pyridine-2-carboxylate) and vanadium (V) alkylperoxidic complexes [16] such as $(Dipic)VO(OOR),L$ **7** [Dipic=pyridine-2,6-dicarboxylate, L=H_2O, HMPT]; both of which are characterized by an X-ray crystal structure.

Table 2 Homolytic oxidations

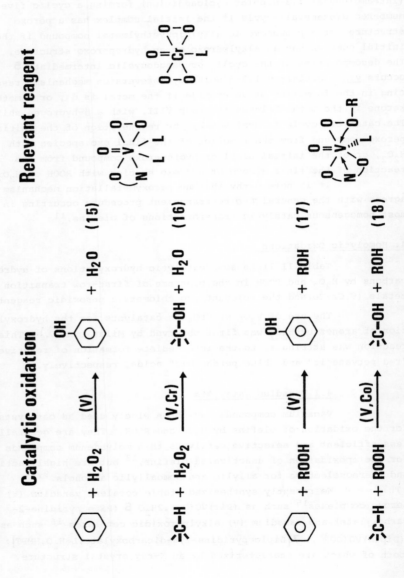

6   7

The red peroxo complex **6** resulted from the reaction of V_2O_5 with picolinic acid in the presence of H_2O_2. This complex is stable in the solid state and in a protic solution (H_2O, MeOH) but rapidly decomposes in a aprotic solution (e.g. CH_3CN) with the evolution of molecular oxygen.

The yellow complex **7** (L=H_2O) was obtained from the reaction of tert-butyl or cumyl hydroperoxide with an aqueous equimolar solution of V_2O_5 and pyridine-2,6- dicarboxylic acid. This complex is the first example of a d° metal alkylperoxidic complex and has a characteristic O,O-triangularly bonded alkylperoxidic structure which is presumably a common characteristic of d° metal alkylperoxidic intermediates involved in the catalytic epoxidation of olefins in Equation 10.

The reactivity of vanadium peroxo and alkylperoxidic complexes is very different from that of the molybdenum analogues.

a) Olefins are epoxidized in a non-stereoselective way. For example cis-2-butene gave a cis + trans epoxide mixture. Furthermore, the reaction is not selective and results in the formation of cleavage and allylic oxidation by-products.

b) Arenes are hydroxylated to phenolic products. While **6** mainly hydroxylates toluene at the ring positions, **7** gives a mixture of phenolic and benzylic oxidation products. The nuclear hydroxylation of toluene occurs with a high NIH shift[17] value (70% for **6** and 62% for **7**) which suggests the transient formation of arene oxide.[17]

c) Alkanes are less readily hydroxylated than arenes. This reaction occurs with a relatively low isotope effect (k_H/k_D= 2.8) and a significant amount of epimerization at the hydroxylated carbon atom. Radical carbon intermediates were revealed in this reaction by trapping experiments with chlorine atoms coming from CCl_4.[15,16]

d) No or little inhibition by basic ligands is observed.

e) In the presence of excess H_2O_2 (Eqs.15-16) or ROOH

(Eqs.17-18), the hydroxylation of alkanes and arenes becomes catalytic.[16,18] The alcohol-ketone or phenol yields (based on H_2O_2) are in the range of 15-20%. Parallel decomposition of H_2O_2 or ROOH occurs together with a consecutive oxidation of the hydroxylated products.

A plausible mechanism for oxygen transfer from vanadium (V) peroxo and alkylperoxidic complexes to olefins, arenes and alkanes is depicted in Scheme 2.

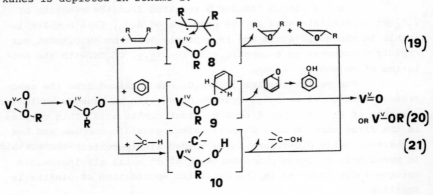

Scheme 2

The homolytic cleavage of V(V) peroxo and alkylperoxidic species results in the formation of biradical $V^{IV}O-O°$ and $V^{IV}OR-O°$ species. These biradical species can:

a) add to double bonds to give a freely rotating free radical intermediate **8** which homolytically decomposes to a <u>cis</u> + <u>trans</u> epoxide mixture and V(V) oxo or alkoxo complex (Eq.19);

b) add to aromatic rings to give a radical intermediate **9** which decomposes to phenol via arene oxide (Eq.20);

c) abstract a hydrogen atom from alkanes to give an intermediate carbon radical **10**, followed by a recombination with the hydroxyl radical to give the alcohol and the V(V) oxo or alkoxo complex.

Concerning the oxidation of olefins (Eq.19), it is noteworthy that the diradical intermediate **8** corresponds to the homolytic cleavage of the metal-carbon bond of the dioxametallocyclic intermediate **5** involved in heterolytic oxidations (Scheme 1). The absence of strained intermediates in homolytic oxidations is probable the cause of the absence of selectivity of these reactions.

4.2. Chromium Catalysts.

Although blue chromium (VI) peroxo complexes $CrO(O_2)_2, L$ were prepared a long time ago,[19] their use as oxidizing reagents has been limited to the oxidation of alcohols[20] and tetracyclone.[21]

We have recently observed that chromium (VI) peroxo complexes such as $CrO(O_2)_2, Ph_3PO$ **11** are able to hydroxylate alkanes[22] Used in CH_2Cl_2-t-BuOH solution under N_2 at room temperature, complex **11** oxidizes cyclohexane to a cyclohexanol-cyclohexanone mixture (9% and 4.5% yield, respectively, based on Cr) and adamantane to 1-adamantanol (8.5%), 2-adamantanol (3%) and adamantanone (3%). Cis-decaline gave cis-9-decalol (2%), trans-9-decalol (3.5%) and 1-and 2-decalone (4.1% and 5%). This important epimerization at the C_9 hydroxylated position suggests a radical carbon intermediate.

Olefins are almost not epoxidized but rather react at the allylic positions. Cyclohexene mainly gave cyclohex-1-one-3 (98%) and cyclohex-1-ol-3 (8%). No epoxide formation was detected from the oxidation of cis or trans 2-butene. Arenes are almost not hydroxylated at the ring positions. No phenol was obtained from benzene, and toluene mainly gave benzylic oxidation products.

It was further shown that Cr(VI) oxo complexes are not the active species since the complex CrO_3, Ph_3PO does not (or very slowly) hydroxylate alkanes under the same conditions.

Although preliminary, these results suggest a radical hydroxylation mechanism, with the active species being the biradical Cr^V-O-O° resulting from the homolytic cleavage of the peroxo moiety. However, in contrast to vanadium, chromium (VI) peroxo complexes do not add to double bonds or arenes, but can only abstract hydrogen atoms from alkanes (Scheme 3).

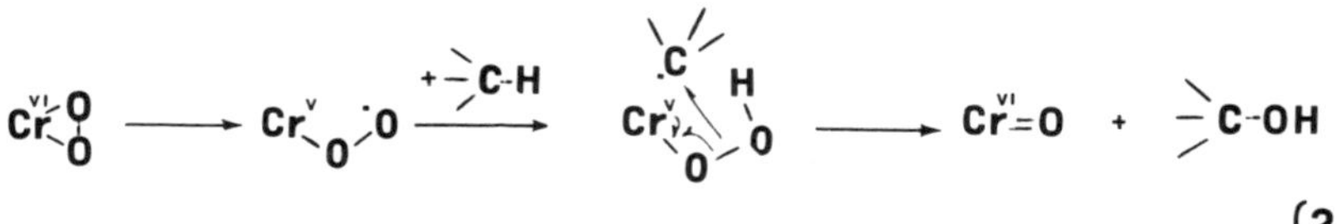

$$Cr^{VI}\underset{O}{\overset{O}{\lessgtr}} \longrightarrow Cr^V\underset{O}{\overset{}{\lessgtr}}O \xrightarrow{+\overset{|}{-}C-H} Cr^V\underset{O}{\overset{\overset{|}{C}\diagdown H}{\lessgtr}}O \longrightarrow Cr^{VI}=O + \overset{|}{-}C-OH$$

(22)

Scheme 3

5. Conclusion.

Transition-metal peroxides therefore, appear as the key reactive intermediates in catalytic oxidation reactions. The heterolytic or homolytic nature of catalytic oxidation seems to be strongly governed by the heterolytic or homolytic dissociation mode of the peroxidic moiety. Heterolytic oxidations require releasable coordination sites on the metal, involve strained metallocyclic reaction intermediates, and are highly selective.

In contrast, homolytic oxidations involve bimolecular radical processes with no metal-substrate association and are less selective. Such homolytic processes are also involved in enzymic hydroxylases which also use first-row transition metals as active centers (e.g. Fe for P_{450}).[23] Further work is needed to control the selectivity of these homolytic hydroxylation processes. This would help for improving the yields of homolytic industrial processes and for understanding the behavior of naturally occurring enzymic oxidations.

REFERENCES

1. a) G. Franz, Ullmanns Encyclopadie der Technishen Chemie, Verlag chemie, Weinheim, 1979, vol 17 p.483
   b) J.E. Lyons, Hydroc.Process., 1980,107

2. R.A. Sheldon and J.K. Kochi, Metal-Catalyzed Oxidation of Organic Compounds, Academic Press, New-York, 1981

3. H.Mimoun, J.Mol.Cat, 1980,7,1

4. H. Mimoun, the Chemistry of Functional Group, Peroxides, S. Patai ed.,John Wiley and Sons Ltd, London, 1982 p.463

5. H. Mimoun, Pure & Appl.Chem., 1981,53, 2389

6. H. Mimoun, Angew.Chem. Int.Ed.Engl, 1982, 21, 734

7. D.P.M. Mingos, Comprehensive Organometallic Chemistry, Pergamon Press, Oxford, 1982, G. Wilkinson, Ed.,vol.3 p.1

8. H. Mimoun, R. Charpentier, J.Fischer, A.Mitschler and R.Weiss, J.Am.Chem.Soc., 1980, 102, 1047

9. K.B. Sharpless and T.R. Verhoeven, Aldrichimica Acta, 1979,12 63

10. T. Katsuki and K.B. Sharpless, J.Am.Chem.Soc.,1980,102,5976

11. M. Tsutsui and A. Courtney, Adv.Organomet.Chem., 1977,16, 241
12. N.A Milas, J.Am.Chem.Soc., 1937,50,2342
13. C.C.Su, J.W. Reed and E.S Gould, Inorg.Chem.,1973,12, 337
14. T. Itoh,K.Jitsukawa, K.Kaneda and S. Teranishi, J.Am.Chem.Soc., 1979, 101,159
15. H.Mimoun, L.Saussine, E.Daire, M.Postel, J.Fischer and R.Weiss J.Am.Chem.Soc,1983, 105, 3101
16. H.Mimoun, P.Chaumette, M.Mignard, L.Saussine, J.Fischer and R.Weiss, Nouv. J.Chim. (in press)
17. D.M. Jerina, Chem.Tech., 1973,120 and references therein
18. H.Mimoun, L.Saussine, E.Daire, A.Robine and J.Guibourd, Fr. Demande, 1982,82/10938
19. O.F.Wiede, Ber.,1897,30, 2178
20. G.W.J. Fleet and W.Little, Tetrahedron Lett., 1977, 42,3749
21. J.E. Baldwin, J.C. Swallow and H.W.S.Chan, J.Chem.Soc.Chem. Commun.,1971,1407
22. E.Daire, A.Pecqueur and H.Mimoun, Manuscript in preparation.
23. J.T. Groves,Metal Ion Activation of Dioxygen, Wiley, New York T.G. Spiro, editor,1980 p.125 and references therein.

Heterogeneous Catalytic Oxidation: A Review of Principles and Practice

By W. J. Thomas
SCHOOL OF CHEMICAL ENGINEERING, UNIVERSITY OF BATH, CLAVERTON DOWN, BATH BA2 7AY, U.K.

1. Introduction

It is the intention in this review to confine attention to the heterogeneous catalytic oxidation of vaporised or gaseous feedstocks. The subject matter covers a resumé of the chemical principles describing oxygen transfer between reactant, catalyst and product, chemical kinetics and the design of hardware for the many industrial processes which have emerged with the passage of time.

Partial oxidation processes using air or oxygen are ubiquitous in practice. Many high tonnage, low cost petrochemical derivatives are manufactured by the partial catalytic oxidation of alkanes, alkenes and aromatics using metal oxide catalysts. For example, oxygen can be directly incorporated into a reactant hydrocarbon to yield an oxygen-containing final product such as in the oxidation of propylene to acrolein or o-xylene to phthalic anhydride. Oxidative dehydrogenation and ammoxidation are also catalytic processes in which oxygen, although a necessary reactant, does not appear in the final product. The reaction of butylene with air to form butadiene and the formation of acrylonitrile from a mixture of propylene, ammonia and air exemplify such processes. The complete catalytic oxidation of hydrocarbons is an effective and practical method of eliminating organic pollutants from gaseous effluents. Turning to the catalytic oxidation of inorganics, the heavy chemical and fertiliser industries have made widespread use of the direct oxidation of sulphur dioxide to manufacture sulphuric acid and the oxidation of ammonia in the manufacture of nitric acid. Oxychlorination, the most prominent example being the Deacon process for the manufacture of chorine, also offers an attractive route for the elimination of hydrogen from, and the incorporation of chlorine into, a reactant. The conversion of ethylene to vinyl chloride is one such catalytic oxychlorination process. These several and various routes for the manufacture of oxidation products are summarised in Figure 1. The chemical mechanisms of a selected number of these processes are briefly discussed in the following section to provide an illustration of some of the important principles.

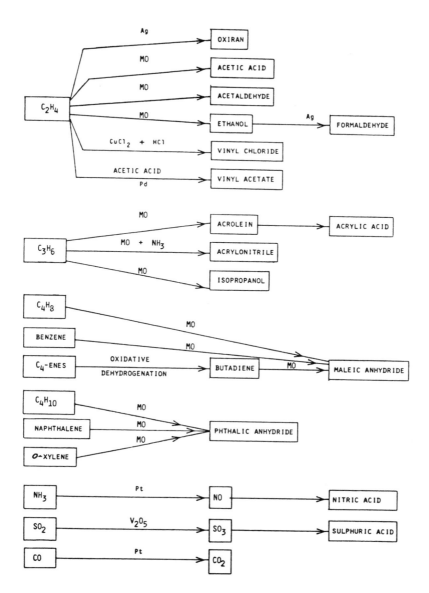

Figure 1   Catalytic Oxidation Process Routes

A phenomenon which has only been observed in comparatively recent years, and which may have potential benefits as well as disadvantages, is the oscillatory nature of certain catalytic oxidation reactions. The principles are not completely understood, but do have their origin in the physical and chemical dynamics of the catalytic oxidation reaction. What little information that currently exists is summarised, paying particular attention to those principles which are most likely to account for the phenomenon.

The important consequences of the highly exothermic nature of catalytic oxidation processes are that the overall reaction rate may be influenced by heat and mass transport and that thermal deactivation of the catalyst may result. A reaction which is controlled by mass transfer effects rather than chemical kinetics is the catalytic oxidation of ammonia. The reaction occurs at a relatively high temperature and the rate of product formation is dependent on the rate at which the reactants are transported from the bulk fluid phase to the catalyst surface. A conventional process design such as a porous catalyst packed in a tubular reactor is therefore inappropriate, so the hardware is configured in such a way that mass and heat transport are not constrained and a non-porous catalyst in the form of a gauze or pad is utilised. Because heat liberated during a catalytic reaction can have deleterious effects on catalyst activity, not only must catalytic reactors be designed to effect rapid removal of heat, but process operating conditions may have to be progressively changed to overcome loss of activity. These principles, which are an important consideration for the successful operation of a reactor effecting catalytic oxidation are also discussed in this review.

2. Chemistry of Catalytic Oxidation

2.1 Catalysis by Metal Oxides

Crucial to the chemistry and catalysis of oxidation reactions involving metal oxides is the mechanism and route by which oxygen is transferred from the gas phase to the catalyst and thence incorporated in the reactant to form the final product.

2.1.1 <u>Allylic Intermediates</u> Evidence has accumulated which suggests that the synthesis of many important petrochemical derivatives from C_3 to C_{10} hydrocarbons involves the formation of allylic intermediates. Isotopic tracer techniques[1] have contributed extensively to the understanding of the elementary steps in the sequence of surface reactions which occur. Catalytic oxidations which involve allylic intermediates propagate as a series of consecutive reactions in which the hydrocarbon reactant is activated by the catalyst. Cullis[2] has reviewed the various mechanistic possibilities. The first step in either the oxidation or ammoxidation of propylene is dissociative adsorption with abstraction of an H atom to form an adsorbed symmetric allyl radical. This allylic species subsequently loses a further H atom prior to either oxygen insertion to form acrolein or further hydrogen abstraction and nitrogen insertion to form acrylonitrile. The probable sequential steps are illustrated in Figure 2. Metal cationic sites are the most plausible points of surface attachment for the allyl radical. The allyl radical may also dimerise to form a diene which by further H abstraction yields cyclohexa-1,3-diene and ultimately benzene.

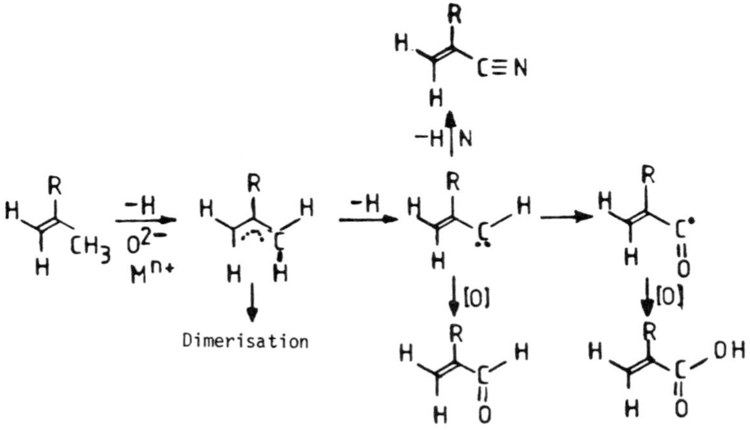

Figure 2   Hydrocarbon Oxidation *via* an Allylic Intermediate

Burrington et al.[3] used allyl alcohol as a source of allyl radicals to investigate propene oxidation over bismuth molybdate. These workers proposed that O insertion occurred prior to the second H atom abstraction which is facilitated by the C-O bond formed between adsorbate and adsorbent. Figure 3 illustrates the proposed mechanism. Although the details of this mechanism differ from the previous scheme, both mechanisms involve the formation of allylic surface species from which H is abstracted and O or N inserted.

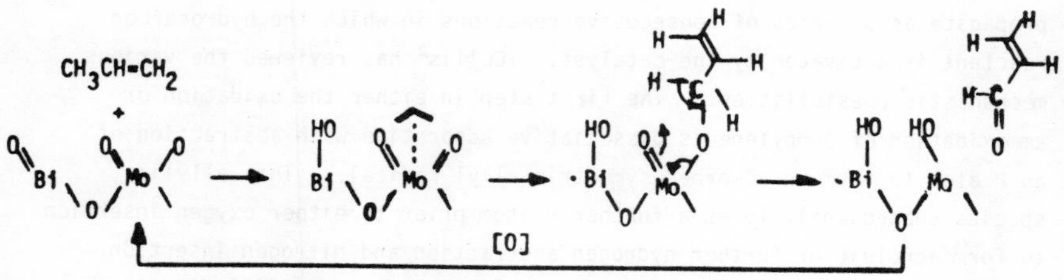

Figure 3   Propene Oxidation over Bismuth Molybdate

2.1.2 Acid-Base Properties   The nature of the product obtained as a result of catalytic oxidation also depends, of course, on the type of catalyst employed.   Insertion of O into adsorbed allylic species appears to require the presence of transition metal cations.   Indeed other classes of oxidation process also require transition metal cations which apparently facilitate the ease with which the original surface is regenerated by a reduction-oxidation sequence (see Figure 3).   It has been demonstrated,[4] however, that there is a correlation between the acid-base properties of a particular catalyst and the product selectivity.   Thus, acidic type products, such as maleic acid formed by the catalytic oxidation of butene or benzene, are preferentially synthesised by catalysts with high acidity.   Excessive acidity, however, causes complete degradation to CO_2.   Catalysts with basic properties give rise to basic products, such as the dienes, by oxidative dehydrogenation.   The role of surface acid centres was examined by Forzatti et al.[5] who studied the oxidation of but-1-ene.   It was proposed that Brønsted sites (surface OH groups) were responsible for diene formation, whereas Lewis sites encouraged O insertion and extensive oxidation.   The acid-base characteristics of admixtures of transition metal oxides depend on factors such as the specific metal oxides used, the relative composition and calcination temperature.   The scheme proposed by Forzatti et al.[5] is summarised in Figure 4, which shows how Brønsted sites, responsible for butene isomerisation, can be dehydroxylated at increased temperatures to form Lewis sites capable of O insertion and extensive oxidation involving bond rupture.   Brønsted sites, on the other hand, could be produced by H_2O vapour present during catalyst calcination.

2.1.3 Oxygen Transfer   Oxygen, as an active species during catalytic oxidation, plays rather different roles depending on the electronic properties of the catalyst.   Bielenski and Haber[6] and also Margolis[7]

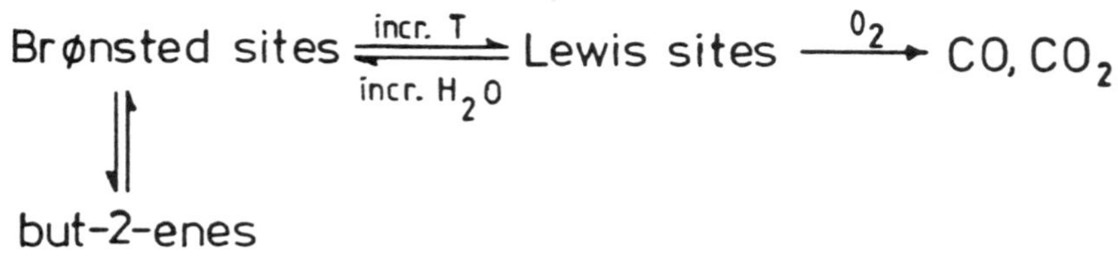

$$\text{Brønsted sites} \underset{\text{incr. } H_2O}{\overset{\text{incr. T}}{\rightleftharpoons}} \text{Lewis sites} \xrightarrow{O_2} CO, CO_2$$

$$\updownarrow$$

but-2-enes

Figure 4   Acid-Base Properties of an Oxidation Catalyst

assert that adsorbed oxygen species may undergo surface transformations in which the surface oxygen gradually becomes more rich in electrons.

$$O_2(ads) \longrightarrow O_2^-(ads) \longrightarrow O^-(ads) \longrightarrow O^{2-}(lattice)$$

Which form is predominant during a specific catalytic oxidation depends on the properties of the reactant-catalyst system and the experimental or process conditions.

It is convenient and rational to classify metal oxides into three groups according to the manner in which oxygen is adsorbed. Thus there are p-type oxides (*e.g.*, NiO, MnO, CoO, Co_3O_4) which adsorb oxygen as an electron-rich species. The cations of these p-type oxides have a tendency to increase their oxidation state and supply adsorbed oxygen with electrons, the consequence of which is that relatively simple molecules such as H_2, CO and CH_4 are oxidised by interaction with adsorbed O^- or lattice oxygen.

The second group of oxides are those which have a tendency to adsorb oxygen in the form O_2^-. Oxides such as ZnO, TiO_2, and V_2O_5, non-stoichiometric n-type semiconductors, fall into this category. The common feature of this group of oxides is the low concentration of electron donating centres. The O_2^-(ads) species may be considered as an electrophilic reactant and will oxidise H_2, CO, CH_4 and SO_2 as well as causing extensive oxidation of hydrocarbons.

Minimal oxygen adsorption characterises the third group of oxides. These oxides are mainly oxysalts in which oxygen is present in the form of well defined oxyanions with the transition metal ion in its highest oxidation state. No chemisorption of oxygen is observed on such unreduced catalysts. An electrophilic hydrocarbon molecule, however, is activated by the cations of the lattice which act as oxidising agents, and consequently becomes adsorbed, the electrons accepted from the oxide becoming delocalised. The formation of adsorbed allyl radicals on bismuth molybdate, for example, can be explained in these terms. The oxide lattice has the ability to insert lattice oxygen into a hydrocarbon. After desorption of the oxygenated product, oxygen vacancies appear in the oxide lattice and these may be filled by oxygen from the gas phase simultaneously reoxidising the reduced cations. This reduction-oxidation cycle (redox mechanism) is capable of accounting for many of the kinetic observations during hydrocarbon oxidation.

2.2 Oxidation by Metals

There are some notable examples of industrially important oxidation reactions which are apparently catalysed by metals. They are

$$C_2H_4 + \tfrac{1}{2}O_2 \xrightarrow{Ag} C_2H_4O; \qquad \Delta H_{298} = 146 \text{ kJ mol}^{-1}$$

$$4NH_3 + 5O_2 \xrightarrow{Pt} 4NO + 6H_2O; \qquad \Delta H_{298} = 227 \text{ kJ mol}^{-1}$$

$$CH_3OH + \tfrac{1}{2}O_2 \xrightarrow{Ag} HCHO + H_2O; \qquad \Delta H_{298} = 158 \text{ kJ mol}^{-1}$$

The catalytic oxidation of ethylene by silver to form ethylene oxide is unique. No other metal is as selective as silver in effecting this reaction of ethylene with air or oxygen at about 250° C. The higher homologues propylene and butylene cannot be oxidised by an analogous reaction, possibly because they may be adsorbed as allylic intermediates prior to extensive oxidation. Carbon dioxide and water, as well as ethylene oxide, are the sole products of combustion and to obtain maximum selectivity (*ca.* > 70%) ethylene dichloride, or a similar chlorinated compound, is added continuously to the feed. This has the effect of partially poisoning the silver surface. The formation of ethylene oxide is believed to occur by an Eley-Rideal mechanism whereby ethylene is adsorbed onto a preadsorbed O_2^-.

The first step in the reaction is therefore most likely to be:

$$Ag + O_2(g) \longrightarrow O_2^-(ads) + Ag^+$$

There is also the possibility that oxygen may be dissociatively adsorbed:

$$4Ag + O_2 \longrightarrow 4Ag^+ + 2O^{2-}(ads)$$

which is supposed to occur on clusters of Ag atoms by an activated process and which becomes more important at higher temperatures. The addition of chlorinated compounds in small amounts is assumed to inhibit this latter reaction. The silver catalyst is normally supported on low area α-alumina. Excessive diffusion limitation can be avoided by employing a support with large pores.

Because the catalytic oxidation of ammonia occurs in practice at elevated temperatures (*ca.* 850° C at atmospheric pressure or about 960° C at 0.8 MPa) and the process rate is governed by the transport of reactants from the fluid phase to the catalyst surface, there is little evidence available to suggest what the chemical mechanism might be. All that can be said is that intermediates such as NH, HNO and NH_2OH have been postulated, but not confirmed. At much lower temperatures in the range 150-200° C, infra-red studies[8] have indicated the strong chemisorption of molecular NH_3.

The catalytic oxidation of methanol by silver in a fixed catalyst bed at 400-600° C is also strongly mass transfer controlled. However, as the reaction temperature is considerably lower than the ammonia oxidation, kinetic experiments deliberately designed to eliminate mass transfer effects can be relied upon with greater certainty.[9] Wachs and Madix,[10] who studied methanol oxidation of a single crystal of Ag and employed the technique of temperature programmed reaction spectroscopy, proposed that methanol reacts with adsorbed oxygen and that hydrogen, as well as the main products formaldehyde and water, arise from a surface interaction between adsorbed species.

$$CH_3OH + O(ads) \longrightarrow CH_3O(ads) + OH(ads)$$

$$CH_3OH + OH(ads) \longrightarrow CH_3O(ads) + H_2O$$

$$CH_3O(ads) \longrightarrow HCHO + H(ads)$$

$$2H(ads) \longrightarrow H_2$$

This mechanism does not, however, account explicitly for observed inhibition of the oxidation rate by methanol, oxygen and water,[9,11] results which can be more easily explained by a redox mechanism.

At a somewhat lower temperature (*ca*. 350-400° C) an iron molybdate catalyst will oxidise methanol to formaldehyde. The catalyst is only active when excess air is employed for the process. It is believed that a redox type mechanism, not dissimilar to the schemes proposed for bismuth molybdate, explains experimental observations.[12]

2.3 Oxychlorination

A brief mention of oxychlorination is apposite, especially as it is so important as an industrial process for making vinyl chloride. Modern technological practice is to react hydrochloric acid vapour and air with ethylene over a potassium chloride promoted copper chloride catalyst supported on porous alumina. Kinetic studies[13] have provided some indication that the Deacon reaction

$$4HCl + O_2 \longrightarrow 2Cl_2 + 2H_2O$$

is rate determining. The product is then formed by chlorination of ethylene.

$$CH_2=CH_2 + Cl_2 \longrightarrow CH_2ClCH_2Cl$$

The precise mechanism is unclear, although adsorption of HCl and O_2 are believed to occur. The desired product, vinyl chloride, is formed successively by pyrolysis:

$$CH_2ClCH_2Cl \longrightarrow CH_2=CHCl + HCl$$

More recent studies[14] of the Deacon reaction support the view that the catalyst is involved in a redox reaction with HCl.

2.4 The Redox Mechanism

A chemical mechanism which explains many of the features of catalytic oxidations by metal oxides and other catalysts involves distinct steps in which the catalyst oxidises the reactant, thereby becoming reduced, and susbequently is reoxidised by oxygen either from the catalyst lattice or by chemisorbed oxygen. This reduction-oxidation (redox) mechanism has already been inferred as a possible route for the oxidation by mixed oxides of some hydrocarbons involving allylic intermediates, the oxidation of methanol by iron molybdate and oxychlorination by copper chloride. The oxidation of SO_2 by V_2O_5 has also been cited[15] as an example of a redox reaction. The reduction step is considered to involve reaction between the catalyst in an oxidised form (Cat-O) and the reactant (R), the catalyst becoming reduced:

$$\text{Cat-O} + \text{R} \xrightarrow{k} \text{RO} + \text{Cat}$$

The subsequent oxidation step is represented:

$$\text{Cat} + \tfrac{1}{2}O_2 \xrightarrow{k_{-1}} \text{Cat-O}$$

More specifically, Mars and van Krevelen[16] explained the kinetics of the partial oxidation of several aromatic hydrocarbons and of sulphur dioxide by means of a redox mechanism. Ascribing the rate constants k_1 and k_{-1} to the reactant oxidation and catalyst reoxidation steps respectively, the steady state overall rate is

$$r = kp_R(1-\theta) = \frac{k_{-1}}{\beta} p_{O_2}^n \theta$$

where p_R and p_{O_2} signify partial pressures of reactant and oxygen respectively, θ the fraction of active sites in the reduced form and β the number of moles of O_2 consumed per mole of reactant. Solving for θ and re-substituting immediately yields

$$r = \frac{1}{(1/kp_R) + (\beta/k_{-1}p_{O_2}^n)}$$

Two limiting forms of the above equation can be written. Thus if $kp_R \gg k_{-1}p_{O_2}^n \theta/\beta$, then reoxidation of the catalyst is rate limiting and

$$r = \frac{k_{-1}}{\beta} p_{O_2}^n$$

In this form of the equation, the observed rate of reactant oxidation equals the rate of oxidation of the catalyst surface and is independent of the reactant partial pressure. Mars and van Krevelen[16] found this to be true for several aromatic hydrocarbon oxidations over a mixed oxide catalyst composed of V_2O_5 and MoO_3. The oxidation of o-xylene to phthalic anhydride was similarly found to conform to such a description.[17] On the other hand, when the reaction between the reactant and oxidised form of catalyst is rate limiting $k_{-1}p_{O_2}^n/\beta \gg kp_R$ and then

$$r = kp_R$$

which is the case for the ammoxidation of propylene over a bismuth molybdate catalyst.

Although the redox mechanism is capable of explaining the kinetics of many catalytic oxidations, there are limitations in adopting the concept too generally. For example, some reactions requiring chemisorbed oxygen to drive the reaction, such as the oxidation of ethylene on Ag and the extensive oxidation of some hydrocarbons, conform to Langmuir-Hinshelwood or Eley-Rideal kinetics, rather than redox kinetics. The redox mechanism also does not explain why intermediate products in some partial oxidation reactions are preferred products. For high selectivity in a reaction in which other undesired oxidation products may also be formed by consecutive steps, the rate of desorption of the desired intermediate should be large relative to the rate of further oxidation. This depends, of course, on the strength of adsorption of the desired intermediate and is a function of catalyst structure.

3. Instabilities and Transient Phenomena

It is natural that the acquisition of experimental data from heterogeneous catalytic reacting species is sought, normally, from systems which are in a stable steady state, for it is only by this means that we can be sure that experiments are reproducible and that data are free from spurious errors. There are, of course, occasions when a deliberate perturbation from the steady state (such as a step change in concentration or temperature) is advantageous in securing kinetic information about the approach to a steady operating state. There are, however, some circumstances which give rise to unstable conditions and which encourage the reacting system to converge upon a more stable operating state. In rather different circumstances a perturbation may lead to oscillation between steady states of the system.

Heterogeneous catalytic oxidations which display more than a single steady state, at least one of which is metastable, are exothermic processes in which heat and mass transport between the fluid phase and the solid catalyst play a dominant role. A multiplicity of steady states can, however, arise by virtue of the chemical kinetics of the catalytic system and is certainly not confined to that class of problems identified by the effects of physical transport processes. The observed phenomena are a result of system instabilities and result in transient states.

3.1 Multiple Steady States

To illustrate how more than one steady state can arise in a reacting catalytic system, we consider a very exothermic reaction such as, for example, ammonia oxidation. Whether the reaction occurs on a metal gauze or in a porous material supporting the active metal, heat is generated by virtue of reaction at the catalyst surface and is dissipated by convection, conduction and (for some conditions) radiation processes. A given portion of gauze or single individual catalyst pellet will be in a steady state of thermal equilibrium if the rate of heat generation Q_g in the pellet volume v_p is balanced by the rate of heat loss Q_ℓ from its exterior surface area a. In the simplest case, when one can ignore intraparticle diffusion effects, the rate of heat generation due to, say, a first order reaction is

$$Q_g = v_p(-\Delta H)kc_i$$

where k is the chemical rate constant, $(-\Delta H)$ is the heat of reaction and c_i the interface concentration. For a steady state the rate of chemical reaction must be balanced by the mass flux of reactant from gas phase to particle, so

$$v_p k c_i = h_D a (c_g - c_i)$$

where h_D is the mass transfer coefficient and c_g is the bulk gas concentration of reactant. Q_g may therefore be expressed as a function of temperature and gas concentration only by elimination of c_i. As a function of temperature, $Q_g(T)$ is a sigmoid curve, as shown in Figure 5. One may superimpose upon such a curve (for a given gas concentration) the straight line representing the rate of heat loss from the particle

$$Q_\ell = ha(T_i - T_g)$$

where h is the particle to gas heat transfer coefficient.

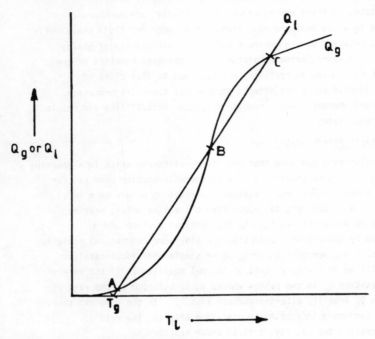

Figure 5   Thermal Instabilities

It is evident from Figure 5 that as many as three steady states A, B and C are possible. States A and C are stable steady states. The operating point B is a metastable state. It is not difficult to imagine the violent temperature perturbations which catalyst particles, located at various positions in a packed tubular reactor, might undergo as a reactor assembly is either started up or shut down. Each particle in the bed is likely to have a different temperature history because the extent of reaction will vary along the bed length, as also will the gas phase temperature.

Different states of chemical stability might also exist for a given catalytic system. Thus Beusch et al.[18] not only reported thermal instabilities during the platinum catalysed oxidation of hydrogen (which were of the kind described above and attributed to mass and heat transfer effects between catalyst particle and gas phase), but also concentration instabilities for isothermal conditions during the palladium catalysed oxidation of carbon monoxide. Now the iso-

Heterogeneous Catalytic Oxidation

thermal reaction rate for carbon monoxide oxidation can be represented by an equation of the form

$$r = \frac{kc_i}{(1 + kc_i)^2}$$

carbon monoxide actually inhibiting the oxidation at relatively high interface concentrations c_i of carbon monoxide. For a steady state this must be balanced by the rate of mass transport from gas to catalyst particle

$$r = h_D a (c_g - c_i)$$

where c_g is the bulk gas phase concentration of carbon monoxide and $(h_D a)$ is the product of the mass transfer coefficient and the particle external surface area. In Figure 6 the curve displaying a maximum represents the reaction rate as a function of bulk gas concentration and the straight line (of slope $h_D a$) represents the mass transfer rate from bulk gas to interface. At steady state conditions these rates will be equal. Again it is noted that there are three possible steady states, the point B being metastable and points A and C stable. Clearly then, concentration instabilities may also exist in heterogeneous reacting systems, but have only been reported for a limited number of systems. It

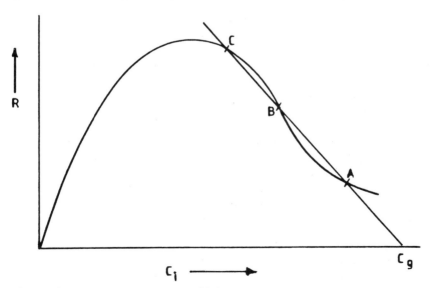

Figure 6   Concentration Instabilities

is also possible for multiple steady states to arise by virtue of a balance between intraparticle diffusive fluxes and a reaction rate described by Langmuir-Hinshelwood kinetics.

A proposal that purely chemical kinetic effects might account for the multiplicity of carbon monoxide oxidation in the presence of platinum has recently been confirmed by Goodman et al.[19] They employed sufficiently small catalyst particle sizes to eliminate intraparticle diffusion and, furthermore, used a recycle reactor to ensure that fluid to particle mass and heat transfer effects were absent. The experimental results obtained display multiple steady state phenomena which are explained solely on the basis of adsorption, surface reaction and desorption rates. Apparently the interaction between these three chemical rate processes is sufficient to lead to a multiplicity of steady states under isothermal conditions.

3.2 Transient Phenomena

It has just been shown that it is possible for some catalytically reacting systems to switch from one steady state to another. In adjusting to such a change, the path followed by the system depends on the direction of the perturbation and leads to the phenomena of hysteresis. For example, consider the heating of a catalyst particle which catalyses an exothermic oxidation reaction. One may superimpose upon the sigmoid heat generation curve a number of heat loss lines, each one of which represents the rate of heat loss for a given gas phase temperature. Figure 7 illustrates how the steady state interface temperature of the particle changes as the gas phase temperature is increased through the sequence T_1, T_2, T_3, T_4 to T_5. The states of stable equilibrium corresponding to these temperatures are at the points of intersection A, B, F, G, H and I of the straight lines with the heat generation curve. On the other hand, if the temperature were decreased from T_5 to T_1 a different path I, H, D, C, B, A would be traced. This is evident from Figure 8, which is a plot of the steady state interface temperature as a function of gas temperature; it is clear that a different steady state path is followed when increasing the temperature to that followed when decreasing the temperature.

Evidence that such hysteresis effects occur during the platinum catalysed oxidation of hydrogen was provided by Beusch et al.[18] Figure 9 shows the stepwise transition between lower and upper steady states which were recorded on increasing the concentration of hydrogen.

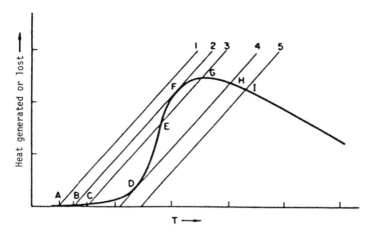

Figure 7   Interface Temperature Changes with Change in Gas Temperature

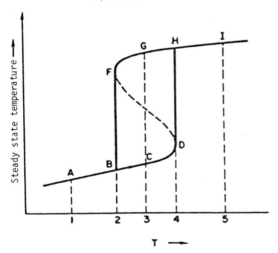

Figure 8   Thermal Hysteresis

Decreasing the concentration from a stable upper steady state operating condition, however, did not have an identical effect and the system displayed hysteresis, the lower stable steady states being approached along a gradually decreasing concentration path.

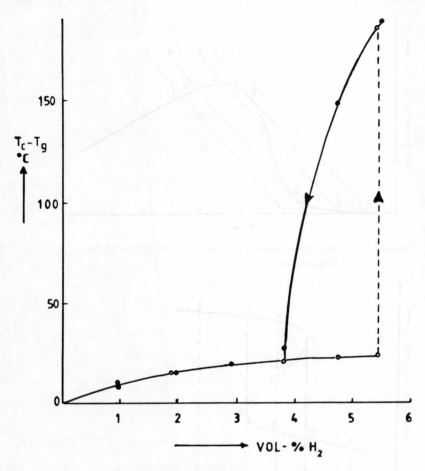

Figure 9   Stepwise Transition between Upper and Lower Steady States during Oxidation of H_2 over Pt.

During the platinum catalysed oxidation of hydrogen sustained oscillations in both the product concentration and the catalyst pellet temperature were observed.[18]   These oscillations occurred at a relatively low value of the gas temperature and had a rather long time period (ca. 1 h).   The coincidence of the temperature and concentration maxima and the long, but regular, time interval between maxima suggests that the oscillatory behaviour does not arise as a result of transitions between stable thermal states of the catalyst particle and gas phase.

Oscillatory behaviour was also noted[20] during the catalysed oxidation of hydrogen over nickel foil for isothermal conditions (Figure 10). Both the amplitude and frequency of oscillation increase with increase in reaction temperature. McCarthy et al.[21] observed little difference (ca. 0.001° C) between gas and catalyst temperature at the maxima and minima of reaction rate oscillations during the platinum catalysed oxidation of carbon monoxide. Contact potential difference measurements at the surface of a nickel foil catalyst during hydrogen oxidation indicate that the composition of the catalyst surface must also be periodically changing. All of these observations point to the probability that chemical, rather than physical, phenomena are the principal causes of oscillatory behaviour.

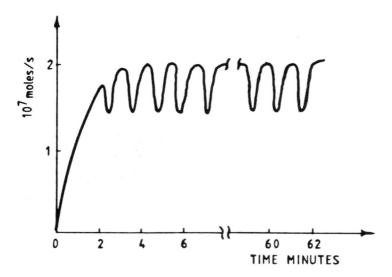

Figure 10   Concentration Oscillations observed during Oxidation of H_2 over Ni Foil

Sheintuch[22] studied the development of oscillatory states occurring when carbon monoxide is oxidised in the presence of platinum foil in a well mixed spinning basket reactor. Simultaneous with the observation of the oscillatory behaviour (Figure 11a) he measured the instantaneous derivative of the time smoothed detector output (conversion versus time) and was therefore able to construct a phase plane portrait, represented in Figure 11b, depicting the approach to the oscillatory state. This

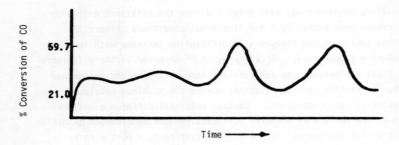

Figure 11(a) Oscillations observed during Oxidation of CO over Pt Foil

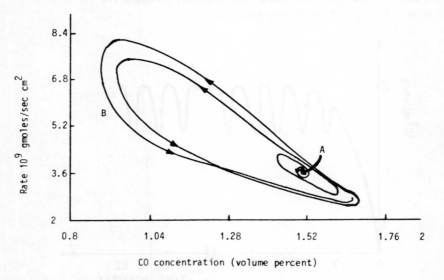

Figure 11(b) Phase Plane Portrait Indicating a Limit Cycle during Oxidation of CO over Pt Foil

so-called limit cycle is typical of the behaviour of systems with a multiplicity of steady states. The limit cycle diagram provides a useful indication of how a system re-establishes either steady or oscillatory behaviour when the system is perturbed. In the case cited, oscillations in the concentration of product grow in amplitude until regular concentration maxima and minima are established: the trajectory therefore commences from the internal point A (Figure 11b) and spirals outwards toward the stable limit cycle B which becomes established after the elapse of a sufficient period of time (Figure 11a).

It should be noted that phase portraits of limit cycles may also be presented as diagrams of the concentration of any appropriate system variable as a function of another variable. Figure 12a, for example, shows the phase plane portrait of a hypothetical catalytic reaction in which gaseous reactants A and B form a product AB by the catalytic action of a surface S. The portrait shows how the surface concentrations of both reactants (x and y respectively) vary and approach, from various starting locations, a stable limit cycle. The corresponding time dependence of the surface concentrations in the stable limit cycle is shown in Figure 12b.

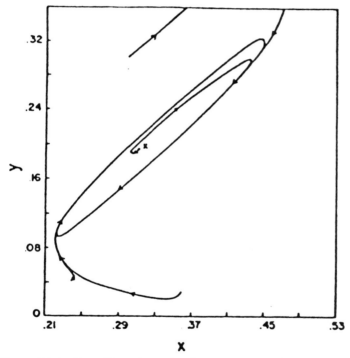

Figure 12(a) Phase Plane Portrait for a Hypothetical Surface Reaction

$A(g) + S \rightarrow A\text{-}S;$
$B(g) + S \rightarrow B\text{-}S;$
$A\text{-}S + B\text{-}S \rightarrow AB(g) + 2S$

Surface concentrations of A and B designated by x and y respectively.

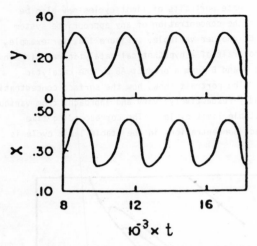

Figure 12(b) Sustained Oscillation in the Surface Concentrations of A (designated by x) and B (designated by y)

Pikios and Luss[23] computed the phase plane portrait described (Figure 12a) by assigning estimated parameter values to the dynamic interaction of gaseous A and B with the surface S. The catalysis was supposed to occur by the following overall reaction steps:

$$A(g) + * \rightleftharpoons \overset{A}{\underset{*}{|}}$$

$$B(g) + * \rightleftharpoons \overset{B}{\underset{*}{|}}$$

$$\overset{A}{\underset{*}{|}} + \overset{B}{\underset{*}{|}} \longrightarrow AB(g) + * + *$$

where the symbol * denotes an active site at the catalyst surface. The surface was assumed to be energetically heterogeneous, the activation energy of the surface reaction varying linearly with the fraction of surface occupied by one of the reactants. A mathematical stability analysis of the two ordinary differential equations describing the dynamics of the reacting system showed that only a single unstable steady state existed which led to sustained oscillations in the system.

Clearly then, a chemical description of the catalytic system is
sufficient in this case to account for sustained oscillations [Figure 12(b)].
Catalytic oxidations in which both the reactant and oxygen compete for
sites fall into this cetegory.

Hlavacek and Rathousky[24] have reported both regular and chaotic
oscillations in the extent of carbon monoxide oxidation in a multichannel catalytic monolith whose tubular channels were coated with
alumina and impregnated with palladium. These oscillations appear to
occur about a unique unstable steady state, the character of the oscillations depending on the physical conditions in the reacting system
and the initial conditions. The experimental observations are
qualtiatively explained by a surface memory effect which can be interpreted as a time lag between the condition of the surface at any
moment and the response of the reactants to the surface condition.

4. Industrial Practice

4.1 Operating Problems

To overcome or circumvent some of the intrinsic operating
difficulties associated with processes involving catalytic oxidations, three types of reactor hardware are adapted for use. These
are the fixed bed, multitubular-shell assembly and fluidised bed
reactors. Operating problems arise principally because of the extreme
exothermicity and rapidity of reaction. A consequence of a high
rate of chemical reaction is that the actual rate of product formation
may well depend on the relative rate at which inter- and intra-particle
transport processes occur. Intraparticle diffusion may be eliminated
or substantially reduced by choosing a sufficiently small catalyst
particle size, but a compromise usually has to be made because, in
fixed bed reactors, particle diameters which are too small lead to
excessive pressure drops and therefore vastly increased pumping costs.
When, because of intraparticle diffusion limitations, there is little
chance of avoiding small catalyst particles, a fluidised bed configuration offers a means of satisfactory operation. If, as in
ammonia oxidation by a platinum catalyst, the rate is limited by mass
transfer from gas phase to catalyst interface, there is little point
in employing a porous support and the catalyst is arranged in the form
of a gauze of intermeshing fine metal wire.

Large quantities of heat also have to be removed from reactor
assemblies in which very exothermic catalytic oxidation reactions occur.

Except in a few instances, the principal mechanisms by which heat is removed from the reaction environment are convection and conduction. In ammonia oxidation and the oxidation of exhaust gases from automobiles the reaction temperature is very high and a not insignificant amount of heat is dissipated by radiation. The choice of operation mode which is usually available is adiabatic, non-isothermal or isothermal. In adiabatic operation heat can only be dissipated by the convective flow of the reactant and product gases through the reactor. Under these circumstances the problem is to ensure that the catalyst bed is sufficiently restricted in length to avoid too large a temperature rise. An example is the catalytic oxidation of sulphur dioxide. Three or four successive trays of catalyst, with intercooling between each tray, are used specifically to ensure that the catalyst temperature does not exceed a particular limit which would cause catalyst deactivation. Isothermal operation of a fixed bed reactor in which a heterogeneous catalytic reaction occurs is not a practical alternative for avoiding high temperatures: catalytic oxidation reactions usually occur so rapidly that the reactor volume requirements for isothermal operation are not compatible with the heat transfer demands. On the other hand, fluidised beds operate in a substantially isothermal mode on account of their excellent heat transfer properties and are therefore employed for many hydrocarbon partial oxidation and ammoxidation reactions. A multi-tubular-shell assembly in which catalyst is placed in the tubes and a coolant is passed, usually countercurrent to the inlet gases, through the shell is another type of reactor frequently employed for catalytic oxidation. This configuration results in non-isothermal operation and excessive temperature rise along the tube length must be avoided by careful design and operation of the reactor: the oxychlorination on supported cupric chloride of ethylene and the oxidation of methanol by an iron molybdate catalyst are reactions which are effected in a cooled multitubular-shell assembly. The reaction of oxygen with a mixture of ethylene and acetic acid vapour over a silica supported palladium catalyst to form vinyl acetate is also effected in a multi-tubular-shell assembly, but in this case boiling water in the shell removes the heat of reaction and the tubes containing the catalyst operate essentially in an isothermal mode.

The examples which follow are intended to illustrate the principles of operation and design of the three classes of reactor. One type of reactor which falls outside these categories and which is increasingly

coming into use for the control of automobile exhaust gas emission is
the multichannel catalytic monolith: its characteristics will also
be briefly described.

4.2 Adiabatic Catalytic Reactors

Adiabatic reactors are frequently encountered in practice. Many
catalytic oxidation reactions are effected by this operational mode.
Because there is no exchange of heat with the environment, radial
concentration and temperature gradients are absent. All of the heat
generated by reaction manifests itself by a change in enthalpy of
the gaseous stream. Furthermore, heat transfer between catalyst and
gas is sufficiently rapid to assume that all the heat generated at any
point along the catalyst bed is transmitted instantaneously to the gas.
Provided one may neglect the longitudinal dispersion of mass and heat,
then the conservation equations for mass and heat within the reactor
lead to a simple relationship (the adiabatic reaction path) between gas
temperature T and concentration c of product:

$$T = T_0 + \frac{(-\Delta H)}{\bar{c}_p} \cdot c$$

where T_0 is the inlet gas temperature, $(-\Delta H)$ the heat of reaction and
$\bar{c}_p$ the average molar heat capacity of the gas stream. A reasonable
approximation is to assume that the ratio $(-\Delta H)/\bar{c}_p$ is temperature independent so that the above equation becomes linear. If this relation
is then substituted into an elementary mass conservation equation for the
reactor, it is possible to estimate conversion as a function of reactor
length. This is a convenient basis for predicting the necessary reactor
size and quantity of catalyst to effect a required conversion of reactant.
If the kinetic reaction data are available in the form of reaction rate
as a discrete function of temperature and conversion, then the reaction
rate information may be displayed as contours of equal reaction rate in the
temperature plane, as is shown in Figure 13 for an equilibrium reaction
such as the SO_2 oxidation. Superimposed on these contours is an
adiabatic reaction path of slope $(-\Delta H)/\bar{c}_p$ and intercept T_0 on the abscissa.
The various values of reaction rate $r(c, T_0)$ along the reactor length
are then simply those points at which the adiabatic reaction path inter-
sects the continuous but bounded family of contours.

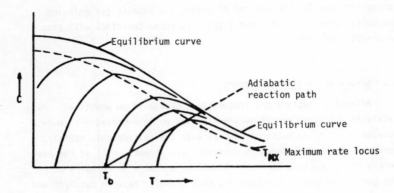

Figure 13   Adiabatic Reaction Path for SO_2 Oxidation

It is often necessary to employ more than one adiabatic reactor to achieve a desired conversion. The catalytic oxidation of SO_2 to SO_3 is a case in point. In the first place chemical equilibrium may have been established in the first reactor and it would be necessary to cool and/or remove the product before entering the second reactor. This, of course, is one good reason for choosing a catalyst which will function at the lowest possible temperature. Secondly, for an exothermic reaction, the temperature may rise to a point at whicn it is deleterious to the catalyst activity. At this point the products from the first reactor are cooled prior to entering a second adiabatic reactor. To design such a system it is only necessary to superimpose on the rate contours the adiabatic temperature paths for each of the reactors. The volume requirements for each reactor can then be computed from the rate contours in the same way as for a single reactor. It is necessary, however, to consider carefully how many reactors in series it is economic to operate. Figure 14 illustrates the adiabatic reaction paths that would be followed by a series of three successive catalyst beds with interstage cooling between each catalyst tray. Variations of such an operational mode exist in practice and Figure 15 is a sketch of an industrial SO_2 converter.

A further example of an industrial catalytic oxidation reaction which occurs adiabatically is the oxidation of ammonia. Industrial plants producing nitric acid employ platinum or platinum alloy gauze as a catalyst to oxidise ammonia. Rather less than the stoichiometric ratio (14.4%) of ammonia to air is fed to the converter to avoid

Heterogeneous Catalytic Oxidation

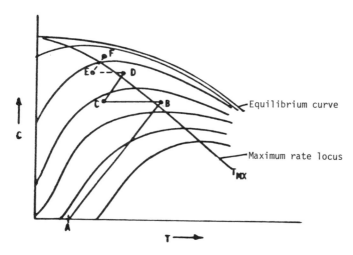

Figure 14   Three Successive Adiatbatic Reaction Paths for SO_2 Oxidation with Intermediate Cooling

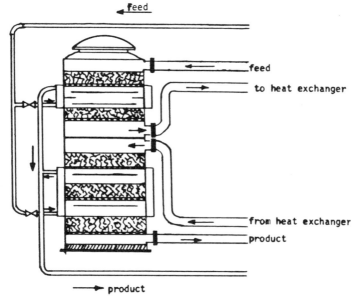

Figure 15   Industrial SO_2 Converter with Heat Exchange between Catalyst Trays

operating near the explosion limit of ammonia in air. The operating conditions for the catalytic converter will depend on the downstream process conditions which are arranged so that the nitric oxide produced is oxidised in the gas phase to nitrogen dioxide by the homogeneous reaction

$$NO + \tfrac{1}{2}O_2 \longrightarrow NO_2$$

As this latter reaction occurs rapidly under compression, it is not unusual for the ammonia-air mixture fed to the catalytic converter to be at about 0.5 MPa or above. The reactant gases are passed downward through the reactor which operates at about 900° C.

The ammonia oxidation is exceedingly exothermic and is very fast. Consequently the relatively slower transport of reactant through the gas phase and the stagnant boundary layer adjacent to the solid catalyst surface (and the corresponding return of product to the gas phase) determines the overall rate of conversion. Such mass transfer limiting conditions do not require so large an extent of surface area as a reaction controlled by surface kinetics. Therefore, rather than disperse platinum catalyst over a porous inlet support (which would cause an unwelcome pressure drop) the catalyst is normally in the form of platinum finely woven gauze, about 30 or 40 layers of which constitute a 4 m diameter pad of catalyst. Only the first two or three layers of the gauze act catalytically until such time as the platinum metal has become inactive due to surface oxidation and platinum loss through vaporisation of the relatively more volatile PtO_2 formed during reaction. The reaction zone consequently moves to successive layers of gauze when the first few layers have lost their activity. Gauzes are then removed from the top of the pad and fresh ones inserted at the bottom of the pad. The total platinum loss amounts to about 200 mg per 1000 kg of nitric acid produced in a plant.

Because the pressure drop through the pad is so low, it is necessary to pay careful attention to the uniform distribution of gas flow over the whole cross section of a gauze layer and throughout the entire pad. Poor gas distribution through the pad would, for example, cause any unreacted ammonia to react with product nitric oxide to form nitrogen, thus reducing the overall conversion and efficiency. Because so much heat is generated during reaction and because of the rapid gas flow through the catalytic converter (less than 1 ms contact time is adequate), there is a delicate balance between convective heat loss and heat generation by reaction. Figure 11 (Section 3.2) shows the kind of balance required (a heat loss line superimposed upon a heat generation curve). Clearly, to achieve the stable operating conditions at C where the conversion is high at a high reaction temperature, the incoming reactant gases must be ignited, otherwise little or no conversion would occur as at A (the intersection at B represents an unstable condition). Such ignition is usually initiated by means of a flame: once the gas temperature is above B, sufficient heat will be generated by reaction to overcome the loss of heat by convective flow and for the reaction to stabilise and be self-sustaining at the operating point C.

4.3 Cooled Fixed Catalyst Beds

Some catalytic oxidation reactions are effected in cooled fixed catalyst beds to avoid an otherwise too large temperature rise. The configuration is usually a shell and tube assembly, as illustrated in Figure 16, in which the catalyst is contained in the tubes and a coolant in the shell. Cooling is accomplished either by allowing the heat produced by reaction to boil a suitable liquid (*e.g.*, water under pressure) or by circulating a molten salt. Alternatively, cooling can be achieved by extracting the heat with a countercurrent flowing liquid. A process for manufacturing vinyl acetate by the catalytic reaction between ethylene, oxygen and acetic acid vapour (oxidative dehydrogenation using a

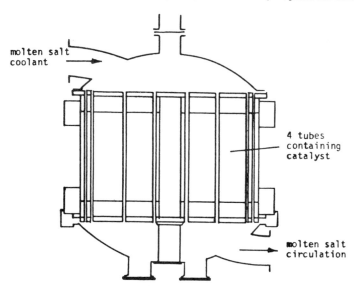

Figure 16   Cooled Fixed Catalytic Bed as Employed for Vinyl Acetate Manufacture

supported palladium catalyst) employs such a shell and tube assembly, the heat of reaction being removed by water boiling in the reactor shell. Examples of shell and tube assemblies cooled by countercurrent flowing liquids are to be found in catalytic processes for the manufacture of formaldehyde (by methanol oxidation on an iron molybdate catalyst), acrylonitrile (by propylene ammoxidation over mixed oxides containing bismuth molybdate) and ethylene oxidation to ethylene oxide on a silver catalyst.

Two inherent operating problems, which must be avoided as far as possible, arise because of the extreme exothermicity of catalytic oxidation reactions. The first of these is the sharp temperature increase of the reactant fluid along the length of the reactor. Figure 17 is a sketch of a typical fluid temperature profile in a cooled fixed bed tube. Near the tube inlet the temperature increases slowly at first, but as more heat is released by reaction, the temperature rises more steeply. At the tube exit where conversion is near completion, much less heat is evolved and the temperature is lower. The axial temperature profile thus reaches a maximum (commonly known as the hot spot) somewhere along the reactor length. The more effective the cooling, the lower is the temperature maximum and a satisfactory reactor design would take account of the heat transfer between reactant and coolant streams and ensure that the temperature maximum is well below any temperature which is deleterious to the catalyst activity. Random hot spots may also occur in packed catalyst tubes, due to uneven catalyst distribution causing inefficient heat transfer: these are impossible to predict and are best avoided by ensuring a uniform catalyst distribution throughout the reactor.

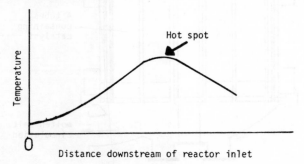

Figure 17 Temperature Rise along a Fixed Bed Catalytic Reactor during an Oxidation Reaction

The second operating difficulty is the extreme sensitivity of some reactor systems to a slight change in coolant temperature. This phenomenon is known as parametric sensitivity and its effect is illustrated in Figure 18 for the catalytic oxidation of naphthalene on a V_2O_5 catalyst contained in a cooled tubular reactor. The predicted sensitiveness is due to the manner in which the heat transfer coefficients from catalyst particle to reactant gas and from the gas phase through the tube wall to the coolant change with temperature and flow conditions. Unwelcome responses to small changes in coolant temperature are best avoided by a careful analysis of the system and a suitable choice of operating policy which circumvents the sensitive operating region.

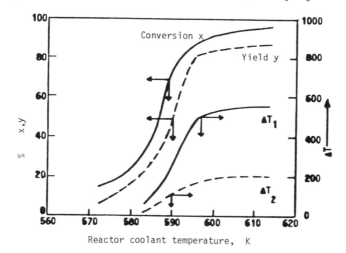

Figure 18   Parametric Sensitivity observed during the Catalytic Oxidation of Naphthalene

ΔT_1 = maximum radial temperature difference between tube centre and wall

ΔT_2 = maximum temperature difference between gas phase and solid catalyst

4.4 Fluidised Beds

The use of fluidised bed reactors for some partial oxidation reactions has certain advantages over tubular type reactors. Apart from the mechanical advantage gained by the ease with which solids may be conveyed, the high wall-to-bed heat transfer coefficient enables heat to be abstracted from the reactor with little difficulty. Furthermore, because of the movement of solid particles within the bed, the whole of the gas in the reactor is substantially at the same temperature and this mode of operation therefore leads to effective temperature control of the reaction environment. If partial and extensive catalytic oxidations are regarded as two consecutive chemical reactions, then the heat generated by chemical reaction as a function of temperature difference between the fluidised bed and its surroundings is a double sigmoid shaped curve, as shown in Figure 19. To restrict conversion in the bed to the desired partial oxidation product, it is necessary to employ not only a highly selective oxidation catalyst, but also to ensure that the bed operates under such conditions that heat is removed from the bed at a rate which is commensurate with the rate at which heat is generated by the partial oxidation step. This, in turn, means choosing the appropriate temperature difference between bed and surroundings and, by correct design procedures, the bed size and materials of construction. Figure 19 illustrates schematically a correct choice of heat removal rate by the intersection of the heat removal line with the heat generation curve.

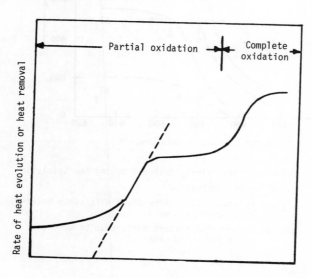

Figure 19   Rate of Heat Generation for Successive Partial Oxidation and Complete Oxidation

Three well known examples of processes employing fluidised bed operations are the oxidations of naphthalene and xylene to phthalic anhydride using a supported V_2O_5 catalyst and the ammoxidation of propylene utilising a mixed oxide composition containing bismuth molybdate. Typically this latter reaction is executed by passing a mixture of ammonia, air and propylene to a fluidised bed operating at about 0.2 MPa pressure, 400-500° C and a few seconds contact time between gas and fluidised catalyst particles.

4.5 Catalytic Monoliths

A particular type of catalytic reactor which has been used principally for the conversion of automobile exhaust gases to innocuous products is the catalytic monolith. The reactor consists of a single block of extruded material containing within it an array of parallel uniform non-intersecting channels upon the walls of which is coated an active catalyst. This configuration is employed because it offers the facility of effecting a mass transfer controlled catalytic oxidation reaction without suffering the disadvantages of any significant pressure drop between inlet and exit of reactor. Such a pressure drop occasioned by the use of a packed reactor has undesirable effects on engine performance. The geometry of the channels varies with the manufacturer and the purpose for which the monolith is used, but channels of circular or triangular cross section are not uncommon. Figure 20 illustrates some configurations. Yet another type is prepared from crimped metal ribbon and arranged in layers in close proximity, so that when the whole is rolled and contained within an outer former of circular cross section, the net result is a tube containing a multitude of irregular but independent, closely packed, channels extending through the whole length of material.

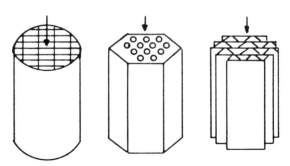

Figure 20   Catalytic Monolith Assemblies

The main object of developing the catalytic multichannel monolith was to respond to legislation for the regulation of automobile exhaust emissions. Thus the requirement is to oxidise completely unburnt hydrocarbons, CO and reduce oxides of nitrogen to nitrogen. This simultaneous oxidation of hydrocarbons and CO and reduction of nitrogen oxides can be achieved by careful control of the fuel-air mixture near the stoichiometric ratio and the use of a supported platinum catalyst containing some rhodium. The catalyst is supported on the walls of the channels and is prepared by introducing a wash coat of alumina impregnated with the platinum-rhodium catalyst or by electrolytic deposition.

Because the reactions occurring are strongly exothermic and have a high activation energy, the reaction rate is fast. Thus the reacting gases diffuse only a small depth into the catalyst layer before reaction is complete, necessitating only a thin layer of catalyst wash coat.

Modelling the behaviour of these catalytic monoliths is difficult. Although intraparticle temperature and concentration gradients are virtually absent, the fast reaction rates imply significant mass and heat transfer effects between bulk gas and channel walls. Furthermore, rapidly changing engine loads introduce sudden perturbations to the system and the dynamic characteristics of these catalytic oxidation reactors also requires understanding. Wei[25] has reviewed some of the difficulties involved and refers to a number of modelling attempts. In a more recent paper Bruno *et al.*[26] describe the use of laser based diagnostics to measure the temperature and concentration profiles of gases in a single channel catalytic monolith. They conclude that at 400-500° C most of the hydrocarbon fuel is converted to CO_2 and H_2O at the catalyst monolith surface, although the importance of homogeneous gas phase oxidation becomes increasingly important at higher temperatures. The rate of oxidation is strongly mass transfer controlled. Using a two dimensional model to describe the continuity of mass and heat (taking into account the mass transfer limitation from fluid to solid) and recognising that the fuel may be converted to product by both catalytic and homogeneous oxidation, their model at least agrees qualitatively with the observed experimental results.

References

[1] W.M.H. Sachtler, *Rec. trav. chim.*, 1970, **89**, 460.
[2] C.F. Cullis and D.J. Hucknall, in 'Catalysis', *Chem. Soc. Special Report* Vol.5, p.273.
[3] J.D. Burrington and R.K. Graselli, *J. Catal.*, 1979, **59**, 79.
[4] D. Vanhove, S.R. Op, A. Fernandez and M. Blanchard, *J. Catal.*, 1979, **57**, 253.
[5] P. Forzatti, F. Trifiro and P.L. Villa, *J. Catal.*, 1978, **52**, 389.
[6] A. Bielanski and J. Haber, *Catal. Rev. Sci. Eng.*, 1979, **19(1)**, 1.
[7] L.Ya. Margolis, *Catal. Rev. Sc.*, 1973, **8(2)**, 241.
[8] D.W.L. Griffiths, H.E. Hallam and W.J. Thomas, *Trans. Faraday Soc.*, 1968, **64**, 3361.
[9] C. Bazilio and W.J. Thomas, unpublished results, University of Bath, 1983.
[10] I.E. Wachs and R.J. Madix, *J. Catal.*, 1978, **53**, 208.
[11] S.K. Bhattacharyye, N.K. Nag and N.K. Gangoly, *J. Catal.*, 1971, **23**, 158.
[12] M. Ai, *J. Catal.*, 1978, **54**, 426.
[13] R.G. Carrubba and J.L. Spencer, *Ind. Eng. Chem. (Process Des. Dev.)*, 1970, **9**, 414.
[14] C.N. Kenney and Y.N. Rojas, University of Cambridge, private communication.
[15] C.N. Kenney, *Catal. Rev. Sc. Eng.*, 1975, **11**, 197.
[16] P. Mars and D.W. van Krevelen, *Chem. Eng. Sc.*, 1954, **3**, 41.
[17] P.H. Calderbank and A.D. Caldwell, *Adv. Chem. Ser.*, 1972, **109**, 38.
[18] H. Beusch, P. Fieguth and E. Wicke, *Adv. Chem. Ser.*, 1972, **109**, 615.
[19] M. Goodman, M.B. Cutlip, C.N. Kenney, W. Morton and D. Mukesh, *Surface Sc.*, 1982, L453.
[20] V.D. Belyaev, M.M. Slin'ko, M.G. Slin'ko and V.T. Timoshenko, *Kinet. Katal.*, 1973, **14**, 810.
[21] E. McCarthy, J. Zahradnik, G.C. Kuczynski, and J.J. Carberry, *J. Catal.*, 1975, **39**, 29.
[22] M. Sheintuch, *A.I.Ch.E. J.*, 1981, **27**, 20.
[23] C.A. Pikios and D. Luss, *Chem. Eng. Sc.*, 1977, **32**, 191.
[24] V. Hlavacek and J. Rathousky, *Chem. Eng. Sc.*, 1982, **37**, 375.
[25] J. Wei, *Adv. Chem. Ser.*, 1975, **148**, 1.
[26] C. Bruno, P.M. Walsh, D.A. Santavicca, N. Sinha, Y. Yaw and F.V. Brucco, *Combustion Sc. Tech.*, 1983, **31**, 43.

Ethylene and Ether from Ethanol

By P. L. Yue*, O. Olaofe, and R. H. Birk
SCHOOL OF CHEMICAL ENGINEERING, UNIVERSITY OF BATH, CLAVERTON DOWN, BATH
BA2 7AY, U.K.

Economics of Ethylene from Ethanol

In a recent review[1] entitled 'Technology and economics of fermentation alcohol', the rapid development and implementation of plans for significant production of fuel alcohol in Brazil was reported. The project included the expansion of distilleries associated with existing sugar mills, the creation of new agricultural and manufacturing complexes to grow and process sugar cane into alcohol. Another programme was initiated to produce alcohol from mandioca, a tropical root crop which is also known as cassava, yuca and tapioca. It was emphasised in the review that new process improvements are essential to make fermentation alcohol an economically competitive fuel.

When examining the economics of biomass conversion, the potential of using fermentation alcohol as a chemical feedstock should not be overlooked. Industrial processes for the catalytic dehydration of ethanol to ethylene have been in existence even before World War I. A number of dehydration plants were built in the 1950's and 1960's in Asia and South America. Some of these are still in operation to-day, but others have given way to large olefin plants using naphtha as feedstock. Winter and Eng[2] have surveyed the status of ethanol-based ethylene plants in India and Brazil. They concluded that although not competitive when large-scale petroleum-based ethylene is available, ethanol-based ethylene has a definite place in the economics of developing countries where agricultural waste products are in abundance.

While ethanol has been dehydrated over a variety of catalysts, industrial methods have mostly been based on activated alumina or on phosphoric acid on a suitable support. Recently a commercially successful dehydration catalyst was reported,[3] but no details of the catalyst formulation were given. Although the vapour-phase dehydration is usually performed in fixed beds, the use of fluidised bed reactors has been studied.[4-6] The study by Irani $et\ al.$[4] showed that when alumina was mixed with solid inerts for a given set of operating conditions, product selectivity depended on the choice of reactor and catalyst dilution ratio. Besides the obvious importance of the price of the raw material (ethanol), proper choices of catalyst and reactor system are therefore critical to the cost of ethylene production. Capital requirements are also significantly affected by the grade of ethylene produced.

Reaction Fundamentals

Apart from water, ether has been found to be the other principal product of ethanol dehydration. Table 1 gives a summary of the kinetics of the reactions and product selectivity reported by various investigators. It can be seen form this table that there is no universal agreement on many important aspects of the apparently simple reaction of ethanol dehydration.

Table 1 Studies of catalytic dehydration of ethanol

Catalyst	Temperature (T/K)	Selectivity (r_o/r_E)	Reaction Pathway	Kinetic Model	Ref.
Alumina	588-668	high	consecutive	-	2
Alumina	580	high	parallel	H-W	7
Linde NaA	543-621	low	-	-	8
Linde Ca, MgA		high			
Linde NaX		low			
Linde Ca, MgX		low			
Norton NaZ		high			
Linde 5A, 10X, 13X	573-648	high	-	P-L	9
Dowex 50X8	353-396	zero	-	H-W	10

A variety of reaction pathways has been proposed to account for the formation of ethylene and ether. The three principal schemes are: (i) consecutive (ii) parallel and (iii) combinations of (i) and (ii). The pros and cons of different reaction pathways have been discussed elsewhere.[11,12] Whichever scheme one prefers, the thermodynamics of the reactions are such that ethylene formation tends to be irreversible while the ether formation reaction can be reversible. Different reaction mechanisms based on surface intermediates such as a carbenium ion or an 'aluminium alcoholate' have been suggested[13,14] to explain the primary reaction steps.

In studies where reaction kinetics were examined, rate data were represented either by power law or Hougen-Watson type of rate models. For power law expresssions, the reaction orders varied between zero to one with respect to alcohol. For the Hougen-Watson models, surface reaction was usually assumed to be rate-controlling. Both single-site and dual-site mechanisms have been tested and are supported equally well by experimental data.

Besides alumina, synthetic zeolites have been found to be effective in ethanol dehydration. Bryant and Kranich[8] used Linde A, X and Z type zeolites and conducted an interesting study on the effect of cations in the zeolites on product selectivity. Some of their results, however, appear to differ from those of others.[9] Nevertheless, the molecular sieving properties of zeolites are expected to influence product selectivity, thus offering some significant advantages over alumina. Another type of dehydration catalyst that is of interest is the sulfonic acid ion exchange resins. Despite the lack of rigid pores and true internal surface in the resins, high catalytic activity can be obtained even at relatively low temperatures due to the presence of the sulfonic acid groups.[10]

Experimental

The present paper presents the results of a laboratory investigation of the dehydration of ethanol. The reaction was conducted in the vapour phase over three types of commercial zeolites and a synthetic ion exchange resin. The kinetics of the product formation reactions as well as the effects of the choice of catalyst and reactor conditions on product selectivity were studied.

Catalysts and Reactor. The three commercial zeolites used were: Laporte 4A, 13X and Norton Zeolon, a NaZ modernite. They cover a varied range of cage structures, pore sizes and acidity. The Laporte zeolites are spherical beads while the Norton Zeolon is in the form of cylindrical extrudates. The uniform and well defined intracrystalline structure of the zeolites is, of course, the basis for their shape-selective properties. The cation exchange resin used was the commercially available Dowex 50X-8 which was a cross linked styrene-divinylbenzene polymer with sulfonic acid groups introduced. It has a matrix framework consisting of an irregular macro-molecular three-dimensional network of hydrocarbon chains. The physical and chemical properties of these four catalysts are given in Table 2.[12,15]

A stirred gas-solid reactor (SGSR) with internal recycle was used. Details of the reactor have been presented elsewhere.[12,15,16] Internal recycle in the SGSR was achieved by spinning the catalyst which was housed in baskets made of wire gauze. The spinning speed was varied to establish the conditions at which gas-solid interphase mass transfer was not rate-controlling.

Table 2 Properties of Catalysts

Property	4A	NaZ	13X	Dowex 50X8
Chemical formula	$Na_{12}Al_{12}Si_6O_{36} \cdot 27H_2O$	$Na_8Al_8Si_{40}O_{96} \cdot 24H_2O$	$Na_{41}Al_{41}Si_{106}O_{294} \cdot 264H_2O$	(styrene-divinylbenzene sulfonic acid resin)
SiO_2/Al_2O_2	2.0	10.0	2.6	–
Nominal aperture, nm	0.42	0.57 x 0.70	1.0	2
Surface area, $m^2 g^{-1}$	10	103	359	850
Bulk density, $kg\ m^{-3}$	740	680	670	–
Framework density, $kg\ m^{-3}$	1270	1700	1310	0.56
Particle size, mm	1-2	1.6 x 3.5	1-2	
Binder	natural clay	self-binding	natural clay	self-binding
Crystalline structure	simple cubic	elliptical	body centre cubic	cross-linking
Physical form	spherical beads	cylindrical extrudates	spherical beads	spherical beads

Procedures. Absolute alcohol of purity higher than 99.5% was delivered by a metering pump to an evaporation unit where it was mixed with dried high purity nitrogen diluent. The mixed feed was heated to nearly the chosen reaction temperature before entering the reactor. The reactor was heated in a temperature-controlled chamber. The range of temperature used was between 270-375° C for the zeolites and 120-142° C for the cation exchange resin. The ethanol feed rate was varied to achieve a range of alcohol partial pressures between 10-80 kPa. Reactor effluent was analysed by a gas chromatograph using a Porapak Q column and a flame ionisation detector. Details of the analyses can be found in references 12 and 15. Mean values of the measurements were obtained from three or more reproducible analyses. Error analysis gave a deviation of no more than 6% from the mean for each component.

Catalyst activity was monitored to determine the extent of deactivation if present. The experiments using Laporte 4A, Norton Zeolon and Dowex 50X-8 were performed within time periods in which catalyst deactivation was negligible or not measurable. For the experiments with Laporte 13X, the catalyst was put on stream for over 40 days. The activity of the catalyst was evaluated as a function of time on stream. From duplicate experiments at reference conditions, it was therefore possible to account for catalyst deactivation in the processing of rate data.

For the zeolites, internal mass transport limitations, apart from intracrystalline diffusion, were evaluated using Laporte 13X by varying the particle size using 0.1-0.3 mm powders and pellets up to a diameter of 8-9 mm. A fixed-bed reactor was used for these experiments because the baskets in the SGSR were not suitable for holding powders.

Results and Discussion

Side Reactions and Transport Limitations. Small quantities of acetaldehyde were formed in some cases. The aldehyde was a product of the ethanol dehydrogenation reaction which was catalysed by the nickel and copper present in the reactor and the associated piping materials rather than by the catalysts. The rate of side-product formation was accounted for in the material balances to obtain the product rates of the dehydration reactions.

External mass transfer at the gas-solid interface was not rate-limiting in the temperature range studied when the SGSR stirred speed was at 2000 revolutions per minute or higher. The experiments performed with different catalyst particle sizes showed that the rates obtained from the 1-2 mm zeolite beads were not controlled by internal mass transfer, with the possible exception of intracrystalline diffusion. Possible influence of intracrystalline diffusion on the reaction rates is a subject to be discussed elsewhere,[17] but will be noted wherever appropriate.

<u>Catalyst Activity and Selectivity.</u> The Laporte 4A zeolites have a nominal cage aperture size of 0.42 nm. The critical diameters of ethylene, ethanol and ether molecules are 0.39, 0.41 and 0.41 nm respectively. Despite the proximity of the dimensions of the molecules to the cage aperture size, both ethylene and ether were formed over the 4A zeolites. However, it should be noted that the catalytic sites are not exclusively intracrystalline, $i.e.$ within the zeolite crystalline structure. There can be a significant number of intercrystalline sites, $i.e.$ between the crystals, which contribute to catalysing the reactions. Increasing the reaction temperature favours the formation of olefin. The equilibrium constant for the ether formation reaction decreases with rising temperature; thus the reverse hydration reaction becomes significant at higher temperatures. It can be seen from Figure 1 that product selectivity, defined as the ratio of the rate of formation of ethylene to that of ether, rises sharply with increasing temperature. The effect of reactant concentration, expressed as partial pressure of ethanol, on product selectivity is also illustrated in Figure 1. Selectivity for ethylene is reduced at higher concentrations of alcohol.

The Norton Zeolon NaZ mordenite consists of tubes with 0.57 x 0.70 nm elliptical sections. These channels should offer less restriction to the movement of the molecules of the reactions. The mordenite has a SiO_2/Al_2O_3 ratio of 10 which is much higher than that of Laporte 4A, and is therefore more acidic than the latter. The surface area of the former is also ten times that of the latter. In spite of these factors, ethanol dehydrates at a much slower rate over the Norton Zeolon than the Laporte 4A. Ether was not always found in the products. The mordenite is more selective towards the formation of ethylene. Of the three zeolites studied, it showed the least dehydration activity. The low activity of NaZ suggests that the dehydration reactions take place via an acid-base mechanism. The NaZ lacks basic sites that are needed for the abstraction of hydrogen atoms from the β carbon.

The Laporte 13X possesses the largest cages among the three zeolites, having a nominal aperture size of 1.0 nm. It has a highly developed surface area of 359 m^2 g^{-1}. These are factors which make the 13X the most active dehydration catalyst. Both acid and basic sites are present in the 13X and its Si/Al ratio is 2.6 which is intermediate between that of Laporte 4A and

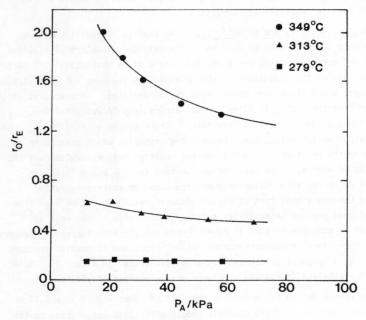

Figure 1 Selectivity over Laporte 4A

the Norton Zeolon. Catalyst deactivation was significant in the extended runs using the SGSR. The catalyst was losing about 1/4 to 1/2% of its initial activity per hour. The rate of deactivation was fairly well represented by an exponential decay model.[18] The rate data were treated with the effect of catalyst deactivation fully accounted for. The variation of the product formation rates with the partial pressure of ethanol at different reaction temperatures is illustrated in Figures 2 and 3. Again ethylene to ether selectivity was found to increase as temperature was raised and as alcohol partial pressure was reduced; see Figure 4.

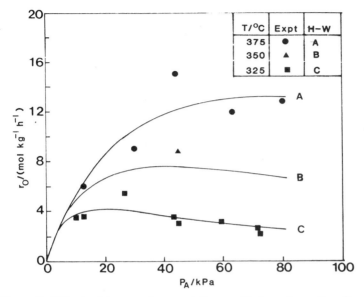

Figure 2 Effect of temperature and P_A on ethylene formation (Laporte 13X)

The reaction temperature required for ethanol dehydration over the synthetic cation exchange resin Dowex 50X8 is much lower. Indeed, temperatures beyond 160° C would soften the resin and render it useless. The catalyst does not have micropores or true internal catalytic sites. The overall dehydration rates obtained from the exchange resin are comparable with that from Laporte 4A, but the resin favours the production of ether. The ethylene formation rate rises sharply at low alcohol partial pressures and then drops or levels off beyond 10-20 kPa. The ether rate, however, increases progressively with increasing partial pressure of ethanol. In the case of the zeolites, production rates of both products slow down at higher alcohol concentrations. This suggests that for the cation exchange

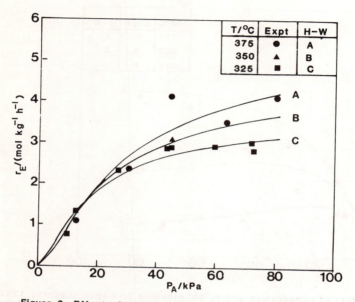

Figure 3 Effect of temperature and P_A on ether formation (Laporte 13X)

resin, the saturation partial pressure of ethanol for ether production is higher than that for ethylene. It can also be seen from Figure 5 that at temperatures around 120° C, nearly exclusive production of ether can be achieved.

Kinetics of Ethylene Formation. All the rate data have been modelled according to both empirical power law (P-L) and Hougen-Watson Langmuir-Hinshelwood (H-W) type of kinetic expressions. Preliminary discrimination of all plausible rate expressions reduced the number of reasonable models to the following:

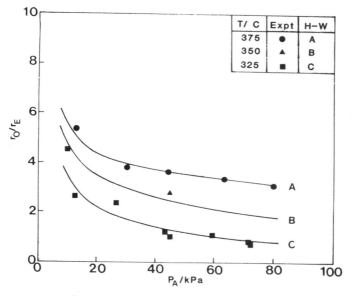

Figure 4 Selectivity over Laporte 13X

P-L $\qquad r_0 = \bar{k}_0 p_A{}^{n_0}$ (1)

H-W Dual Site $\qquad r_0 = \dfrac{k_0 K_{AO} p_A}{(1+K_{AO} p_A)^2}$ (2)

H-W Single Site $\qquad r_0 = \dfrac{k_0' K_{AO}' p_A}{(1+K_{AO}' p_A)}$ (3)

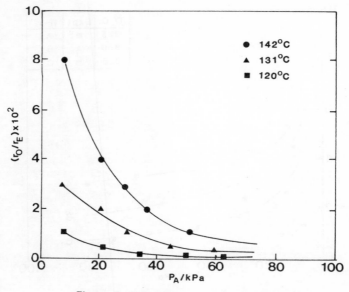

Figure 5 Selectivity over Dowex 50-X-8

In the latter models, surface reaction is assumed to be rate-controlling. Product adsorption-desorption terms are negligible and have been excluded from the denominator of equations (2) and (3). Similarly, product-desorption controlling has been ruled out. Parameters in the models were estimated according to the methods proposed by Kittrell.[19] Both multi-linear and non-linear regression analyses were applied to the experimental data. Results from linear regression were used as first estimates for non-linear regression. Values of the model parameters and details of the kinetic analyses will be presented elsewhere.[17,20]

The empirical power law can adequately represent ethylene formation over Laporte 4A and Norton Zeolon, but not the data for Laporte 13X and Dowex 50X8. The H-W single site surface reaction mechanism also gives satisfactory correlation of data for the two former catalysts. Only the H-W dual site mechanism, *i.e.* equation (2), succeeds in representing the kinetic data over all four catalysts. The solid lines in Figure 2 represent model predictions for ethylene formation over Laporte 13X. There is excellent agreement between the values of the model parameters obtained from both linear and non-linear techniques. The kinetic parameters at different reaction temperatures follow the Arrhenius equation. The temperature variation of the adsorption-desorption equilibrium constants is in accordance with the van't Hoff type of equation. It should be noted that although the dual site surface reaction model fits the rate data well, in the case of ethylene production over Laporte 13X, intracrystalline diffusion is becoming the rate-controlling step at 375° C. The shift in rate-controlling step is reflected in the activation energy of the reaction.[17]

<u>Kinetics of Ether Formation.</u> Ether formation, of course, is a bimolecular reaction. The reverse reaction may also be significant, especially at higher temperatures. Preliminary screening of rate expressions led to the following candidates for modelling:

P-L
$$r_E = \bar{k}_E p_A^{n_E} \qquad (4)$$

L-H
$$r_E = \frac{k_E K_{AE}^2 (p_A^2 - p_W p_E / K^*)}{(1 + K_{AE} p_A)^2} \qquad (5)$$

R-E
$$r_E = \frac{k_E' K_{AE}' (p_A^2 - p_W p_E / K^*)}{(1 + K_{AE}' p_A)} \qquad (6)$$

The Langmuir-Hinshelwood (L-H) model assumes that surface reaction takes place between two adjacently chemisorbed alcohol molecules to form the chemisorbed products. The Rideal-Eley (R-E) model assumes that an alcohol molecule in the gas phase reacts with another chemisorbed alcohol molecule, but not in the presence of an adjacently vacant site. The product ether goes directly into the gas phase.

Again the power law fails to represent the data for Laporte 13X and Dowex 50X8. Neither does the R-E model give very successful correlation. Satisfactory fitting of rata data for all four catalysts is achieved only by the L-H model. The predicted rates of ether formation over 13X by the L-H model have been plotted as solid lines in Figure 3. Good agreement between model prediction and experimental rates over Dowex 50X8 can also be seen in Figure 6. Once more, it should be mentioned that a satisfactory statistical test of models does not imply proof of reaction mechanism. Values of activation energy[17] for ether formation over 13X show that a transition of the rate-controlling step from surface reaction to intracrystalline diffusion occurs at about 350° C.

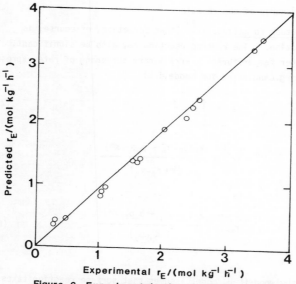

Figure 6 Experimental ether formation rate over Dowex 50-X-8 compared with model prediction

Conclusions

Ethanol can dehydrate to ethylene and ether over zeolites Laporte 4A, 13X, Norton Zeolon NaZ and cation exchange resin Dowex 50X8. The activity of the zeolites follows the order

Laporte 13X > Laporte 4A > Norton Zeolon NaZ

In the case of the Laporte zeolites, selectivity of ethylene to ether decreases as the zeolite cage aperture size increases. High temperature and low alcohol partial pressure favour the formation of ethylene. The cation exchange resin is active at much lower reaction temperatures and produces nearly exclusively ether at around 120° C.

Kinetic analysis shows that the Hougen-Watson type of rate expression based on a dual-site mechanism gives the best representation of rate data. However, intracrystalline diffusion can limit the rates at certain temperatures.

References

[1] C.R. Keim, *Enzyme Microb. Technol.*, 1983, 5, 103.
[2] O. Winter and M-T. Eng, *Hydrocarbon Processing*, Nov. 1976, 125.
[3] N.K. Kochar, R. Merims and A.S. Padia, *Chem. Eng. Prog.*, 1981, 77, No.6, 66.
[4] R.K. Irani, B.D. Kulkarni, L.K. Doraiswamy and S.Z. Hussain, *Ind. Eng. Chem. Process Des. Dev.*, 1982, 21, 192.
[5] T.R.A. Magee, PhD Thesis, 1975, Queen's University, Belfast.
[6] I.A. Zenkovich and K.V. Topchieva, *Int. Chem. Eng.*, 1971, 11, 393.
[7] J.H. de Boer, R.B. Fahim, B.G. Linsen, W.J. Visseren and W.F.N.M. de Vleesschauwer, *J. Catal.*, 1967, 7, 163.
[8] D.E. Bryant and W.L. Kranich, *J. Catal.*, 1967, 8, 8.
[9] T.N. Krishnaprasad and M. Ravindram, *Hungarian J. Ind. Chem.*, 1977, 5, 13.
[10] R.L. Kabel and L.N. Johanson, *J. Chem. Eng. Data*, 1961, 6, 496.
[11] M.E. Winfield, "Catalysis", P.H. Emmett (ed.), Reinhold, New York, 1960, Vol.7, p.93.
[12] R.H. Birk, PhD Thesis, 1983, University of Bath.
[13] W.S. Brey and K.A. Krieger, *J. Amer. Chem. Soc.*, 1949, 71, 6637.
[14] K.V. Topchieva and K.Yun-Pin, *Zh. Fiz. Khim.*, 1955, 29, 1854.
[15] O. Olaofe, PhD Thesis, 1982, University of Bath.
[16] M.L. Brisk, R.L. Day, M. Jones and M.A. Warren, *Trans. I. Chem. E.*, 1968, 46, T3.
[17] R.H. Birk, P.L. Yue and W.J. Thomas, paper to be presented at 8th Int. Symp. Chem. Reaction Eng., Edinburgh, 1984.
[18] S. Szepe and O. Levenspiel, *Proc. 4th European Symp. Chem. Reaction Eng.*, Brussels, 1968, p.265.
[19] J.R. Kitterell, "Adv. Chem. Eng.", T.B. Drew, G.R. Cokelet, J.W. Hoopes and T. Vermeulen (eds), Academic Press, New York, London, 1970, Vol.8, p.97.
[20] O. Olaofe and P.L. Yue, paper to be published.

Nomenclature

$\bar{k}_E$, $\bar{k}_O$	P-L kinetic rate constants for ether and ethylene formation respectively, in mol kg^{-1} h^{-1} kPa^{-n}.
k_E, k_O	H-W kinetic rate constants based on dual site mechanism for ether and ethylene formation respectively, in mol kg^{-1} h^{-1}.
k_E'	R-E kinetic rate constant based on single site mechanism for ether formation, in mol kg^{-1} h^{-1} kPa^{-1}.
k_O'	H-W kinetic rate constant based on single site mechanism for ethylene formation, in mol kg^{-1} h^{-1}.
K^*	thermodynamic equilibrium constant for ether formation.
K_{AE}, K_{AO}	kinetic adsorption coefficients based on dual site mechanism for ether and ethylene respectively, in kPa^{-1}.
K_{AE}, K_{AO}'	kinetic adsorption coefficients based on single site mechanism for ether and ethylene respectively, in kPa^{-1}.
n_E, n_O	order of ether and ethylene formation reactions respectively.
p_A	partial pressure of alcohol, in kPa.
p_E	partial pressure of ether, in kPa.
p_W	partial pressure of water, in kPa.
r_E, r_O	rate of ether and ethylene formation respectively, in mol kg^{-1} h^{-1}.
T	reaction temperature, in K or °C.

Acknowledgement

The catalysts were provided *gratis* by Laporte Industries, Norton Chemicals and Dow Chemical. Helpful discussions with Mr. C.W. Roberts of Laporte Industries Ltd. and Professor W.J. Thomas of the University of Bath are much appreciated. O. Olaofe was supported by a British Commonwealth Scholarship, and R.H. Birk by a Rotary International Foundation Scholarship and a University of Bath Studentship.

Novel Pathways to Enhance Selectivity in Fischer–Tropsch Chemistry

By P. A. Jacobs
KATHOLIEKE UNIVERSITEIT LEUVEN, KARDINAL MERCIERLAAN 92, B-3030 LEUVEN, BELGIUM

Introduction

The catalytic conversion of synthesis gas, i.e. carbon monoxide and hydrogen mixtures has been the subject of a renewed and concentrated research mainly during the past five years. This effort has been based on two key assumptions[1] :
- the cost of crude oil and oil-derived basic chemicals and fuels will continue to escalate rapidly
- the gasification of coal and of biomass will become economically attractive. Consequently, a very large number of molecules, as hydrocarbons and oxygenates, has been catalytically prepared from synthesis gas.

Typical product molecules from synthesis gas are given in Table 1, together with the synthesis gas composition required and the usage of the raw material. Processes in which all carbon and oxygen of the feed gas is retained in the product molecule are more efficient than processes which produce hydrocarbons and waste water or carbon dioxide. The first class of processes is used to produce oxygen-containing basic feedstocks for the chemical industry, while the hydrocarbon-producing processes are situated mainly in the fuel industry or consist of so-called alternative processes for the production of non-oxygen-containing chemical raw materials, such as light olefins and BTX. In the latter class of processes decreased feed efficiencies are obtained when oxygen is eliminated as carbon dioxide instead of water. On the contrary a lower H_2/CO ratio is required in such cases, which results in a less severe catalytic adjustment of the synthesis gas composition via the water-gas-shift (WGS) reaction :

$$CO + H_2O \rightleftharpoons CO_2 + H_2 \qquad \Delta H_R = -40 \text{ kJ} \qquad (1)$$

Table 1. Theoretical reaction stoichiometry of synthesis gas conversions

Product	H_2/CO ratio	feed usage in g feed per g of product[a]
methanol	2	1.0
ethanol	2	1.4
acetic acid	1	1.0
ethylene glycol	1.5	1.0
acetaldehyde	1.5	1.0
methane	3.0 (1)[b]	2.1 (3.7)[b]
ethylene	2.0 (0.5)[b]	2.3 (4.1)[b]
benzene	1.5 (0.3)[b]	2.4 (4.4)[b]
isooctane	2.1 (0.6)[b]	2.3 (4.1)[b]
hexadecane	2.1 (0.6)[b]	2.3 (4.1)[b]

a, with water as eventual co-product; b, with CO_2 as co-product

These hydrocarbon synthesis reactions have been known as the Fischer-Tropsch (FT) synthesis. A modern view[2] claims that FT-chemistry can be limited to the following basic reaction :

$$CO + 2H_2 \longrightarrow (-CH_2-) + H_2O \qquad \Delta H_R = -165.5 \text{ kJ} \qquad (2)$$

The net overall reaction of (1) and (2) is :

$$2CO + H_2 \longrightarrow (-CH_2-) + CO_2 \qquad \Delta H_R = -205.4 \text{ kJ} \qquad (3)$$

Catalytic manufacture of hydrocarbons which appear the least adventageous group of processes from synthesis gas will be considered in the present paper. Only a few years ago these routes from coals to chemicals and fuels were projected to become economically attractive by 1990 since crude oil prices were predicted to reach $ 100 per barrel[3]. Current projections claim that crude oil prices in 1990 will only amount to half of that value, as a result of a decreased demand for crude due to oversupply and an increased emphasis on natural gas and natural gas liquids[1].

In view of all this, the focus of this article is on this of the technical factors that is decisive for the success of future FT-type chemistry : its selectivity. It will include a discussion on selectivity using Sasol and Mobil technology and some recently published results in key areas, related to selectivity enhancement.

FT Product Selectivity using SASOL technology

Product Distribution in Synthol and Arge Reactors. The price ratio of oil versus coal, which was at about 10:1 in Europe in the period 1920-1940, was a strong incentive for the production of liquid fuels from coal-derived synthesis gas. The cost ratio of oil to coal today in South Africa is even higher[3]. As a result three FT plants (SASOL I, II and III) are in operation, using fixed bed reactors (ARGE technology) or entrained fluidized beds (Synthol reactors), using promoted iron catalysts. Typical carbon-atom selectivites in terms of fuel compositions for the two types of commercially operated reactors are given in Figure 1. The Arge reactors produce mainly long-chain hydrocarbons, the Synthol

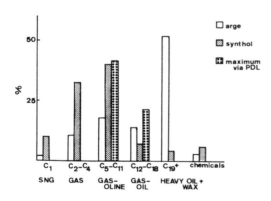

Figure 1. Carbon atom selectivity using SASOL technology (partially after ref. 5)

reactors mainly gasoline and gases rich in olefins[5]. The gasoline product cut form the Synthol reactor contains mainly olefins and has a relatively high lead-free research octane number (RON), while the gasoline cut straight from the fixed bed reactor has a low octane number since it is mainly n-paraffinic in nature.

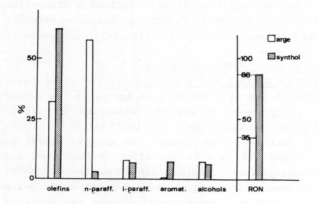

Figure 2. Straight SASOL gasoline composition (after ref. 4)

On the other hand, the diesel fraction from the Arge reactor has a high cetane number since it is very rich in n-paraffins (Fig. 3). The diesel product cut from the fluidized bed reactor is less interesting for straight use since it contains mainly olefins and a non-negligible amount of aromatics. Therefore, subsequent upgrading processes will be needed in order that the product compositions meet the required specifications as fuels for internal combustion engines.

Secondary Separate Upgrading of Product Cuts from Sasol Reactors. The secondary upgrading of the low octane and cetane fuels occurs via classical refinery procedures (Table 2), dehydration and isomerization reactions. The diesel and gasoline yield is increased via oligomerization of light olefins, while tailgas is either regenerated by steam reforming or transformed into hydrogen by a subsequent water-gas-shift process and used for the synthesis of ammonia. Wax can be hydrocracked to extinction into diesel fuel. These secondary upgrading procedures give the whole FT plant a high flexibility with respect to diesel and gasoline production. Variations in the diesl to gasoline production ratio between 1 and

10 can be obtained[5,6].

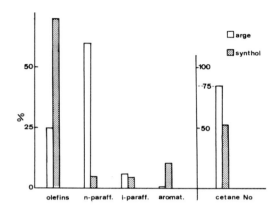

Figure 3. Straight SASOL gasoil composition (after ref. 4)

Table 2. Secondary catalytic upgrading of Sasol FT products (4,5)

F.T. Product	Catalyst	Conversion from/to	reaction
Synthol gasoline	(unspecified)	acid oxygenates/olefins	dehydration
C_3+C_4 gas	H_3PO_4/kieselguhr	olefins/isoolefins	skeletal isomerization
tailgas	nickel	CH_4/synthesis gas	steam reforming
oils	unspecified	oils/high quality fuels	hydrofining reforming isomerization dewaxing
wax	unspecified	wax/diesel	hydrocracking

Other Fischer-Tropsch Technology. Besides the fixed and fluidized bed reactors a three-phase slurry reactor was developed by Rheinpruessen-Koppers[14,15] as an alternative to overcome the disadvantages of the fixed bed reactors :
- insufficient heat removal
- need for hydrogen-rich syn-gas,

- low conversion,
- nonuniform catalyst loading.

This slurry reactor was a 8.6 m bubble column, with a diameter of 1.5 m. The slurry consisted of molten wax, which is a product fraction from the synthesis with a promoted iron catalyst[14].

Other types of reactors have been used by the Pittsburgh department of Energy (PETC)[13] as oil circulation, wet and hot gas recycle and tube wall reactors. Typical operation conditions and product cuts are given in Table 3 for these reactors. Their mode of

Table 3. FT process selectivity investigated at the Pittsburgh Energy Center (after ref. 13)

reactor	oil circulation	cold gas recycle	hot gas recycle	tube-wall
catalyst	fused iron	steel lathe turnings	lathe turnings	flame sprayed taconite
H_2/CO	1:1	1.3	1.5	2.1
catalyst age / h	458	442	488	517
space velocity / h^{-1}	602	1002	607	-
temperature K	517	575	581	613
pressure / bar	21	28	28	46
(H_2 + CO)conv. / %	67	89	80	93
Product distribution / wt%				
gasoline (C_3-477K)	59	44	59	18
gasoil (977-590K)	13	4	9	2
heavy distillate (590-723K)	19	2	6	-
wax (723K)	10	1	3	-
alcohols	-	10	-	-

operation is summarized in Table 4. The product distribution slightly differs for the different processes although modification at the catalyst level is able to overcome these differences. Indeed, the high gas yield obtained in the tube-wall reactor with taconite can be decreased substantially in favor of the gasoline and gasoil yield by promoting this catalyst with K_2O.[13] It seems that the use of this type of reactor would be a major factor to increase overall plant thermal efficiency[13].

Table 4. Mode of operation of FT reactor used at PETC (after ref. 13)

Reactor	Mode of operation	Characteristics
Oil circulation	- syngas + recycle oil through a fixed bed - oil is externally cooled	- minimum pressure drop - good catalyst life with steel lathe turnings
Slurry	- syngas through a suspension of catalyst in oil - heat adsorbed by oil	- nitrided iron produces high amount of oxygenates
Gas recycle	- large volumes of recycle gas through fixed bed remove heat	- excellent temperature control
Tube wall	- catalyst flame-sprayed to outer wall of heat exchanger tube	- long catalyst life with flame sprayed-taconite - regeneration possible

Kinetic Limitations upon FT Selectivity
The Polymerization Distribution Model. Dry[6] from Sasol experimental data reported an interdependency of FT product fractions. It follows that when the concentration of one component is determined by catalyst nature or operational conditions that all others are fixed. It was already recognized at a very early state of FT research that the product distribution could reasonably well be described by a chain growth polymerization mechanism[7,8] or by the so-called Flory-Schulz polymerization distribution law[9,10] (PDL) :

$$W_n = n\alpha^{n-1}(1-\alpha)^2 \qquad (4)$$

W_n represents the weight fraction of carbon number n, α the growth factor and n the product carbon number. This model implies that :
(a) chain growth of the products occurs via a stepwise addition of C_1-intermediates
(b) the growth factor α is constant and independent of chain length and is the only parameter which determines carbon number product selectivity

(c) chain growth termination occurs via desorption and as a result 1-olefins are primary products.

It requires that a semi-logarithmic plot of the molar product distribution against their carbon number is linear.

A high number of published FT product distributions can be fairly accurately described by this formalism,[11] including as well results from laboratory reactors, pilot plants and large scale units. Theoretical variations of the carbon number distributions with the growth factor α are shown in Figure 4. The higher this

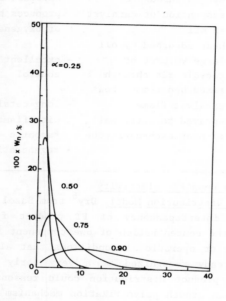

Figure 4. Theoretical FT distribution according to the polymerization distribution law or Schulz-Flory kinetics : weight % of different product carbon numbers at different values for the growth factor α.

factor becomes, the wider is the range of observed carbon numbers and the lower the selectivity of the individual products. For any carbon number an optimal polymerization degree exists with a corresponding maximum selectivity. A typical product distribution in terms of these carbon numbers is given in Figure 5 together with the maximum yields of certain product cuts : SNG, gasoline, gasoil and waxes. These maxima for gasoline and gasoil are relatively close to the experimentally observed yields obtained in Synthol and Arge reactors, respectively (Figure 1). It should be noted that minor deviations from the PDL-behaviour occur often at the C_1 and C_2 yields : the C_1 content being too high, the C_2 level being too low. Possible reasons for this behaviour have been discussed earlier[12], but are not relevant in the present discussion.

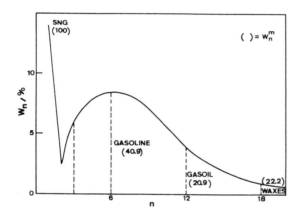

Figure 5. Practical FT product distribution against their carbon numbers, observed for a multitude of catalysts together with the theoretical maximum values for certain product cuts which can be obtained according to SF kinetics.

Factors determining α. The growth factor α is dependent on the catalyst composition, i.e. the nature of the catalytic element and the presence of basic promotors. Physical parameters as reaction temperature and synthesis gas composition also influence this growth factor. Their qualitative interrelation is given in Table 5.

Improved Selectivity using New Catalyst Materials
Carbon Number Distributions which deviate from Schulz-Flory behaviour. The previous section clearly indicates that FT selectivity on a given catalyst can be optimized by changing the physical operational parameters till certain critical values.

Table 5. Parameters determining the growth factor of FT carbon numbers on catalysts following a PDL-formalism[6,12]

---

1. Nature of catalyst : - changes in the order : Ru > Co > Fe

2. Presence of basic promotors or CO_2 increases α : higher basicity gives higher α values

3. Synthesis gas composition : α is increased for decreasing CO/H_2 ratios

4. Reaction temperature : α decreases at increasing reaction temperatures

---

Further improvement of selectivity for fuels, light olefins aromatics or C_{15}-C_{20} olefins (for detergent use) will require catalysts which circumvent SF-kinetics. Efforts are reported by shifting the primary reaction routes. Non PDL-behaviour has been reported for metals on special supports such as zeolites and microporous aluminas. A selection of literature data showing deviations from a polymerization growth scheme is available.[12] It seems that the supports mentioned are able to cut the carbon number distribution of the FT products at a value which apparently is determined by the pore size of the support or the particle size of the metal. In other words, the growth probablility α remains constant up to a certain value for n and then drops considerably to zero or to a lower value. Examples for representative catalytic systems are given in Figure 6. The variation of α with carbon number or the non-PDL-behaviour was explained more or less satisfactorily by the respective authors who reported the data. Rationalization of these data was done along the the following lines[12] :
- in zeolites, the existence of cage effects and size effects of the metal phase were used to explain the enhanced selectivity;
- on other supports, the existence of diffusion effects were thought to be responsible for this behaviour.

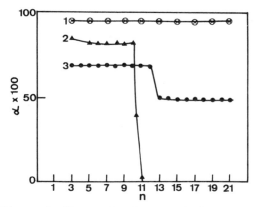

Figure 6. Change in the growth factor with carbon number for FT products obtained with 1, Ru on silica (after ref. 12); 2, Ru on NaY zeolite (after ref. 16); and 3, Co on alumina (after ref. 17).

Besides these other possible reasons for non-PDL-behaviour are reported as[12] :
- wax deposition in the catalyst bed
- operation in transient conditions
- artifacts in sampling
- occurrence of secondary reactions
- presence of discrete types of sites
- mechanistic effects : e.g. chain growth via 2 discrete intermediates.

Although this kind of FT chemistry is essential in the preparation of tailor-made catalysts, the effects are not clearly established and difficult to reproduce from one research group to another[17]. However, the investigation and understanding of this kind of effects seems timely and economically relevant.

Chemically Modified Iron Catalysts for Light Olefin Production from Syngas. During the past several years many modified catalysts for light olefin production have been reported. The literature on this has been reviewed[12,24]. A majority of the data has been obtained in non-reliable conditions, mainly at very low CO conversions, sometimes lower than 1 %, and after short reaction times so that steady-state conditions are not reached and accurate carbon mass balances are impossible to make.

Data on catalysts which do not seem to suffer from these drawbacks are collected in Table 5. Relatively high yields of light olefins can be obtained this way with ethylene yields which are definetely higher than those predicted by the polymerization distribution law. The carbon number distribution however follows the PDL-formalism[12] with a characteristic drop in α around C_5. Common to these approaches is the presence of very basic manganese oxide in the catalyst formulation, at least under reaction conditions. Although the detailed chemistry is not understood, the presence of basic oxides as MnO is known to result in a decreased hydrogen chemisorption capacity on the metal. This phenomenon is rationalized in terms of Strong-Metal-Support-Interaction (SMSI)[28].

<u>Improved FT Selectivity via Secondary Effects</u>. It was shown earlier that FT product upgrading can be done <u>via</u> classical technology. With the advent of the new family of shape-selective catalysts (ZSM-5 or ZSM-11) used by Mobil Company in the methanol to gasoline conversion[18], a more efficient upgrading is possible. The shape

Table 5. Chemically modified iron catalysts for light olefin production from syngas

Catalyst	alloy of iron and electrolytically pure manganese	100 Fe : 100 Mn : 10 ZnO : 4 K_2O	Ru on MnO
Temperature/K	356	593	623
Pressure / bar	14	10	10
H_2/CO	0.8	1.0	0.5
GHSV/h^{-1}	353	500	350
C_2-C_4 / wt %	50.2	86.6	79.0
% olefins	80	82	88
$\dfrac{C_2^=}{C_2^=+C_3^=}$ ×100	31	59	48
ref.	25	26	27

selective features of these zeolite supports restrict the size of the product molecules. In this way whole FT effluents or fractions of it can be converted either into a high-octane gasoline, into a mixture of light olefins and BTX or into diesel or an olefinic naphtha fraction. This can be achieved in separate reactors or in a single reactor containing a dual component catalyst consisting of a classical FT catalyst and a Mobil methanol cracking catalyst. The degree of intimacy in the mixing of the 2 components is an important parameter in the selectivity of the final catalyst. Since recently modified catalysts have been presented by other companies, different processes will be discussed separately.

Upgrading to High Octane Gasoline using a Mobil Methanol Cracking Catalyst

Using model hydrocarbon-type feedstocks, it has been shown that the same product selectivity (i.e. light olefins and aromatics) can be obtained over a HZSM-5 catalyst as with methanol as feed[29]. A high octane gasoline can also be obtained when the whole FT-effluent or fractions of it are passed over such a Mobil Methanol Cracking catalyst. This is illustrated in Table 6 for different

Table 6. Octane number (R+O) of FT feeds treated by a Mobil methanol cracking catalyst

Feed type	Feed	Product	Reference
Sasol FT effluent	55	92	19
C_5-477K liquid fraction	59	90.6	20
Aqueous FT phase + lower boiling cut (below 477K)	-	92	21
Synthol light oil	54	92.7	22

FT product cuts. It is striking that a single fixed bed operation is almost able to double the octane number. Table 7 shows the changes a FT product cut undergoes upon passing it over ZSM-5 type zeolites. The major part of the olefin content (the primary FT products) are transformed into a mixture of aromatics, naphthenes and to a minor extent to paraffins. This change in product composition explains the increased octane numbers of this kind of product.

Table 7. Change in composition of a FT (C_5-477K) liquid fraction after treatment with a HZSM-5 catalyst (after ref. 20) at 602 K and WHSV of $2h^{-1}$

	Charge	Product
R + O	59	90.6
Composition/ vol. %		
Paraffins	31	40
Olefins	56	20
Naphthenes	4	17
Aromatics	9	23

It results that the exit of a FT reactor can be fed directly to a methanol-cracking-type reactor. This two stage syngas conversion using a dual reactor gives product compositions similar to those of Table 8. It is striking in this respect that the temperature of the second reactor determines the product composition and properties. At elevated temperatures a high yield in aromatics and LPG-type molecules is obtained. At lower temperatures the product is acyclic in nature, with a good yield of light olefins. It seems therefore that cyclization and aromatization of light olefins is a major conversion at higher reaction temperatures over this cracking catalyst.

Syngas Conversion into Hydrocarbons Rich in Aromatics using Dual Component Catalysis. Knowing all that precedes, it is logical to grind together a FT and ZSM-5-type catalyst, expecting that in carefully controlled conditons a product spectrum imposed by the cracking catalyst will be obtained. From the results of Table 9, it can already be derived that the reaction temperature will determine the product properties. On the other hand, this temperature should allow the FT catalyst to work with a high growth factor in order to minimize the C_1 content. A selection of literature data on this type of conversion is shown in Table 9. Since aromatic-rich mixtures are only delivered by the ZSM-5 catalyst at high temperatures, admixing to a FT catalyst which gives low C_1 yields in these conditions is a prerequisite. This can be achieved with oxides known as isosynthesis catalysts.[23] The Ru-catalyst operates at temperatures where the effect of the ZSM-5 is only slightly visible. The iron dual composite represents an intermediate situation.

Table 8. Two-stage syngas conversion using a dual reactor (after ref. 23)

Temp./K. F.T. reactor	544	544	544
ZSM-5 reactor	544	616	727
CO conv./%	66	79	95
Effluent / wt %			
H_2	9.6	7.8	4.5
CO	29.1	17.9	4.5
CO_2	31.2	37.9	30.1
H_2O	11.9	14.7	34.6
hydrocarbon	18.2	21.8	26.2
Hydrocarbon distribution / wt %			
C_1	14.9	11.4	17.3
$C_2^=$	6.1	4.2	7.4
C_2	10.4	-	-
$C_3^=$	2.5	10.5	18.0
C_3	5.1	-	-
$i-C_4$	0.3	12.8	10.7
$n-C_4$	2.8	9.1	6.9
$C_4^=$	3.0	-	-
$i-C_5$	1.7	9.6	5.2
$n-C_5$	1.4	5.0	1.4
C_6^+ non arom.	51.9	14.0	2.9
C_6^+ arom.	-	23.5	30.1

A method using a dual reactor and a tri- and bi-composite catalyst has been presented recently by researchers from Shell[34]. It is schematically shown in Fig. 7 and constitutes a very flexible arrangement for the synthesis of different hydrocarbon cuts, including those in the gasoil and C_{20}^+ range.

Table 9. Syngas conversion into hydrocarbons rich in aromatics using dual component catalysts consisting of a FT catalyst and a HZSM-5 zeolite

Catalyst	ZrO_2 + HZSM-5	0.05 RuO_2 + 0.95 HZSM-5	0.22 Fe + 0.78 HZSM-5	HZSM-5[a]
Ref.	30	31	31	32
Temp./K	700	567	590	644
Pressure/bar	83	77	14	1
H_2/CO	1	2	2	-
CO conv./%	20	90	83	-
Product distribution/wt %				
C_1	2	26	38	1
C_2-C_4	13	14	29	44
C_5^+	85	60	33	55
% arom. in C_5^+	99.8	19	28	75

a, using methanol as feed

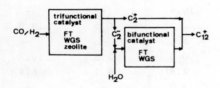

Figure 7. Schematic represenation of a two-step Shell process for synthesis of C_{12}^+ hydrocarbons out of synthesis gas (after ref. 34).

High diesel yields are reported for a new Gulf-Badger process using an unspecified catalyst in a fluidized bed reactor. The product distribution is of the SF-type and the process is obviously less flexible.

Synthesis of Olefinic Naphtha over Dual Component Catalysts via Mobil Technology. Syngas can also be converted directly to a high octane olefinic naphtha using a dual component catalyst consisting of a FT and a special type of methanol cracking catalyst[35]. Fig. 8 shows that the key parameter determining whether aromatic gasoil or olefinic naphtha is produced is the Al content or acid site density of the zeolite. In any case the CO conversion and octane number of the C_5^+ product remains the same. When the acid site density decreases, the boiling point of the liquid product decreases, as well as its content in aromatics. The olefin yield increases at the same time as illustrated by the olefin content of the C_5 fraction.

Dual Component Catalysts for the synthesis of Light Olefins. In a recent DOE approach,[36,37] iron precipitated on the surface of silicalite promoted with K produces up to 36 % light olefins. With HZSM-5 in the same conditions[37], less olefins and more gasoline hydrocarbons are formed. The difference in behaviour is consistent with the difference in acid site density of the materials.

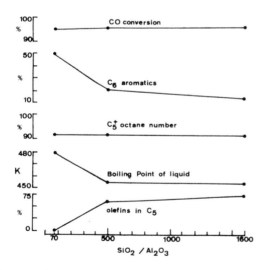

Figure 8. Synthesis of olefinic naphtha over dual component catalysts : influence of the acid site density of the HZSM-5 component (after ref. 35)

Conclusion

It is shown that using the established FT technology and secondary upgrading, flexibility in the products can be obtained. Increase in the FT selectivity by chemical modification of the catalyst is also possible and can be achieved in one step. Combining the classical FT catalyst (or reactor) with the new Mobil Methanol Cracking process gives, depending on the detailed nature of this catalyst, either olefins, high octane gasoline or olefinic naphtha. Combination of these 2 processes can be done in several ways and so even more flexibility for the production of certain hydrocarbon cuts is reached.

Acknowledgment

The author acknowledges the Belgian National Fund of Scientific Research for a permanent research position as Senior Research Associate.

References

[1] R.L. Pruett, Preprints Div. Fuel Chem. A.C.S., 1983, 28(2), 163.
[2] J. Haggin, C&EN, 1981, Oct. 26, 22.
[3] A. Aquilo, J.S. Alder, D.N. Freeman and R.J.H. Voorhoeve, Hydroc. Process, 1983, March, 57.
[4] M.E. Dry, Catalysis 1, J.R. Anderson and M. Boudart, eds., Springer Verlag, Berlin, Heidelber, New York, 1981, p. 159.
[5] M.E. Dry, J. Molec. Catal., 1982, 17, 133.
[6] M.E. Dry, Ind. Eng. Chem. Prod. Res. Dev., 1976, 15, 282.
[7] E.F.G. Herington, Chem. Ind., 1964, 347.
[8] R.A. Friedel and R.B. Anderson, J. Am. Chem. Soc., 1950, 72, 1212 and 2397.
[9] G. Henrici-Olivé and S. Olivé, Ad. Polym. Sci., 1974, 15, 1.
[10] G. Henrici-Olivé and S. Olivé, Angew. Chem., 1976, 88, 144.
[11] P.A. Jacobs, Catalysis by Zeolites, B. Imelik et al., eds., Elsevier Scientific, Amsterdam, 1980, p. 293.
[12] P.A. Jacobs and D. Van Wouwe, J. Molec. Catal., 1982, 17, 195.
[13] M.J. Baird, R.R. Schehl, W.P. Haynes and J.T. Cobb, Ind. Eng. Chem. Prod. Res. Dev., 1980, 19, 175.
[14] W.D. Deckwer, Oil Gas J., 1980, Nov.10, 198.
[15] H. Kölbel and M. Ralek, Catal. Rev. Sci. Eng., 1980, 21, 225.
[16] H.H.Nijs and P.A. Jacobs, J. Catal., 1980, 65, 328.

17 P. Struyf, P.A. Jacobs and J.B. Uytterhoeven, New Ways to Save Energy, D. Reidel, Dordrecht, Boston, Lancaster, 1983, p. 334.
18 for a review see C.D. Chang, Catal. Rev. Sci. Eng., 1983, 25, 1.
19 J.C. Kuo, U.S.P. 4,046,830 (1977).
20 H.R. Ireland, A.W. Peters and T.R. Stein, U.S.P. 4,045,505 (1977).
21 J.C. Kuo, U.S.P. 4,046,831 (1977).
22 J.C. Kuo, C.D. Prater, J.J. Wise, U.S.P. 4,049,741 (1977).
23 T.J.-H. Huang and W.O. Haag, E.P.A. 0,029,140 (1980).
24 Y.C. Hu, Hydroc. Process., 1983, May, 88.
25 H. Kölbel and D. Schneldt, Erdöl u. Kohle, 1977, 30, 139.
26 B. Bussemeier, C.D. Frohning and B. Cornelis, Hydroc. Process., 1976, Nov.
27 E.L. Kugler, A.C.S. Meeting, 1980, San Francisco.
28 S.J. Tauster, S.C. Fung and R.L. Garten, J. Am. Chem. Soc., 1978, 100, 170.
29 R.A. Stone and C.B. Murchison, Hydrocarb. Process., 1982, Jan., 147.
30 C.D. Chang, W.H. Lang and A.J. Silvestri, J. Catal., 1979, 56, 268.
31 W.O. Haag and T.J. Huang, U.SP. 4,157,338 (1979).
32 C.D. Chang and A.J. Silvestri, J. Catal., 1977, 47, 249.
33 H.H. Storch, N. Columbic and R.B. Anderson, The Fischer-Tropsch and related syntheses, J. Wiley, New York, 1951,
34 L. Schaper and S.T. Sie, U.S.P. 4,338,089 (1982).
35 F.G. Dwyer and W.E. Garwood, U.S.P. 4,361,503 (1982).
36 V.U.S. Rao and R.J. Gormley, U.S.P. R,340,503 (1982).
37 V.U.S. Rao and R.J. Gormley, Hydroc. Proc., 1980, Nov., 142.

In situ Electron Microscopy Studies of Catalysed Gasification of Carbon

By R. T. K. Baker
CORPORATE RESEARCH SCIENCE LABORATORIES, EXXON RESEARCH AND ENGINEERING COMPANY, PO BOX 45, LINDEN, N.J. 07036, U.S.A.

The gasification of carbonaceous solids is an area of tremendous industrial importance, having impact on several diverse processes. It has been known for many years that a small amount of an inorganic imparity can have a profound effect on the rate of gasification of carbon. This feature is taken advantage of in gasification of coal and chars, and removal of carbon from coked catalyst particles and cracker tubes. The gasification of carbon by oxidizing gases is often the limiting factor in the performance of graphite structures at high temperatures. In nuclear reactors catalytic poisons are often added to the gas coolant to overcome the deleterious effects of impurities present in the high-temperature graphite components.

Catalytic gasification is a unique example of a heterogeneous catalyst system, because the carbonaceous support is also a reactant and is consumed during the reaction. It is a fascinating reaction since one can directly observe, with the aid of a microscope, the catalyst in action. The most important gasification reactions of carbon are with O_2, CO_2, H_2 and steam, and catalysts are used extensively to increase the rates of these processes.

This paper describes the use of controlled atmosphere electron microscopy[1] (CAEM) to investigate a number of aspects exhibited by catalyst particles during the gasification of graphite.

Experimental Methods

The CAEM technique, which is shown schematically in Figure 1, enables one to study reactions between gases and solids at very high magnifications, while they are taking place, under realistic conditions of temperature, pressure and reaction time. The key design feature in the technique is the ability to operate at high gas pressure in the specimen region while maintaining very low pressure in the rest of the microscope. This situation is achieved by incorporation of a gas-reaction cell in the specimen chamber of a JEM-120 electron microscope. This device allows for differential pumping around the specimen and chamber regions.

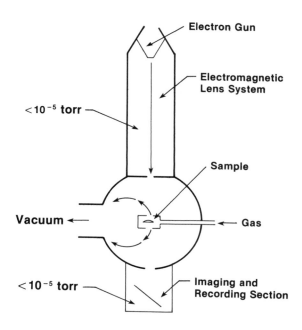

Figure 1 Schematic representation of the controlled atmosphere electron microscope.

In order to capture the dynamic information generated in these studies, sophisticated continuous recording qualities have been introduced into the system. Part of the transmission image is viewed by a Phillips Plubicon TV camera equipped with a fiber optics lens system. The magnified image is displayed on a monitor and continuously recorded on video tape. From this type of investigation, it is possible to obtain not only qualitative information on the behavior of a surface, but also quantitative estimates of kinetic parameters from the rates of motion of small particles and other features.

All of the work discussed in this article was performed with natural single crystal graphite obtained from Ticonderoga, New York State. The crystals were released from the pyroxene mineral in which they were embedded by repeated leaching in concentrated hydrochloric acid, followed by treatment in hydrofluoric acid, and finally the crystals were washed in deionized water. Prolonged interaction with hydrochloric acid was necessary to ensure complete removal of calcium salts. If this step was not performed proficiently, then an addition of hydrofluoric acid calcium in the form of a fluoride was left as a residue on the crystals and could function as a catalyst in subsequent reactions. After separation, the crystals were first cleaved between glass slides coated with a polystyrene cement and then released and washed with acetone. The crystals were then mounted on glass slides with aqueous polyvinylpyrrolidone (PVP) adhesive. The PVP was allowed to harden and the crystals further cleaved with Scotch tape. Successive layers were removed until an optically transparent portion of the crystal remained affixed to the slide. Such crystals, which varied in thickness from 15 to 100 nm, were found to be excellent transmission specimens. The cleaved crystals were freed from the slide onto a clean water surface and eventually mounted on coated specimen holders.

Additives were introduced onto the graphite specimens by one of two procedures: (a) By evaporation of high purity metal wires from a tungsten filament at a residual pressure of 10^{-6} Torr, or (b) as an atomized spray of a soluble salt. The former was the preferred method as it afforded more precise control of the metal loading on the support. The reactant gases used in this work, oxygen, hydrogen and argon were 99.99% minimum purity (Scientific Gas Products Inc.) and were used without further purification. Steam was introduced by allowing argon to flow through a bubbler containing deionized water maintained at 0°C prior to entering the gas reaction cell, conditions which produced an argon/water ratio of about 40:1.

Results and Discussion

The factors which control the mode of action of a catalyst in graphite-gasification reactions can be divided into chemical and physical aspects. Since electron microscopy has, as yet, contributed very little to the understanding of the chemical aspects, this area will only be mentioned superficially. Comprehensive treatments of the various proposed theories of the mechanistic features of the gasification reactions have been presented previously.[2-4] Some general requirements of the catalyst include the ability to dissociate a reactant gas molecule and a capacity to dissolve carbon atoms from the graphite and allow these species to diffuse through or over the catalyst particle surface.

In this paper the emphasis is placed on information obtained from CAEM studies concerning the physical factors which determine whether a particular particle will act as a catalyst for graphite gasification and the mode by which it will function. The most likely condition for a particle to become catalytically active is if it is located at a graphite edge or step site. Detailed examination of the particles in these regions shows that prior to catalytic attack they undergo distinct changes in morphology, resulting in a change in contact angle from obtuse to acute, <u>i.e.</u> a transformation from a non-wetting to a wetting condition.

A more fundamental understanding of the mechanisms of adherence and the factors influencing the contact of metal particles on graphite can be obtained from a treatment of the surface forces affecting the metal melt-solid substrate system. For a metal particle resting on the edge site of a graphite surface, as shown in Figure 2, the contact angle, θ, at equilibrium is determined by the surface energy of the support, γ_{gs}, the surface energy of the metal, γ_{mg}, and the metal-support interfacial energy, γ_{ms}, and is expressed in terms of Young's equation:

$$\gamma_{gs} = \gamma_{ms} + \gamma_{mg} \cos \theta \qquad [1]$$

Equation [1] can be written in the form

$$\cos \theta = \frac{\gamma_{gs} - \gamma_{ms}}{\gamma_{mg}} \qquad [2]$$

If γ_{mg} is larger than γ_{gs} the contact angle is larger than 90° and the particle is in a non-wetting state; if the reverse is true, θ is less than 90° then wetting occurs, and if $\gamma_{gs} = \gamma_{ms} + \gamma_{mg}$, then the particle will spread out over the support surface. The ability of particles to undergo this transformation suggests that a significant degree of atomic mobility exists, particularly in the surface layers. Provided that the chemical requirements are satisfied, the mode of catalytic attack is governed by the degree of wettability of the graphite.

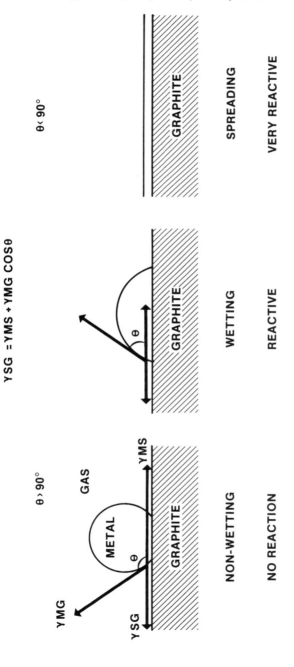

Figure 2 Changes in shape of particles as a function of the wettability of the catalyst on graphite.

For the intermediate wetted state where the catalyst is present as a discrete cap-shaped particle, then the mode of attack would be expected to proceed by a channeling action. An example of this form of catalysis is presented in Figure 3. Throughout the reaction, the active particles maintain contact with the graphite interface and as a consequence always remain at the leading face of the channel. Once this form of attack has been initiated, the particles will undergo frequent changes in shape, but the leading faces will always have hexagonal facets, characteristics of the crystalline graphite structure. If the catalyst material preferentially wets one of the graphite faces, then the channels will tend to be straight with occasional 60° and 120° bends. If no preferential wetting occurs, then channels will follow random pathways.

For the situation where catalyst particles spread along the graphite edges in the form of a thin film, then the subsequent catalytic mode of attack is by edge recession. An example of this type of behavior is shown in Figure 4. Edge recession, like channel propagation, may exhibit a very ordered motion if the catalyst undergoes preferential wetting of a particular graphite face. Spreading of the catalyst in this manner results in the most efficient use of the additive in that the contact area between catalyst and carbon atoms is maximized.

In this paper examples have been selected which demonstrate the influence of various catalysts on gasification of graphite in steam, oxygen and hydrogen environments.

(a) <u>Potassium/Graphite-Steam</u>--When graphite specimens containing a monolayer of potassium carbonate were heated in 2.0 Torr wet argon, particles were observed to form on the edge sites at 550°C; however, no particle nucleation was seen on the basal surfaces of

Figure 3 Catalytic channeling by nickel particles on graphite at 950°C in 1.0 Torr hydrogen.

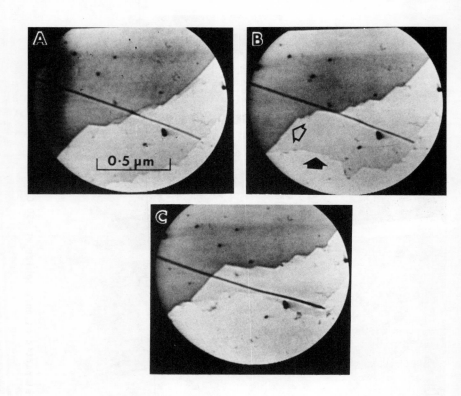

Figure 4 Catalytic attack of graphite by the edge recession mode; the receding edges are indicated by the arrows.

the specimen. As the temperature was gradually raised above 550°C, particles located on the graphite edges disappeared, and almost simultaneously gasification of these regions took place by the edge recession mode. The edges began to recede at many locations, at first as irregular shaped notches and later acquiring a hexagonal facetted appearance similar to the situation presented in Figure 4. In the absence of added potassium species, no change in graphite appearance was observed under the conditions used here. We therefore believe that the disappearance of the particles reflects spreading of the salt to a thin film on the graphite surface rather than evaporation.

On continued heating, the reaction became more widespread and it was possible to establish that the recession was occurring along directions parallel to the $\langle 11\bar{2}0\rangle$ crystallographic orientations. This property is determined by noting the orientation of edges with respect to twin bands present in the graphite specimens, which are aligned along the $\langle 10\bar{1}0\rangle$ crystallographic directions. The potassium-catalyzed reaction thus exposes the $\langle 11\bar{2}0\rangle$ or "zig-zag" /\/\/\/\ presentation of the surface, which must in this case be less reactive than the $\langle 10\bar{1}0\rangle$ or "armchair" configuration. This is perhaps not too surprising since the $\langle 11\bar{2}0\rangle$ orientation presents <u>one</u> uncondensed carbon atom per ring exposed at the edge whereas the $\langle 10\bar{1}0\rangle$ orientation presents <u>two</u> uncondensed carbons together.

Although the edge recession reaction followed the expected trend of an increase in rate with increasing temperature, there were some edge regions which exhibited anomalous behavior, recession coming to an abrupt halt and not being restarted by either raising or lowering the temperature. At present we do not understand the reasons for this phenomenon; however, loss of catalyst can be discounted for two reasons. (i) Deactivation is a sudden event and the maintenance of constant activity until it happens argues against gradual removal of catalyst. Moreover, areas immediately adjacent to spent regions continue to react. (ii) the spent regions can be reactivated by treatment in oxygen indicating that catalyst material is still present after deactivation.

In summary, K_2CO_3 is the prime example of a catalyst which in its active mode spreads across the entire graphite surface. The morphology of the K_2CO_3 catalyst during gasification is striking confirmation of a strong interaction between the catalyst and the edges of the graphite lattice. The interfacial bonding is strong enough to compete with the cohesive bonding within the bulk salt and effectively results in dispersion of the catalyst along the active edge, thus accounting for the high carbon conversion activity.[5,6]

(b) <u>Nickel/Graphite-Hydrogen</u>--Treatment of nickel/graphite specimens in 1.0 Torr hydrogen resulted in particle nucleation at 755°C. Catalytic attack of the graphite commenced at 845°C and this action was seen as the development of fine channels, which emanated from edges and steps on the graphite surface. The typical appearance of channels produced after reaction of nickel/graphite with hydrogen at 950°C is shown in Figure 3. The channels are up to 150 nm in width and are predominantly straight and aligned along the $\langle 11\bar{2}0 \rangle$ crystallographic orientations of the graphite.

As the temperature was increased, it became apparent that active nickel particles were gradually disappearing from the heads of the channels, either due to deposition of material on the sides of the channels or for some reason, volatilization. This behavior has also been reported by other workers[7] who attributed the disappearance of nickel as being due to diffusion of metal into the graphite structure. Figure 5 is a schematic representation showing the gradual depletion in catalyst material as the channel increases in length. It was apparent that the channel width was reduced in a step-wise fashion, suggesting that nickel was being lost from the particle by a gradual process and that the particle was periodically undergoing a reorganization in its shape. Detailed examination of many experiments demonstrated

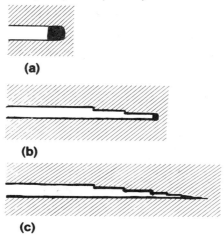

Figure 5 Schematic representation of loss of catalytic gasification activity by nickel in hydrogen.

that the larger the initial size of the catalyst particle, then the longer the period that channeling action was maintained before the particle disappeared. Continuous observation also showed that active particles became thinner and that channel depth decreased towards the end of the track. This phenomenon which started at 980°C was essentially complete by the time the temperature had been raised to 1100°C. Continued heating in hydrogen up to 1250°C produced no further catalytic action or restoration of the original particles. A detailed quantitative analysis of these events demonstrates that nickel is progressively laid down as a near monolayer film on the walls of the channels while the particles move forward.[8]

If hydrogen was replaced by oxygen and the specimen reheated, then at 850°C small particles, 2.5 nm diam., started to reform along the edges of the original channels, which were in the process of undergoing expansion due to uncatalyzed oxidation. Eventually at 1065°C these particles proceeded to create very fine channels emanating from the edges of the original channels, Figure 6a. Substitution of oxygen by hydrogen at this stage resulted in a temporary loss of activity, which was restored after a period of 20 minutes at 900°C in the form of straight channels with similar characteristics to those found during the initial hydrogen step.

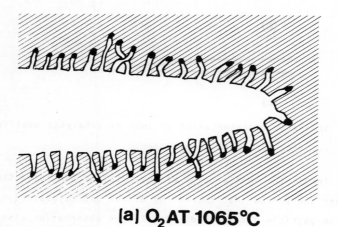

(a) O$_2$ AT 1065°C

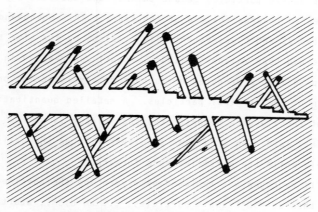

(b) STEAM AT 935°C

Figure 6 (a) Schematic representation of reactivation of nickel in oxygen; (b) Reactivation of the metal in steam.

In a second series of experiments the oxygen cycle was replaced by a steam treatment. Introduction of steam (1.0 Torr argon saturated with water at 0°C) resulted in "break-up" of the wetted nickel film at 830°C and on continued reaction fresh channels were propagated from the edges of the original ones and this effect became more pronounced at 935°C, Figure 6b. In this case, the rate of the uncatalyzed reaction was insignificant and as a consequence all channels remained parallel-sided throughout the reaction. The re-introduction of 1.0 Torr hydrogen resulted in the continued formation of channels which were formed almost instantly when the temperature was raised to 850°C.

It is probable that the delay in onset of activity in the final hydrogen treatment in the hydrogen-oxygen cycling experiments was due to the time required to reduce nickel oxide back to the metallic state. In the presence of steam the bulk of the particle remains in the metallic state and therefore catalytic channel activity would be expected to proceed in an uninterrupted fashion in the final hydrogen step.

There is little doubt that in the deactivated state nickel remained as a thin film along the walls of the channel. Evidence for this conclusion is seen in observations that during regeneration in either oxygen or steam the nickel always nucleated to re-form particles at the site of the initial channel edge. If diffusion of metal into the graphite structure had taken place,[7] then it is difficult to see how re-formation of particles could occur in such an ordered manner.

Finally, the perplexing issue is why nickel wets and spreads on graphite in an irreversible mode when heated in hydrogen at 980°C, when it does not do so at lower temperatures. It is possible that the material actually spreading onto the channel walls was not pure nickel but carbidic nickel, having a higher surface tension than the metal. Upon cooling, the nickel carbide could decompose to form metallic nickel with a graphite monolayer coating. This "sandwich" model would thus explain the observed loss of catalytic activity and the decrease in hydrogen chemisorption capacity.[9]

(c) <u>Influence of Tungsten, Rhenium, and Tungsten-Rhenium on the Graphite-Oxygen and Graphite-Hydrogen Reactions</u>--
Controlled atmosphere electron microscopy has been used as a microanalytical probe of tungsten-rhenium particles on graphite. The test reactions for this probe were the catalytic gasification of graphite in 5.0 Torr oxygen and 1.0 Torr hydrogen. Both of these reactions are extremely sensitive to the nature of the catalyst surface. From a comparison of the qualitative and quantitative behavior of the mixed system with that of its pure constituents, it has been possible to determine in the bimetallic system which component preferentially segregated to the particle surface during reaction in oxygen and hydrogen.

When the single component/graphite systems were heated in 5.0 Torr oxygen, the observed patterns of behavior were very similar, differing only in the temperatures where the various events occurred, Table 1.

In both cases, initial catalytic attack took place by the channeling mode. As the temperature was raised there was an increasing tendency for active particles to become depleted in size as material was deposited on the walls of the channels. As a consequence the width of the channel was reduced and channel walls receded at a faster rate than uncontaminated edges, causing channels to acquire a fluted appearance. This sequence of events

TABLE 1

A Comparison of the Behavior of W, Re and W-Re on the Graphite-O_2 and Graphite-H_2 Reactions

CATALYST	GRAPHITE-OXYGEN	GRAPHITE-HYDROGEN
W	Pitting at 705°C Channeling at 770°C Spreading of active material to produce edge recession at 910°C. E_a(channeling) = 58.7 kcal mole^{-1}	NO REACTION
Re	Channeling at 550°C Spreading of active material to produce edge recession at 635°C E_a(channeling) = 47.7 kcal mole^{-1}	Channeling at 715°C E_a(channeling = 35.9 kcal mole^{-1}
W-Re	Channeling at 635°C No Spreading Action E_a(channeling) = 47.7 kcal mole^{-1}	NO REACTION

is shown schematically in Figure 7. When all the material from a given particle had been deposited on the walls, linear propagation of the channel ceased. This type of behavior is indicative of a strong interaction between the catalytic material, probably in the form of WO_3 and ReO_2 with the graphite.

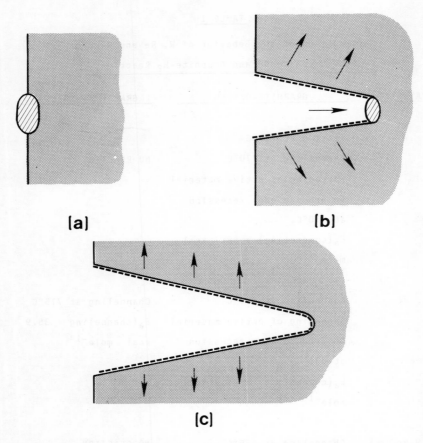

Figure 7 Effect of wetting and spreading action on catalytic channeling mode.

Detailed quantitative kinetic analysis of many reaction sequences showed that both tungsten and rhenium were active catalysts for the graphite-oxygen reaction, although the former exhibited only marginal activity.

When the reduced mixed component system on graphite was heated rapidly to 900°C in oxygen, then on subsequent temperature cycling only channeling activity was observed, there being no

Electron Microscopy Studies of Catalysed Gasification of Carbon

tendency for the catalyst particles to spread along graphite edges. This observation indicates that there is strong bonding between tungsten and rhenium, which is reflected in a reduction in the strength of the interaction with the graphite substrate.

The Arrhenius plots of tungsten-rhenium catalyzed rate and that of its pure constituents for the graphite-oxygen reaction are presented in Figure 8. The similarity in the derived values of the apparent activation energies for the mixed system and the

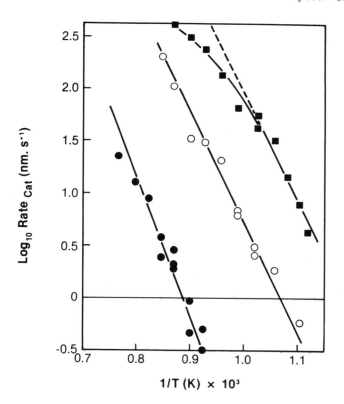

Figure 8 Arrhenius plots of catalyzed rate for tungsten-rhenium, ■; tungsten, ●; and rhenium, O; for the graphite-oxygen reaction.

rhenium catalyzed reaction indicates that under oxidizing conditions the bimetallic particle surface is enriched with rhenium.

When rhenium/graphite specimens were heated in 1.0 Torr hydrogen, catalytic attack of the graphite by the channeling mode commenced at 715°C. The channels were predominantly straight and exhibited a preferred orientation, being parallel to the $\langle 11\bar{2}0 \rangle$ crystallographic directions. The channeling action remained the exclusive form of attack up to 1050°C, the highest temperature studied.

In sharp contrast, no catalytic attack was observed when either the tungsten/graphite or tungsten-rhenium/graphite samples were treated in hydrogen up to 1050°C. It is therefore concluded that following reduction, the surfaces of the bimetallic particles are enriched with tungsten.

Finally, no article discussing the use of microscopy techniques for the study of catalytic gasification of carbon would be complete without reference to the pioneering efforts of Hennig[10] and Thomas,[11] which combined with the early work of Long and Sykes[12] has provided the basis for the majority of investigations in this area.

References

1. R. T. K. Baker and P. S. Harris, J. Sci. Instrum., 1972, 5, 793.
2. P. L. Walker, Jr., F. Rusinko, Jr. and L. G. Austin, in "Advances in Catalysis," (D. D. Eley, P. W. Selwood, and P. B. Weisz, Eds.) Academic Press, New York, 1959, Vol. 11, p. 133.

3. J. B. Lewis, "Modern Aspects of Graphite Technology," (L. G. F. Blackman, Ed.) Academic Press, New York, 1970, p. 129.
4. D. W. McKee, "Chemistry and Physics of Carbon," (P. L. Walker, Jr., and P. A. Thrower, Eds.) Marcel Dekker, New York, 1981, Vol. 16, p. 1.
5. K. Otto and M. Shelef, 6th Int. Congr. Catalysis, London, 1976, paper B47.
6. C. A. Mims, J. J. Chludzinski, Jr., J. K. Pabst and R. T. K. Baker, submitted to J. Catalysis.
7. C. W. Keep, S. Terry and M. Wells, J. Catalysis, 1980, 66, 451.
8. R. T. K. Baker, R. D. Sherwood and E. G. Derouane, J. Catalysis, 1982, 75, 382.
9. A. J. Simoens, E. G. Derouane and R. T. K. Baker, J. Catalysis, 1982, 73, 175.
10. G. R. Hennig, "Chemistry and Physics of Carbon," (P. L. Walker, Jr., Ed.) Marcel Dekker, New York, 1966, Vol. 2, p. 1.
11. J. M. Thomas, "Chemistry and Physics of Carbon," (P. L. Walker, Jr. Ed.) Marcel Dekker, New York, Vol. 1, p. 122.
12. F. J. Long and K. W. Sykes, Proc. Roy. Soc. (London), 1948, A193, 377.

Ethylene Glycol from Synthesis Gas

By J. B. Saunby
UNION CARBIDE CORPORATION, RESEARCH AND DEVELOPMENT DEPARTMENT, PO BOX 8361, SOUTH CHARLESTOWN, W.V. 25303, U.S.A.

Introduction

While the "oil shocks" of the 1970's stimulated widespread interest in synthesis gas chemistry, particularly in industrial research laboratories, a research program was initiated by Union Carbide (UCC) in 1969 to explore routes to industrial chemicals from syn gas. The driving force for this program was the possibility of eliminating the expensive and somewhat inefficient cracking of naphtha etc. to ethylene, thus reducing investment and manufacturing cost for a range of petrochemical products. One of the early successes in this program was the discovery, in 1971, of the rhodium-catalyzed, direct conversion of synthesis gas to ethylene glycol and the first patent was issued to UCC on this technology in 1973. Extensive research by UCC and many other industrial and academic research laboratories has been focused on this direct conversion. A wealth of new chemistry has been discovered but the high pressures, high catalyst concentration and problems of catalyst stability have hindered the commercialization of a one-step route.

Recognition of the possible intermediacy of formaldehyde in the direct conversion stimulated research on its use as a feedstock. Processes based on its conversion to glycolaldehyde and hydrogenation to ethylene glycol have been proposed by several researchers. The coupling of carbon monoxide via alkyl nitrites to give the corresponding dialkyl oxalates has been studied extensively, particularly by UBE Industries in Japan, who have built and operated an oxalic acid plant based on this technology. It also forms the basis of an attractive route to ethylene glycol by the hydrogenation of the dialkyl oxalate and this process has been developed in a joint program by UBE Industries and UCC. This appears to be the best process available today for the commercial production of ethylene glycol from syn gas.

Single-Step Routes to Ethylene Glycol

The direct conversion of synthesis gas to ethylene glycol has attracted much research attention by many companies, in addition to UCC, in recent years.[1] The first demonstration of this reaction was reported in patents to DuPont,[2,3] and involved the use of soluble cobalt catalysts under H_2/CO pressure on the order of 3000 atm and temperatures up to 290°C. Products in addition to ethylene glycol were methanol, glycerine, ethanol, and other alcohols. Further studies of the cobalt catalyst system have been reported more recently by several groups.[4-7] This research has shown that the same products are obtained under lower pressures of synthesis gas (e.g. 200 atm), but the rates are very low and selectivity to ethylene glycol suffers. Chemical and spectroscopic evidence supports the identification of the active catalyst species as a mononuclear cobalt hydride complex, $HCo(CO)_4$.

Rhodium Catalyzed Reactions

Research at Union Carbide showed that rhodium catalysts are more active and selective than cobalt catalysts for the production of ethylene glycol.[8-10] By-products include methanol and glycerine, as well as minor amounts of other compounds. Reaction conditions generally studied range from 400 to 1000 atm and 220 to 290°C. This catalytic system appears to be much more complex than the cobalt system, and some of the factors affecting catalyst activity, selectivity, and stability have been investigated.

The addition of Lewis base "promoters" to this catalytic system was found to have very significant effects. Useful Lewis bases include amines and various salts, and combinations of bases from these two classes give optimum results. Shown in Figure 1 is a series of experiments in which the anion in cesium salts is varied.[11] Figure 2 illustrates the effects of changing the cation in acetate salts.[11] The amount of Lewis base promoter used is also critical in determining the activity and selectivity of the catalytic system, as illustrated by Figure 3.[11] Here the optimum ethylene glycol rate and selectivity are observed at a CsO_2CH/Rh ratio of 1:6, suggesting that perhaps the active species involved in ethylene glycol formation contains cesium and rhodium in this ratio.

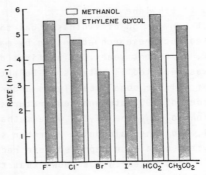

Figure 1. Effect of cesium salts on rhodium-catalyzed CO hydrogenation. Reaction conditions: 3 mmol Rh, 10 mmol 2-hydroxypyridine, 0.45 to 0.50 mmol of cesium salt, 75 ml tetraglyme solvent, 544 atm, 220°C.

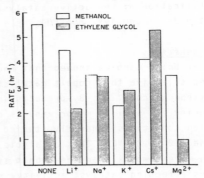

Figure 2. Effect of acetate salts on rhodium-catalyzed CO hydrogenation. Reaction conditions same as in Figure 1.

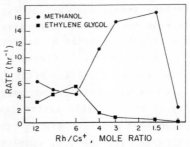

Figure 3. Plot of rates as a function of Rh/Cs^+ ratio. Cesium formate promoter; reaction conditions same as in Figure 1.

Studies of catalyst solutions have confirmed that anionic rhodium complexes are present during catalysis. Observation by high-pressure infrared spectroscopy under catalytic conditions provides evidence for the presence of $[Rh_5(CO)_{15}]^-$ and $[Rh_{13}(CO)_{24}H_3]^{2-}$ (Figure 4). The Cs^+/Rh ratio cited above as

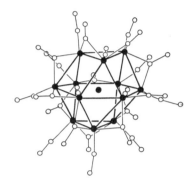

Figure 4. Structure of $[Rh_{13}(CO)_{24}H_3]^{2-}$.

providing optimum ethylene glycol rates and selectivities would be consistent with the involvement of these complexes in catalysis.

Possible roles for such metal clusters in catalytic processes have been the subject of much speculation, but there is no clear evidence that these clusters are the active species in this catalytic system. On the other hand, the key to selectivity could well be the redox potential of the catalyst, with highly reduced clusters (low Rh/Cs^+ ratio) being capable of reducing a $>Rh-CH_2OH$ species to methanol faster than CO insertion occurs which leads to the ethylene glycol precursor

$$Rh-\overset{O}{\underset{\|}{C}}-CH_2OH$$

The presence of anionic metal complexes in catalytic solutions has led to studies of the effects of ion pairing between these negatively charged complexes and their positively charged counterions.[13,14] Evidently, ion pairing decreases the activity of the system for ethylene glycol production, and any factor which minimizes ion pairing promotes ethylene glycol formation. Properties of the reaction solvent are very important in determining the extent of anion-cation interaction.[15] Solvents which can complex the cation and minimize its interaction with the metal anion are found to be good reaction solvents; examples are tetraglyme and crown ethers. Solvents with high dielectric constants, such as sulfolane and N-methylpyrrolidone, also can effectively separate the ions and thus increase the catalytic activity. Mixtures of solvents from these two categories provide the best results.

Higher reaction pressures bring about large increases in product formation rates, as shown in Figure 5.[16] Pressure dependences for both methanol and ethylene glycol formation are greater than second-order, and the ethylene glycol/methanol selectivity is seen to improve slightly at higher pressures.

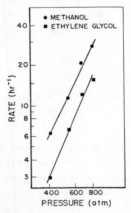

Figure 5. Effect of pressure on rhodium-catalyzed CO hydrogenation. Reaction conditions: 3 mmol Rh, 1.25 mmol pyridine, 75 mL sulfolane, 240°C. $H_2:CO = 1$

Increased reaction temperatures also give higher rates to products, as shown in Figure 6.[11] Effective activation energies for the reactions producing ethylene glycol and methanol are both approximately 18 kcal/mole, as estimated from Figure 6. The reaction rates at higher temperatures deviate from the relationship, however, giving lower values than predicted by extrapolation. This presumably occurs because of catalyst instability at higher temperatures. The temperature at which such instability becomes noticeable is determined largely by the reaction pressure and the solvent-promoter combination employed. Catalyst instability is manifested by

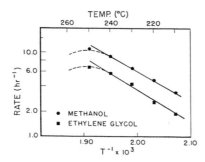

Figure 6. Effect of temperature on rhodium-catalyzed CO hydrogenation. Reaction conditions: 3 mmol Rh, 10.5 mmol 2-hydroxypyridine, 0.5 mmol cesium 2-pyridinolate, 75 mL butyrolactone, 544 atm.

loss of soluble rhodium species from the catalyst mixture and formation of precipitates which appear to include large rhodium carbonyl clusters.

Results described above were obtained in batch experiments over relatively short time periods (1-6 hr). This catalytic system has also been investigated in a continuous reaction unit with catalyst solution recycle, as shown in Figure 7.

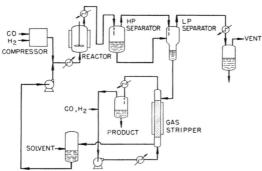

Figure 7. Reaction unit for continuous operation of homogeneous catalytic CO hydrogenation.

The synthesis gas is allowed to react with the soluble catalyst in the stirred tank reactor shown. A liquid stream from the reactor is depressurized and separated into gas and liquid fractions. The liquid fraction is fed into a product stripping unit, which removes reaction products from the catalyst solution by countercurrent stripping with CO or H_2/CO. The stripped catalyst solution is then recycled to the reactor under pressure. Long-term catalyst stability can be studied in such reactions, and an example of results from one of these runs is presented in Figure 8.

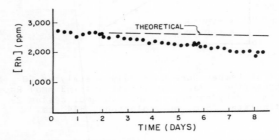

Figure 8. Rhodium concentration in the catalyst solution during a period of continuous operation at 850 atm, 240°C. The theoretical level shown accounts for losses due to sampling.

A gradual decline in the rhodium catalyst concentration is seen over a period of several days. Much of the rhodium lost from solution may be resolubilized by periodically increasing the CO/H_2 ratio in the reactor vessel[17] or by lowering the reaction temperature.[18]

To summarize results obtained with the rhodium catalyst, selectivity to alcohols (including ethylene glycol, methanol, and glycerine) is very high under a great variety of conditions. Ethylene glycol is a major product constituent under most conditions of temperature and pressure examined. However, the activity of the catalyst is relatively low except at high temperatures, and elevated temperatures adversely affect catalyst stability. Long-term continuous experiments exhibit some loss of catalyst from solution. Although much of this loss is reversible, its existence and the lower-than-desirable activity of the catalyst are concerns because of the scarcity and high price of rhodium.

Ruthenium Catalyzed Reactions

Ruthenium catalysts have also been investigated as catalysts for the conversion of synthesis gas to ethylene glycol. In our research program it was found that although unpromoted ruthenium catalysts produce methanol and no ethylene glycol, the addition of ionic promoters has a large effect on activity and selectivity.[19,20] Figure 9 shows the effects of

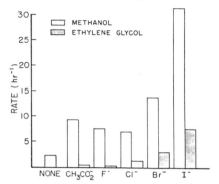

Figure 9. Effect of potassium salts on ruthenium-catalyzed CO hydrogenation. Reaction conditions: 6 mmol Ru, 18 mmol potassium salt, 75 mL sulfolane, 850 atm, 230°C.

several ionic promoters, and it may be seen that the halides, especially iodide, are particularly effective for promoting the formation of ethylene glycol. As shown in Figure 10, under most conditions the identity of the cation in the iodide salt has little effect on catalyst activity and selectivity. It thus appears that ion pairing effects are not as large a factor in determining the catalytic activity in this system as in the rhodium system previously described. Solvents found to be the best for the rhodium catalyst are, however, also useful for the

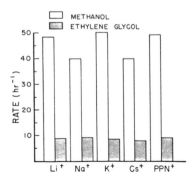

Figure 10. Effect of iodide salts on ruthenium-catalyzed CO hydrogenation. Reaction conditions same as in Figure 9 except that 3 mmol of Ru were used.

ruthenium catalyst; those employed generally are sulfolane, 18-crown-6, and N-methylpyrrolidone.

Process variables such as temperature, pressure, and catalyst concentration are very important in determining the ratio of ethylene glycol to methanol produced, in contrast to the rhodium system. For example, Figure 11 shows that the rate

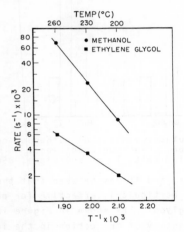

Figure 11. Effect of temperature on ruthenium-catalyzed CO hydrogenation. Reaction conditions: 15 mmol Ru, 60 mmol KI, 75 mL 18-Crown-6, 850 atm.

to methanol increases much more rapidly than the rate to ethylene glycol as the reaction temperature is raised. Figure 12 illustrates the effect of increasing pressure in this catalytic system. Both methanol and ethylene glycol rates are highly dependent on reaction pressure (somewhat greater than third-order dependence), and the effect is slightly greater on the ethylene glycol product rate.

An interesting effect is seen with varying catalyst concentrations, as shown in Figure 13. It is evident that the turnover frequency (activity on a metal atom basis) for methanol declines with increasing catalyst concentration, while that for glycol formation increases. This effect may be indicative of intermolecular processes involved in the catalyst chemistry leading to the glycol product.

As can be seen in Figures 10-13, the ruthenium catalyst tends to produce substantially more methanol than ethylene glycol. This high activity for methanol production is also

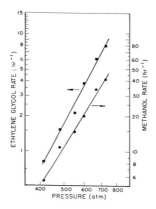

Figure 12. Effect of pressure on ruthenium-catalyzed CO hydrogenation. Reaction conditions: 15 mmol Ru, 45 mmol NaI, 75 mL N-methylpyrrolidone, 230°C.

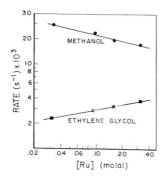

Figure 13. Effect of catalyst concentration on ruthenium-catalyzed CO hydrogenation. Reaction conditions: [KI] = 6 [Ru], 75 mL N-methylpyrrolidone, 850 atm, 230°C.

apparent in Figure 14, which compares the methanol- and ethylene glycol-producing abilities of Co, Rh and Ru catalysts under a standard set of conditions. Under these conditions the Ru and Rh catalysts produce ethylene glycol at a similar rate, but methanol production by the ruthenium catalyst is much higher.

By modifying the reaction parameters described above, i.e. reaction pressure, temperature, and catalyst concentration, it is possible to operate the Ru catalyst so

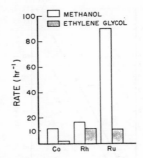

Figure 14. Comparison of activities of cobalt, rhodium, and ruthenium catalysts for homogeneous CO hydrogenation at 850 atm, 230°C.

that the rate to ethylene glycol is greater than the rate to methanol; for example, at 180°C and 850 atm, a 0.4 M solution of Ru promoted by NaI in sulfolane produced ethylene glycol at a rate of 2.05 hr^{-1} and methanol at a rate of 1.71 hr^{-1}. This rate to ethylene glycol is lower than desirable, largely because of the low temperature required to obtain this selectivity.

The stability exhibited by the ruthenium catalyst is excellent. No losses of catalyst are observed in batch runs, and a catalytic system has been operated in the continuous unit with liquid recycle shown in Figure 7. After a period of several weeks of continuous operation, catalyst accountability was essentially quantitative and no deactivation was observed.

To summarize the catalytic behavior of the ruthenium system, excellent stability and good selectivity to alcohols are observed, but more methanol than ethylene glycol is usually produced. Higher temperatures give increased rates to ethylene glycol, but cause even greater increases in methanol rates. As for the rhodium system, high pressures (>600 atm) give the best rates to ethylene glycol.

Mechanism of Ruthenium Catalysis

In view of the high activity of the ruthenium catalyst, an effort was made to understand the mechanisms by which products are formed by this system. Analyses by high-pressure infrared spectroscopy during catalysis indicate the presence of two ruthenium complexes, $[HRu_3(CO)_{11}]^-$ and $[Ru(CO)_3I_3]^-$. These anionic complexes can be prepared in the laboratory and studied individually. When either complex alone is used in a catalytic reaction, relatively low catalytic activity is observed. However, a mixture of the two complexes provides the expected normal activity. This and other observations indicate that both complexes are involved in the catalytic process. A possible scheme showing the involvement of these two complexes in the early steps of the catalytic cycle is shown in simplified form in Figure 15.

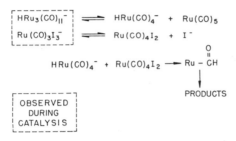

Figure 15. Possible roles of ruthenium complexes in the iodide-promoted catalytic system.

Here the two predominant species observed during catalysis are shown to react with H_2/CO to produce small equilibrium amounts of more reactive intermediates, postulated to be $[HRu(CO)_4]^-$ and $Ru(CO)_4I_2$. Transfer of the reactive hydride ligand on the first complex to a highly electrophilic carbonyl ligand on the second complex could comprise the first actual CO reduction step, producing a ruthenium formyl complex. Related reactions of $[HRu(CO)_4]^-$ support the postulation of this type of hydride transfer.[21] An analysis of the complex stoichiometry obtained in a catalytic cycle based on the scheme of Figure 15 indicates that such a reaction pathway is consistent with the observed kinetics.[20]

Possible steps after initial production of a metal formyl complex are shown in Figure 16.
This is a possible general mechanism which may apply to the cobalt, rhodium, and ruthenium catalytic systems for conversion of synthesis gas to ethylene glycol. Reaction of the formyl complex with hydrogen or a metal hydride may produce a coordinated formaldehyde intermediate, which is then converted to the product methanol or the postulated intermediate glycolaldehyde as shown. Glycolaldehyde may then undergo

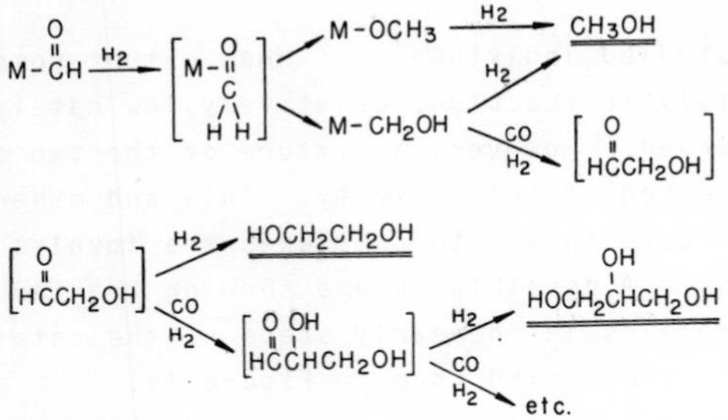

Figure 16. Proposed general mechanism for product formation in homogeneous CO hydrogenation.

hydrogenation to ethylene glycol or chain extension to the three-carbon aldehyde, glyceraldehyde. A continuation of this type of chain growth would lead to the small quantities of higher polyols, such as glycerol and erythritol, which can be obtained as products of the rhodium and ruthenium systems.

If the proposed mechanism of Figure 16 is correct, it may be expected that methanol and the higher polyols will always be co-produced with ethylene glycol, although the selectivities will vary from one catalyst to another. A second point which is raised by the proposed mechanism concerns the involvement of a formaldehyde-containing intermediate. Although production of free formaldehyde in such reactions is thermodynamically unfavorable, the intermediate formation of a coordinated formaldehyde species is plausible. This step probably represents the most difficult portion of the entire synthesis gas conversion process.

As described below, several catalytic systems which closely resemble the cobalt, rhodium, and ruthenium systems have been found to convert formaldehyde to methanol, glycolaldehyde, and/or ethylene glycol by reaction with synthesis gas. Conditions are much milder than those required for direct synthesis gas conversion. Such systems for conversion of formaldehyde and synthesis gas to ethylene glycol have been studied at Union Carbide. Although the mild operating conditions make these processes appear attractive initially, the relatively high cost of the dry formaldehyde usually required by these systems is a strong negative factor in their economics. If aqueous formaldehyde is used, the rate/selectivity advantages disappear and methanol is the major product.

Multi-Step Routes to Ethylene Glycol

Most of the multi-step routes use formaldehyde or methanol derived from synthesis gas as the feed and for this discussion are considered synthesis gas routes.

DuPont Process

In the process used commercially by DuPont until 1968, formaldehyde is carbonylated using an acid catalyst to give glycolic acid.

$$HCHO + H_2O + CO \longrightarrow HOOCCH_2OH$$

The glycolic acid is esterified with methanol and the resulting methyl glycolate hydrogenated to form ethylene glycol and methanol, which is recycled:

$$HOOCCH_2OH + CH_3OH \longrightarrow CH_3OOCCH_2OH + H_2O$$

$$CH_3OOCCH_2OH + H_2 \longrightarrow CH_3OH + HOCH_2CH_2OH$$

The major drawbacks to this process are the need for a strong acid catalyst, e.g. sulfuric acid, in the carbonylation reaction and the need to carry out this reaction at a high pressure (700-1000 atm.) to minimize the water gas shift reaction, which would otherwise lower the yield based on carbon monoxide. Overall, the ethylene glycol yield based on CO is approximately 50 mol percent and on hydrogen 70 mol percent.

Chevron Process

The chemistry of this route is similar to that in the DuPont process except that HF is used as the acid catalyst in the carbonylation step. This allows operation at a lower pressure (70 atm.) and gives improved selectivity. Various process improvements are claimed in patents to Chevron,[22,23] including _in situ_ formation of esters by feeding alcohol to the carbonylation reactor _etc._; however, this process has not been operated commercially.

Monsanto Process

In this process, formaldehyde is hydroformylated to glycolaldehyde at pressures up to 150 atm. using a rhodium catalyst. Recent publications by Chan indicate that he has improved the rates, conversions and selectivities to glycolaldehyde originally reported by Spencer,[24,25,26,27] who reached the following conclusions:

- Only amide solvents are effective.
- Glycolaldehyde is a strong inhibitor, acting as a bidentate ligand.
- Methanol formation increases dramatically at temperatures above 110-120°C and if aqueous formaldehyde rather than paraformaldehyde is used.
- Formaldehyde conversion is low, usually less than 50 percent.

Chan improved the system considerably and found that amines function as promoters in his catalyst system and if amine is present, solvents other than amides can be used. Conversions of 90 percent were obtained and when amines or bases, such as triethylamine or sodium hydroxide, are used the selectivity to glycolaldehyde is 70 percent. A major disadvantage to this process is the need for anhydrous formaldehyde, as with aqueous formaldehyde the major product is methanol. Glycolaldehyde also inhibits the reaction; thus low concentrations must be maintained in the reactor. Significant problems must be overcome if this process is to be commercialized.

PPG Process

In another version of catalyzed carbonylation of formaldehyde, PPG uses strongly acidic cation-exchange resins (e.g. Nafion^R) as the catalyst. The best examples in a recent patent[28] use an acetic acid solvent and acetoxyglycolic acid is formed, but again paraformaldehyde rather than aqueous formaldehyde is used. Under these conditions high reaction rates are obtained at moderate reaction conditions (150 atmos) with good selectivity (95 percent) to acetoxyglycolic acid. This must then be hydrolyzed prior to hydrogenation, adding an extra processing step compared to other routes through glycolic acid. The other major drawback to this process is the need to use paraformaldehyde.

USI Process

In another version of the process based on formaldehyde, USI (then National Distillers) disclosed in a 1978 patent[29] the use of a soluble rhodium carbonyl catalyst in an organic solvent to produce glycolaldehyde from formaldehyde, CO and hydrogen. The advantage is that the glycolaldehyde is more readily hydrogenated to ethylene glycol than glycolic acid. It is also possible that the two steps can be combined. The reaction takes place at pressures from 150-300 atm and although high conversions (50-85 percent) and good selectivity to glycolaldehyde (85 percent) are reported, our experience with similar chemistry would indicate potential catalyst stability problems and again, the presence of water is expected to have a deleterious effect.

UBE/UCC Process

In this process methyl nitrite is formed by the reaction of methanol with nitric oxide:

$$2CH_3OH + 2NO + 1/2O_2 \longrightarrow 2CH_3ONO + H_2O$$

The coupling of methyl nitrite to form dimethyl oxalate is carried out over a palladium catalyst and the nitric oxide released is recycled to the nitrite formation step.

$$2CH_3ONO + 2CO \longrightarrow CH_3O\overset{O}{\overset{\|}{C}}\overset{O}{\overset{\|}{C}}OCH_3 + 2NO$$

The oxalate ester is then hydrogenated over a copper catalyst to form ethylene glycol and methanol which is also recycled to the nitrite formation step:

$$CH_3O\overset{O}{\overset{\|}{C}}\overset{O}{\overset{\|}{C}}OCH_3 + 4H_2 \longrightarrow HOCH_2CH_2OH + 2CH_3OH$$

The overall reaction converts 2:1 synthesis gas to ethylene glycol plus one mole of water:

$$4H_2 + 2CO \longrightarrow HOCH_2CH_2OH + H_2O$$

All of the reaction steps take place at moderate temperatures and pressures, the highest pressure being in the hydrogenation step which is typically operated at 35 atm. The process flow diagram is shown in Figure 17.

High selectivity has been demonstrated in each step of this process. The overall selectivity in the conversion of carbon monoxide to ethylene glycol is over 90 percent and the nitric oxide efficiency to methyl nitrite is 98-100 percent.

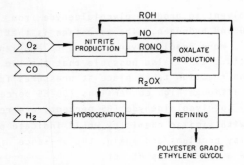

Figure 17. Flow Diagram of UBE/UCC Syn Gas to Glycol Process

Following a joint development program on this process by UBE and UCC, an integrated pilot plant has been built and operated to examine the effects of recycle and to produce ethylene glycol continuously. From operation of the pilot plant it is concluded that there are no technical problems remaining in commercializing this process. Construction and operating costs for a commercial plant are comparable to a conventional ethylene-based plant and the key factor in the cost of producing ethylene glycol is raw material cost. Based on our detailed analysis, this is the best synthesis gas route to ethylene glycol available for commercialization today.

Acknowledgements

Thanks to the many researchers at Union Carbide who participated in these studies, especially Dr. B. D. Dombek who also assisted in the preparation of this paper, and to Union Carbide Corporation for permission to publish this work.

[1] B. D. Dombek, in "Advances in Catalysis," Academic Press, New York, 1983, Vol. 32, 325.
[2] W. F. Gresham and C. E. Schweitzer (DuPont), U.S. Patent 2,534,018 (1950).
[3] W. F. Gresham (DuPont), U.S. Patent 2,636,046 (1953).
[4] J. W. Rathke and H. M. Feder, J. Am. Chem. Soc., 1978, 100, 3623.
[5] H. M. Feder and J. W. Rathke, Ann. N.Y. Acad. Sci., 1980, 333, 45.
[6] W. Keim, M. Berger, and J. Schlupp, J. Catal., 1980, 61, 359.

[7] D. R. Fahey, J. Am. Chem. Soc., 1981, 103, 136.
[8] R. L. Pruett and W. E. Walker (Union Carbide Corp.), U.S. Patent 3,833,634 (1974).
[9] R. L. Pruett and W. E. Walker (Union Carbide Corp.), U.S. Patent 4,133,776 (1979).
[10] R. L. Pruett, Ann. N.Y. Acad. Sci., 1977, 295, 239.
[11] W. E. Walker, D. R. Bryant, and E. S. Brown (Union Carbide Corp.), U.S. Patent 3,952,039 (1976).
[12] J. L. Vidal and W. E. Walker, Inorg. Chem., 1980, 19, 896.
[13] L. Kaplan (Union Carbide Corp.), U.K. Patent 1,565,979 (1980).
[14] L. Kaplan (Union Carbide Corp.), U.K. Patent 1,602,508 (1980).
[15] E. Cesarotti, R. Ugo, and L. Kaplan, Coord. Chem. Rev., 1982, 43, 275.
[16] L. Kaplan (Union Carbide Corp.), U.S. Patent 4,224,237 (1980).
[17] R. W. Beisner and S. C. Winans (Union Carbide Corp.), U.S. Patent 4,225,530 (1980).
[18] R. W. Beisner and S. C. Winans (Union Carbide Corp.), U.S. Patent 4,224,235 (1980).
[19] B. D. Dombek, J. Am. Chem. Soc., 1981, 103, 6508.
[20] B. D. Dombek, J. Organomet. Chem., 1983, 250, 467.
[21] B. D. Dombek and A. M. Harrison, J. Am. Chem. Soc., 1983, 105, 2485.
[22] S. J. Lapporte and W. F. Toland (Chevron Research), U.S. Patent 3,754,028 (1973).
[23] S. Suzuki (Chevron Research), U.S. Patent 3,911,003 (1975).
[24] Two New Routes to Ethylene Glycol from Syn Gas, C&E News 1983 (April) 41.
[25] A. Spencer, J. Organomet. Chem., 1980, 194, 113.
[26] A. Spencer (Monsanto), U.S. Appl. 861,474, 1977.
[27] S. C. Chan, W. E. Willis, and D. E. Mills, 3rd Int. Symposium on Homogeneous Catalysis, "Rhodium Catalyzed Hydroformylation of Formaldehyde," 1982.
[28] D. R. Neilsen (PPG Industries), German Patent 2,617,085 (1976).
[29] R. W. Goetz (National Distillers), German Patent 2,741,589 (1978).

Chemicals from Coal Gasification

By P. McBride[1]* and H. W. Patton[2]
[1]KODAK LTD., CHEMICALS DIVISION RESEARCH DEPARTMENT, KIRKBY, MERSEYSIDE, U.K.
[2]TENNESSEE EASTMAN COMPANY, EASTMAN CHEMICALS DIVISION, EASTMAN KODAK COMPANY, KINGSPORT, TENN. 37662, U.S.A.

Three years ago, Eastman announced a bold new venture that made news around the world. Eastman disclosed its plans to take a new and important turn in the road - to produce acetic anhydride and other chemicals from coal. As far as I know, Eastman is the first manufacturer in the United States to produce a modern generation of industrial chemicals from coal. The symbolic groundbreaking, which was held in 1980, highlighted a decade of work by Eastman men and women to identify, develop, and assemble the numerous technologies necessary for a viable commercial venture.

Eastman Kodak is probably best known for its easy-to-use photographic films and cameras and its high-quality photofinishing services.

This is the primary business of the Photographic Division based in Rochester, New York. The Eastman Chemicals Division is the other operating division of Kodak and is among the 20 largest domestic chemical companies with 1982 sales revenues of $2.2 billion. The Chemicals Division provides many of the chemicals and resins for Kodak's imaging products and is a major trade supplier of over 300 industrial chemicals, fibers, and plastics. The division employs about 19,000 people and has manufacturing locations in Tennessee, Texas, South Carolina, New York, and Arkansas in the U.S. and in Ectona, England. Headquarters for the Chemicals Division is in Kingsport, Tennessee, which is also the site of the new coal gasification plant.

Before reviewing Eastman's gasification project, it is important to know the way Eastman has produced acetic anhydride through the years in order to understand more fully the significance of the decision.

Tennessee Eastman began making acetic anhydride when Eastman Kodak transferred the production of cellulose acetate for film base from Rochester to Kingsport about 50 years ago. Acetic anhydride was originally made from by-product acetic acid from a wood distillation process for methanol. By the 1940s, it became more economical to purchase alternative raw materials derived from petroleum. In 1951, Eastman built a plant in Texas to provide a reliable supply of more basic raw materials made from natural gas liquids. The current process involves the catalytic decomposition of hot acetic acid vapor into ketene and water. This is a very energy-intensive step. The highly reactive ketene readily combines with additional acetic acid to form acetic anhydride.

Today, Eastman has the capacity to produce about one billion pounds per year of acetic anhydride, primarily for internal use in the manufacture of photographic film base, TENITE cellulosic plastics, ESTRON acetate cigarette filter tow and textile yarns, and certain coatings chemicals.

In the late 1960s, Eastman stepped up its program of energy conservation and began a search for lower-cost chemical feedstocks even before the growing national concern caused by the ten-fold increase in petroleum prices during the past decade. An early study, conducted by Eastman in 1970, acknowledged the declining domestic petroleum reserves and projected that coal would become a more attractive energy source and an important chemical feedstock in the longer term.

In 1970, no one really anticipated the actual events which would take place in the Middle East. The run-up in the price of foreign oil from $3 per barrel to $12 per barrel in 1973 following the Arab oil embargo confirmed the pressing need for alternative supplies of non-petroleum feedstocks. It was judged that the development of modern conversion methods to transform coal into more usable fuel products and chemical feedstocks would be the domain of the larger energy companies.

However, Eastman has the ability to develop the downstream processes to utilize coal-based feedstocks to produce chemicals now made from petroleum.

Estimates were prepared which indicated that acetic anhydride from medium-Btu or synthesis gas could eventually be produced from coal gasification processes with coal as energy source at costs substantially below the equivalent amount of ethylene.

Further analysis identified other target molecules of interest to Eastman which might be produced from the mixture of carbon monoxide and hydrogen called synthesis gas or syngas.

As a result, a research program was begun to synthesize acetic anhydride on a bench scale. The main emphasis was the development of suitable catalysts to produce acetic anhydride at lowest cost possible, with a minimum of by-products, and under conditions easily achievable in commercial production. Although initial yields were relatively low, it was determined that acetic anhydride could be made from methyl acetate and carbon monoxide, a component of syngas.

Continuous pilot plants of modest size were built in 1977 to test various materials of construction and to optimize the conversion and yields. Also, it was demonstrated that the rather expensive catalyst system which had been developed could be re-used indefinitely. The data from the pilot operations were used to formulate mathematical models for the conceptual flowsheet design.

The process was scaled up to the full-sized commercial plant, and the mathematical models aided in optimizing the final engineering design.

Now, I'll review Eastman's gasification plans and the important considerations which led to the processes selected. The gasification complex will include a Texaco coal gasification plant for syngas manufacture, raw gas cleanup and separation facilities, a sulfur recovery unit, a coal-fired steam plant with electric power cogeneration, and chemical plants to produce methanol, methyl acetate, and acetic anhydride. Each day, the plant will gasify approximately 900 tons, or eleven carloads, of high-sulfur coal from nearby mines.

The new facilities have been constructed on a 55-acre site across the Holston River from the main plant at Kingsport, Tennessee.

Bechtel Petroleum was responsible for the engineering design, equipment procurement, and project management. The Daniel Construction Company, which has a long history of successful projects for Eastman, served as the general construction contractor. The venture is financed entirely with company earnings and no federal assistance is involved.

Although the principal emphasis was always on coal, many factors had to be considered before selecting the most economical feedstock for the syngas plant. All types of syngas plants should be located close to their raw material source to minimize transportation charges. Location on the Gulf Coast or on a natural gas pipeline will generally favor natural gas. A coal-gas plant should be near the mine mouth to minimize freight charges.

The major portion of the unit cost of syngas from a natural gas-based plant is the feedstock gas itself which is included in the operating cost.

Many experts agree that the price of petroleum and natural gas will rise much faster than coal prices, which will be closely related to its mining and delivery costs.

A more expensive plant is needed for coal due to the increased complexity of gasifying a solid material, product gas clean-up, and sulfur recovery. As a result, the minimum economical plant size for coal is about ten times larger than a natural gas reformer. However, the relatively low annual operating cost of a coal-based plant offsets the higher initial capital cost.

The maximum incorporation of the total weight of the syngas into the product chemical is also important to the cost efficiency of the overall process. In general, syngas will reach its maximum advantage in the production of oxygenated materials where the largest possible portion of the syngas is converted into a saleable product.

The primary feedstocks and their associated processes produce widely different syngas compostions, which may be expressed as the volumetric ratio between hydrogen and carbon monoxide. Natural gas reforming produces the highest ratio of hydrogen in a syngas and is particularly suited to ammonia production where no carbon monoxide is required. At the other extreme, coal gasification provides the maximum concentration of carbon monoxide in a syngas and is favored for carbonylation reactions, such as with methyl acetate to produce acetic anhydride.

The maximum material efficiency will be realized for various organic chemicals by using the feedstock and its associated process which most nearly corresponds with the hydrogen to carbon monoxide ratio of the desired chemical. Although relatively costly, the syngas composition from each process may be adjusted by employing a water-gas shift reactor as required.

As I've stated, choosing an economical syngas feedstock is a complex process which is influenced by long-term raw material price projections, plant size and location, associated conversion processes, and the desired chemical product. When all factors were considered for Eastman's project to produce acetic anhydride, the best alternative was to locate a gasification complex in Kingsport. Coal will be provided for syngas manufacture from the nearby Appalachian coal fields which have proven to be a reliable supply of coal.

In the mid-1970s, the prospect of natural gas curtailments in Kingsport prompted Eastman to investigate alternative fuel sources. Thus, we became familiar with modern coal gasification technologies and the modifications necessary to produce chemical feedstocks. We evaluated three basic types of commercial coal gasifiers including the fixed or slowly moving bed, fluidized bed, and entrained bed units. Since each type of gasifier has relative advantages and disadvantages, no single design will be suitable for all potential uses of the product gas.

When all factors were considered for Eastman's project, the Texaco coal gasification process, involving an entrained bed gasifier, provided the most economical source of syngas from local coals. This conclusion was reached after several internal studies, and confirmed by outside evaluations performed independently by two well-known engineering contractors, Bechtel and Fluor.

Finally, a successful test run was made with the selected coal at the Texaco pilot gasifier in Montebello, California.

Eastman has licensed Ruhrkohle A.G. and Ruhrchemie A.G.'s developments in coal grinding and waste heat recovery made during their four-year operation of a Texaco gasifier in West Germany.

In the entrained bed gasifier, oxygen, water or steam, and finely ground coal are introduced into the reaction vessel through a common feed nozzle and pass downward through the reactor concurrently. The operating temperature is maintained above the melting point of the coal ash so that the unreactive slag will drain from the unit. As a result of the relatively high operating temperature and the unique flow pattern, the gasification reaction rates are very fast. By-product tars are destroyed within the unit and the production of methane is minimized. The fact that the product gas is composed mainly of carbon monoxide and hydrogen makes this unit particularly suitable for chemical reactions.

The Texaco coal gasification process is a highly developed process which soon will be commercialized for several different applications. This coal-based process is a variation of the widely licensed, proprietary partial oxidation process producing syngas from a number of hydrocarbon feedstocks, principally low-valued, high-sulfur crude, and residual oils. The Texaco gasifier, which operates on a wide variety of low-quality coals at pressures up to 1200 psi, has a high throughput and single pass conversion. Since the raw gas is generated at elevated pressures, the size of the gas clean-up equipment is reduced and the need to compress the product gas prior to its use as a chemical feedstock is minimized.

The Texaco coal gasification process, because of its inherent cleanliness, is an environmentally acceptable means of providing gaseous products from coal. Since the coal is handled as a water slurry, wet grinding eliminates the dust problems ordinarily associated with dry-grinding operations.

The relatively high operating temperature within the gasifier leaves only trace amounts of hydrocarbon by-products in the coal gas. The majority of the water needed for slurrying, cooling, and scrubbing operations is re-used within the process. A very small blow-down stream is withdrawn to prevent the accumulation of water-solubles. The coal ash is drained from the reaction area as a molten slag and quenched in a water bath in the bottom of the gasifier. The course ash is removed as glassy pellets through a lockhopper system. The inert ash pellets have shown very low levels of leachability and qualify for non-hazardous waste landfilling.

Eastman's plant contains two gasifiers, each of which is capable of producing the required volume of syngas for chemical manufacture. This equipment duplication permits a reliable on-stream operation as experience is gained with this new process and allows for planned maintenance, including refractory replacement.

Air Products and Chemicals is operating a three-train industrial gas facility to supply oxygen to Eastman's gasification complex and nitrogen to other parts of the main plant. The oxygen is transported at high pressure _via_ pipeline to Eastman's coal gasifier where it reacts with a coal-water slurry to form primarily carbon oxides and hydrogen with trace amounts of methane. The sulfur in the coal is converted mainly to hydrogen sulfide. The hot product gas is first scrubbed with water to cool the gas and remove any ash particles.

A portion of this stream is then sent to a water-gas shift reactor to increase the hydrogen content by the cobalt-molybdenum catalyzed reaction of carbon monoxide and water.

Finally, low-pressure steam is recovered as the product gas is cooled before additional purification. The full recovery of this heat can increase the thermal efficiency of the process by almost 20% and is, in fact, essential for its economical operation.

The raw product gas then enters a Rectisol unit to remove the so-called "acid gases". The carbon dioxide and hydrogen sulfide are absorbed by a cold methanol wash. The purified syngas is then cryogenically separated into a relatively pure carbon monoxide stream for the acetic anhydride plant and a hydrogen stream suitable for methanol production.

The Rectisol unit and the gas separation plant are highly integrated and were provided by Lotepro, a United States subsidiary of Linde A.G. which performed the engineering in West Germany. The by-product carbon dioxide from the Rectisol process is vented after water-scrubbing to remove any residual methanol.

Elemental sulfur is recovered from the hydrogen sulfide in a Claus unit designed by Ford, Bacon, and Davis. A Shell Claus off-gas treating unit removes the last traces of sulfur from the Claus vent gas. One advantage of coal gasification over direct burning is that the sulfur is converted into a form which is more easily removed. Overall, 99.7% of the sulfur originally contained in the coal will be recovered in molten form and solid.

In all areas of Eastman's new complex, discharges are kept within state and federal regulatory standards by the use of advanced environmental control equipment.

Methanol is produced from carbon monoxide, carbon dioxide, and hydrogen using a highly efficient, low-pressure process licensed from Lurgi. The proper gas feed composition for methanol production is formed by combining the hydrogen-enriched syngas from the shift reactor and the hydrogen stream from the gas separation unit. The methanol is reacted with by-product acetic acid from our cellulose esters manufacturing operation to form methyl acetate. This esterification process is also an Eastman innovation which resulted from a successful pilot operation of a novel system to produce refined methyl acetate in a very energy-efficient manner.

Finally, the purified carbon monoxide from the gas separation plant is reacted with methyl acetate to form acetic anhydride using the proprietary catalyst system and process developed by Eastman and described earlier.

Without this key breakthrough in acetic anhydride technology, the entire commercial venture would not be possible. Eastman has assigned the worldwide licensing rights to Halcon.

At its peak, over 2600 workers were employed on the construction of the eight plants in this mammoth complex. The powerhouse began operation in September 1982. We began testing the gasifier in June 1983 . At this point, we have successfully operated each of the chemical plants on a limited basis.

Following further equipment checkout, full-scale commercial production of chemicals will begin later this year. Eastman's use of coal as a raw material will strengthen our ability to supply chemical products to our customers at affordable prices and without disruption.

Eastman's project is a small but important step in the nation's effort to reduce dependence on foreign oil since the chemicals produced in this complex would require the equivalent of one million barrels of oil a year by using conventional technology. Coal-based syngas will provide economical feedstocks for chemical manufacture and will become more attractive as the differential between coal and petroleum prices increases. Eastman is committed to a program of finding new ways to broaden the use of coal as a means of producing other chemicals now made from petroleum.

The coal conversion segment of the synthetic fules industry will gradually develop in other areas. The fully integrated domestic oil companies recognized the declining reserves of petroleum many years ago and obtained extensive rights to lands bearing oil shale, uranium, lignite, and coal.

These oil companies and others have invested hundreds of millions of dollars in pilot plants and demonstration plants to convert coal into more usable fuels and chemical feedstocks. The first and most promising ventures are likely to be coal gasification plants producing medium-Btu fuels or syngas.

Large-scale commercialization, based on the methanation of coal gas to form a truly synthetic natural gas equivalent, will follow much later in the 1990s. New generations of gasifiers employing pressurized operation, catalytic generation, advanced fluidized beds, and improved sulfur removal will be developed. A relatively small portion of the total capital of a coal-based syngas plant is committed to the gasifier area.

Therefore, it will be economically attractive for the gasification complexes already in place to be modified to take advantage of the new, more efficient gasification techniques as they become available.

The petrochemicals industry currently utilizes petroleum and natural gas to manufacture products valued at $81 billion annually. The industry faces a tremendous challenge in the 1980s to keep plants running both reliably and profitably while adjusting to the changing social and political environments and contending with international unrest. Fortunately, the financial rewards for good performance are equally great.

Indeed, the chemical industry is at a turn in the road. The most successful companies of the 1980s will be those which develop advanced technologies to move their operations toward lower-cost feedstocks and fuels.

For its acetic anhydride, Eastman began with wood, converted to petroleum, and is now taking a new and important turn in the road by moving to coal.

Biomass as a Chemical Raw Material

By K. J. Parker
'MURRENS', READING ROAD, CROWMARSH GIFFORD, WALLINGFORD, OXFORDSHIRE
OX10 8EN, U.K. (FORMERLY OF TATE & LYLE)

The potential problem posed to mankind by the finite limits to natural resources has been the focus of considerable interest over the past decade, in particular, with respect to energy. The urgency of the situation has probably been overstated, since the known reserves of oil and natural gas still exceed the quantity of hydrocarbons so far extracted. Deposits of coal are sufficient to meet current rates of extraction for another three hundred years, while fossil carbon reserves as shale oil and tar sands, so far virtually unexploited, are even more substantial.[1]

Nevertheless, the effect of this threat of the ultimate exhaustion of the fossil fuel reserves has been to trigger a search for regenerable sources of energy, often at national level, resulting in rapid progress in the development of alternative energy sources. In particular, the need for transport fuels as a replacement for petrol and diesel fuel has concentrated attention on renewable sources of fixed carbon.

The only energetically replenishable route to fixed carbon available is through the process of photosynthesis, which utilises solar energy to fix atmospheric carbon dioxide as carbohydrate in the chloroplasts of green plants and algae. Further conversion to structural carbohydrates and lignin provides plant tissue, collectively termed 'biomass'.[2] The annual production of biomass world-wide is estimated to be of the order of 2×10^{11} tonnes (dry matter), equivalent to ten times the world's current annual consumption of energy. Approximately one half is in the form of forest trees. Wood is a traditional fuel, around half of the annual production of felled timber being burnt directly.[3]

Organic chemical feedstocks have tended to be derived from by-products of the production of fuels. Coal-tar, important as a source of chemicals up to the 40's, is a by-product of coal-gas production, while the petrochemical industry is dependent for its primary raw materials on the by-products of catalytic cracking and reforming of petroleum hydrocarbons. In fact, compared to fuel production, the organic chemical industry in the US consumes around 8% of the petroleum used, for example.[4]

For the same reason, research into the utilisation of biomass has been directed primarily towards its potential as a source of fuel. Its value as a route to chemical products has not been developed beyond its traditional role as a source of natural products, usually too complex for economic synthesis.

Generally, however, biomass is too thinly dispersed and its energy is too dilute for it to be used directly as an energy source. Some processing involving an energy input, such as collection, drying, compaction and transport, is required before it can be used as a fuel. Still further processing is necessary in order to produce an automotive fuel, such as methanol, ethanol or hydrocarbons, in which the specific energy approaches that of conventional hydrocarbon fuels.

A degree of energy concentration is achieved naturally in many plant species, where the oxidation state of the carbohydrate is further reduced, as in the formation of lignin, an aromatic polymer of coniferyl alcohol, or even in the formation of hydrocarbons, a characteristic, for example, of many species of the family Euphorbiaeceae, which includes the rubber tree Hevea. Energy concentration is also achieved naturally where the plant lays down storage carbohydrates in specific regions, as in the starch crops: cereals, potatoes, yams and the sago palm, for example, and sugar crops: sugar cane, sugar beet and sweet sorghum being the most important.

Even where an advantage is derived from such natural concentration, as in the use of cassava, for example, as a raw material for the production of fermentation ethanol, the energy content of the product may not be significantly greater than the total energy consumed in producing the fuel, resulting in only a small net energy gain.[5] This, of course, does not mean that the process has no value. A liquid fuel is a concentrated source of energy readily stored and transported. In effect, the process provides a means of converting a low-grade source of energy into an automotive fuel, or chemical feedstock.

It may be that the product has a higher value as a chemical intermediate than as a fuel. In such a situation the energy balance becomes largely irrelevant.

This principle is well illustrated by the use of sugar cane as an energy crop. Sugar cane is one of the most efficient species for converting solar energy, achieving a photosynthetic efficiency of 2% compared with the average of around 0.1%. The fixed energy may be recovered by direct burning of the cane as harvested for steam raising and electricity generation. Where fuel alcohol is the required product, sugar cane juice is fermented directly after extraction; the fibrous residue, known as bagasse, is burnt to provide process energy, giving a net yield of just under half of the initial energy content of

Biomass as a Chemical Raw Material

the cane as alcohol. Where the sugar is of greater value than the equivalent as ethanol, raw sugar is produced and the residual impure syrup, molasses, may be fermented to give either potable or fuel alcohol.

Fermentation alcohol may itself have a higher value as a chemical feedstock than as a fuel, the choice depending on economic and geographical factors. Essentially, the technology developed for the utilisation of biomass as a source of fuels is equally applicable to its conversion into chemical feedstocks. Whether the product is used as a food, a fuel or a chemical intermediate is determined by its economic value in each application.

Sources of Biomass

The term biomass covers a wide spectrum of organic materials of natural origin. It comprises forests - which represent 90% of the world's biomass - grassland, agricultural crops and aquatic plants, the latter contributing only around 2%, despite the far higher proportion of the Earth's surface (70%) covered by water. In addition, domestic, agricultural, food processing and industrial wastes provide a collected source of raw materials.[6]

Such resources have been evaluated in detail as energy feedstocks, utilised either by direct combustion or after conversion to suitable fuels by thermal processing or fermentation. The main limitations to the utilisation of biomass are its low bulk density, high water content and seasonal availability. As a chemical feedstock, biomass has the further disadvantage of a widely variable and non-uniform chemical composition.

There are two broad solutions to this problem already available which will largely define the source of biomass required. These are:
> 1. Total utilisation, treating biomass as a source of fixed carbon.
> 2. Selective utilisation, processing biomass to obtain chemically defined intermediates.

Both routes have been used extensively for the conversion of biomass into fluid (liquid and gaseous) fuels, though there may be limitations imposed by cost or energy input, which do not necessarily apply to the production of chemical intermediates.

Thermal Processing. When biomass is available at a low initial water content, for example as wood waste or cereal straw, direct thermal conversion is feasible. The pyrolysis of wood results in the formation of solid (charcoal), liquid and gaseous products, the composition of which represents the sum of the decomposition processes of its three major components, lignin, cellulose and hemicelluloses.

Historically, kilning of wood out of contact with air was used to produce charcoal, the aqueous distillate being a source of methanol and acetic acid. In addition, it also contains the lower aliphatic acids, aldehydes and ketones, furan, furfural and furfuryl alcohol. The tarry distillate contains substituted phenols derived from lignin.[7]

Alternatively, wood, heated with dilute aqueous alkali, liquifies to form a viscous tar, from which as many as 39 phenols have been separated, approximately half of which have been characterised. Under these conditions, the aromatic compounds are not only derived from the lignin but are also formed by the condensation and aromatisation of 2- and 3-carbon decomposition products of cellulose.[8]

Such complex mixtures, in which no one component is present to the extent of more than 2%, do not provide a source of simple chemical intermediates competitive with synthetic routes.

The alternative to pyrolysis is complete gasification, a process developed for coal, for which the technology is well established.[9] As a raw material for gasification, biomass has several advantages over coal. It is generally lower in ash and sulphur; it has a higher hydrogen to carbon ratio; it has a higher oxygen content. Disadvantages are its high moisture content, low density, high transport and storage costs and unsuitable physical form. The low thermal conductivity and fibrous structure do not allow the use of conventional gasifiers, where very rapid heating rates are required to reduce the formation of char, and to give maximum gas yields.

In practice, it is necessary to gasify in the presence of added oxygen and steam at a temperature of around $1000°C$, when a mixture of carbon monoxide and hydrogen, synthesis gas, is obtained. The technology for converting synthesis gas into methanol, or any of a range of products such as ethanol, ethylene or methane, is well established. Methanol can be further converted catalytically into methane, ethanol, acetic acid or hydrocarbons, as in the Mobil process.[10]

The main limitations of thermal gasification of biomass are the high production capacity needed to benefit from economy of scale and hence the high capital investment, and the availability of biomass at the production site, which is determined by the maximum distance over which biomass can be transported economically.

<u>Anaerobic Fermentation</u>. The biological gasification of biomass by anaerobic fermentation is well established and widely practised.[11] The fermentation employs mixed cultures of anaerobic methanogenic bacteria which are capable of utilising carbohydrates, including cellulose and hemicelluloses, proteins, lipids and organic acids, with the production of a methane-rich gas mixture. The gas composition and yield vary according to the substrate, fermentation conditions, temperature and solids concentration. Typically the gas consists of methane (55-80%) and carbon dioxide (20-35%), with small quantities of nitrogen, hydrogen and hydrogen sulphide. Biogas yields range from 300-500 l/kg (dry organic solids), with an energy content of around 23 MJ/m^3. Energy

recovery is between 30 and 60%, depending on ambient temperature and the temperature of operation.

Biogas can be used directly as a fuel, though its energy content is too low for it to be transported economically by pipeline. It has not been considered as a source of methane as a chemical feedstock owing to its low conversion rate and low substrate concentration.

<u>Selective Utilisation of Biomass</u>. The alternative to the total utilisation of biomass is to extract selectively those components which will provide a chemical feedstock competitive with established sources. Most plant species produce a wide range of compounds in addition to their structural components, many of which can be extracted into a suitable solvent. Some plant species have been identified which contain up to 10% hydrocarbons on a dry weight basis, giving rise to the concept of 'energy farming'. Cultivation of <u>Euphorbia lathyris</u> as an energy crop has been proposed by Calvin,[12] for which yields have been claimed equal to that of sugar cane, with the production of 5.3 barrels of crude extract and 2 tons of sugars per acre.

The major fraction extracted from the whole dried plant by hexane comprises a mixture of triterpenoids (85%) and long-chain hydrocarbons and alcohols. The extract can be converted into a conventional liquid fuel by thermal processing using a Zeolite catalyst. The sugars, comprising 20% of the dry weight of the plant, are fermentable to fuel alcohol.

In an extensive screening programme, 14 plant species have been identified having potential dry matter yields of between 11 and 22.4 tonnes/ha/yr, with up to 10% extractable oils, several times the yield of conventional oilseed crops.[13] Depending on the species chosen and the solvent used, a wide variety of sterols, long-chain alcohols, fatty acids, non-glyceride esters and terpenes, in addition to hydrocarbons, can be separated. Many of these products are suitable for industrial uses, for example as plasticisers, or as chemical intermediates.

<u>Carbohydrates as Chemical Feedstocks</u>. The predominant components of biomass are carbohydrate, cellulose being the most abundant. Pure cellulose is produced as cotton, for example, though the major source of cellulose is as wood pulp for the paper industry. Despite the enormous scale on which cellulose is produced, little use has been made of this raw material as a chemical feedstock, except for the production of regenerated cellulose fibre (<u>e.g</u>. Viscose Rayon) and film (cellophane) and chemically modified cellulose derivatives, such as the acetate and the carboxymethylether.

Cellulose is a polymer of D-glucose in which the glucose repeating unit is linked almost exclusively 1 - 4 in the β-configuration. The polymer thus has three free hydroxyl groups (one primary and two secondary) which are chemically reactive. Partial derivatisation of cellulose by esterification or etherification radically modifies its physical properties, including its solubility in water and organic solvents, leading to a wide range of applications.

Hydrolysis of cellulose, either by acid catalysis or by enzymes, gives glucose. Neither process has so far been found to be economic. Nevertheless, since glucose is fermentable to ethanol, wood hydrolysis and fermentation of the product provides an attractive alternative to thermal gasification as a route to a liquid fuel or chemical feedstock. An integrated process in which lignin is recovered and used as a process fuel and the fermentation residues are fermented anaerobically to methane, has been proposed for the production of ethanol from wood, giving a liquid fuel output to input ratio of 9:1. A net yield of ethanol of 633 gal/acre/yr from 10 tonnes oven-dry wood chips is predicted.[14]

No significant use has been identified for the hemicellulosic fraction of wood, which comprises mainly pentosans. Hydrolysis gives pentoses which are not readily fermented to alcohol. Birch xylan is the raw material for the production of xylitol by hydrogenation of xylose produced by acid hydrolysis. Xylitol, a pentahydric alcohol, has been proposed as a non-cariogenic substitute for sugar.

High-temperature digestion of pentosans with dilute sulphuric acid leads to the formation of furfural. In practice, agricultural by-products, such as rice or oat hulls, corn cobs or sugar cane bagasse, are used as the source of hemicelluloses, giving yields of up to 22% on raw material solids.

After cellulose, starch is probably the most widespread carbohydrate in nature. Its annual production in the form of grain, cereal and root crops is estimated to be in excess of 10^9 tonnes worldwide. Starch, like cellulose, is a polymer of glucose in which the glucose units are also linked 1 - 4, but in the -configuration. Starch is not, however, a single chemical entity: it comprises a mixture of the linear polymer amylose (typically 15-30%) and the highly branched amylopectin which has also 1 - 6 α-linkages. It may also carry phosphate ester groups, as, for example, in potato starch. In contrast to cellulose, starch is readily hydrolysed to glucose, either by acid catalysis or enzymically. Produced as a by-product of the extraction of gluten from wheat and maize, it is sufficiently low priced to allow glucose to be sold at below the price of sugar. In North America and Japan, High Fructose Corn Syrup, that is corn syrup in which part of the glucose has undergone enzymic conversion into the sweeter fructose, is rapidly displacing sucrose from the industrial liquid sugar markets.

Glucose is fermentable to ethanol and is the substrate for the production of fuel alcohol in the U.S. under their Gasohol programme.[15] In 1981 the U.S. Department of Agriculture reserved 150m bushels of corn, equivalent to 385m gal ethanol, towards the target production of 500m gal fuel alcohol. At prevailing oil costs fermentation ethanol is not competitive with synthetic ethanol as a chemical feedstock in the US.

As an intermediate for the production of chemical derivatives, glucose finds limited use in the manufacture of sorbitol, mannitol, gluconic acid, ascorbic acid, glucoheptonic acid and methylglucoside. It also provides a cheap substrate for fermentation processes, such as in the production of antibiotics, or Xanthan gum, for example.

Starch gives the expected derivatives by reaction of the free hydroxyl groups to give a range of esters, ethers and oxidation products, which have achieved commercial importance in many applications. Prior to 1950, only the acetate, used as a substitute for gelatin and natural gums, and the nitrate were produced in commercial quantities. Since then, a wide range of modified starches in which a low degree of substitution of hydroxyl groups (typically 0.01-0.1) by ester or ether groups is made. The effect of this limited derivatisation is to modify the physical properties of starch. For example, water solubility, gel temperature, solution viscosity and solubility in organic solvents are modified.

Sucrose as a Chemical Feedstock

As a potential feedstock to the chemical industry, sucrose has several unique advantages. First, it is produced as an anhydrous crystalline solid by an established industry from two major agricultural crops: sugar cane in the tropics and sugar beet in temperate regions. Secondly, it is a pure defined organic compound of low molecular weight (342) with multiple chemical functionality. Thirdly, it is readily fermentable by a wide range of micro-organisms. Fourthly, its production cost is comparable with the price of petroleum-derived chemical feedstocks.

The annual production of sucrose from all sources now exceeds 100 million tonnes: a further 15 million tonnes is obtained as molasses. Most of the raw sugar production is consumed as food, and 85% of the molasses in animal feeds. Probably not more than one million tonnes is used in non-food applications, mainly in fermentation. Recently, however, Brazil has expanded its production of fuel ethanol to meet a production target of 8 billion litres by 1985. The whole of this output will be met by the fermentation of sugar cane juice.[16]

Fermentation. In considering sucrose as a raw material for the chemical industry, its value as a fermentation substrate should not be overlooked. The production of power alcohol is widely practised and the technology highly developed. Sucrose, either in molasses or in sugar cane or beet juice, is particularly suited to continuous fermentation, which gives high conversion rates and efficient utilisation of the substrate.

Alcohol is readily converted into ethylene in almost quantitative yield by catalytic dehydration over an alumina catalyst at 350°C. For this conversion, the ethanol does not require to be specially dried as it does for use as a fuel in admixture with gasoline.[17] Ethanol and ethylene are the starting point for the production of many chemical intermediates by conventional technology, such as ethyl acetate, acetic acid, acetaldehyde, ethylene oxide, vinyl chloride and their derivatives. Ethanol is thus itself an important chemical feedstock which can be competitive with petroleum-derived ethylene and alcohol in particular economic situations, for example, where sugar is produced cheaply and indigenous oil deposits do not exist.[18]

Ethanol is only one product which can be obtained by the fermentation of simple sugars, representing a particular product from a more generalised route. In the presence of oxygen, heterotrophic microbes generally gain energy by the oxidation of organic substrates to carbon dioxide and water. Deprived of oxygen many such organisms are capable of using carbohydrates as sources of energy, for example by converting glucose into pyruvic acid with the formation of reduced pyridine nucleotide and energy in the form of two moles of adenine triphosphate. The cycle is completed by the reoxidation of the NADH with transfer of two protons to pyruvic acid to form lactic acid, or ethanol and carbon dioxide, depending on the organism.

The yield of both lactic acid and ethanol is almost theoretical under optimal conditions. The major limitations of fermentation processes are the low rate of reaction, requiring relatively large volumes and long residence times, and the dilution of the product requiring a high energy input to concentrate and separate the product from water. The reduction of pyruvic acid may follow alternative routes to other fermentation products, including acetone, n-butanol, isopropanol, butan-2,3-diol, glycerol, acetic, propionic and butyric acids or methane, according to the organism and the conditions.

An important group of products of fermentation of simple sugars are the polysaccharide gums, examples of which are elaborated by a large number of micro-organisms. Of the many microbial polysaccharides which have been studied several are currently being produced on a commercial scale.[19]

Xanthan gum is an exopolysaccharide produced by strains of <u>Xanthomonas campestris</u>, which has found wide applications ranging from foods as a thickening and dispersing agent to oil-well drilling and textile printing and dyeing. Other polysaccharides of potential commercial importance produced microbially from sugars are dextran, mainly by <u>Leuconostoc</u> spp., pullulan from <u>Aureobasidium pullulans</u>, alginic acid from <u>Azotabacter vinelandii</u> and curdlan from a strain of <u>Alcaligenes faecalis</u> var. <u>myxogenes</u>.

Some simple products such as citric and itaconic acids are not readily produced synthetically and the preferred route is by fermentation. This is particularly true of the antibiotics, such as penicillin, the tetracyclines, streptomycin and several vitamins, nucleotides and amino-acids, for which purely chemical synthetic routes are not economic.

Thermal Degradation of Carbohydrates. As an alternative to fermentation, the controlled degradation of carbohydrates will, in general, lead to products of simpler chemical structure. Simple sugars give similar products, so that glucose or sucrose would react similarly. Despite extensive research, the yields of the isolated product tend to be low making the processes non-competitive with alternative routes from petroleum-derived starting materials. Apart from the inherently low weight yield consequent upon the loss of elements of water, degradative reactions are not readily controllable, giving rise to a mixture of products. For example, the production of glycerol by the high-temperature hydrogenation of sucrose, is accompanied by the formation of ethylene and propylene glycols, which cannot be re-cycled.

The production of laevulinic acid in the high-temperature acid degradation of sucrose depends on the intermediate formation of 5-hydroxymethylfurfural, which is only formed from the fructose moiety of sucrose under the conditions of the reaction.[20] The yield is further lowered by the alternative polymerisation of hydroxymethylfurfural in competition with the ring opening reaction yielding laevulinic acid.

Under mild conditions, using an ion-exchange resin as a catalyst in the presence of a solvent, sucrose can, however, be converted into 5-hydroxymethylfurfural in 82% yield.[21]

In strongly alkaline solution at high temperatures, sucrose undergoes generalised decomposition with the formation of two-, three- and four-carbon fragments. In the presence of lime, lactic acid is the predominant product, though the process does not offer a viable route to lactic acid production compared with fermentation.[22] In the presence of ammonia, resynthesis leads to the formation of a wide range of nitrogen heterocyclic compounds, including derivatives of imidazole, pyrazine, piperazine and pyridine.[23] With simultaneous high-pressure hydrogenation, 2-methylpiperazine is obtained in 27% yield, though the reaction has not proved economically viable.[24]

Oxidation of sucrose with nitrogen tetroxide gives glucaric acid and, in the presence of vanadium pentoxide, oxalic acid.[25]

None of these reactions has, so far, achieved any economic significance. Sucrose does not have any special advantage over other carbohydrates in degradative reactions and is more costly. The potential of sucrose as a chemical intermediate lies in its synthetic chemistry which is only beginning to be developed.

Sucrose

Sucrose Derivatives. As a non-reducing disaccharide, sucrose will undergo the typical reactions of an octahydric alcohol (three primary and five secondary hydroxyl groups). Being polyfunctional, selective or partial substitution does present problems. The relative reactivity of the hydroxyl groups will depend not only on inherent steric factors, but also on the nature of the reaction, the conditions (temperature, solvent, time, molar proportions of reactants), reagent reactivity, hydrogen bonding and hydroxyl group acidity, for example. It is possible to rank the hydroxyl groups of sucrose in order of relative reactivity. For monosubstitution, this is found to be $6' > 6 > 4 > 1' > 2 > 3,3',4'$. Primary hydroxyl groups will react preferentially to secondary where the formation of the activation complex is sterically hindered by a bulky reactant, giving further opportunity for selective substitution.

As an alcohol, sucrose readily forms esters, ethers, acetals and carbamates, the more accessible and important derivatives. The chlorodeoxy-derivatives are also proving to be of considerable interest, having a number of applications. Furthermore, the hydroxyl groups may be replaced in a nucleophilic displacement reaction of a suitable ester by a range of groups such as bromide, iodide, fluoride, azide, thiocyanate, sulphide, nitrile, etc. considerably extending the synthetic potential of sucrose.[26] Oxidation of selected hydroxyl groups to aldehyde, or carboxylic acid is possible without disruption of the glycosidic ring. Internal anhydride bridges and epoxides are readily introduced into the sucrose molecule.

Numerous potential industrial applications of sucrose derivatives appear in the literature, many of which form the subject of patent specifications, though they have not been widely adopted as industrial chemical intermediates, and only a few esters are commercially available. Some of the more important derivatives of sucrose and their industrial uses are as follows.

Sucrose Esters. Sucrose octa-acetate, an intensely bitter substance, finds use as a denaturant, and as a plasticiser. It is a crystalline material (mp 86° C) which is not entirely satisfactory as a plasticiser for cellulose acetate films. For this application, the mixed ester, sucrose diacetate hexaisobutyrate was found to have optimal properties.[27] It has a high viscosity, low volatility and high stability to hydrolysis and discoloration. It is also non-toxic and is used as a clouding agent, to disperse essential oils, in soft drinks.

Sucrose octabenzoate[28] is also very stable to light and to hydrolysis and is completely soluble in non-hydroxylic organic solvents. It is used as a plasticising agent in nitrocellulose, acrylic and polyvinylchloride-acetate films and lacquers. It may be blended with the acetate-isobutyrate to which it is closely similar. It is also used, like the acetate-isobutyrate, in the manufacture of transparent papers, having a refractive index close to that of cellulose.

Sucrose mono-esters of long-chain fatty acids have surface-active properties and are completely non-toxic. Consequently, they have found numerous applications both in the food and the animal feed industries, as dispersing and emulsifying agents.[29] Sucrose mono-stearate, mono-palmitate, mono-oleate and di-stearate are commercially available, having been manufactured in Japan for several years.

Sucrose esters are readily biodegradable, and can be used in detergents, where environmental problems can be created by persistent detergent residues. For this purpose, however, purified sucrose esters are too costly. The low-cost sucroglyceride mixtures produced by a solventless process can be economically included in detergent formulations, the detergents being competitive in cost and efficacy with those based on conventional linear alkylsulphonates. Markets include those for domestic and industrial detergents, hand cleansers, hard-surface cleaners, machine cutting lubricants, amongst others.

Sucrose esters of stearic, palmitic, lauric, and oleic acid, being completely non-toxic, bland, non-allergenic and non-irritant are suitable for use in foods and are permitted food additives in some countries; for example, in Japan, France, Belgium and Switzerland.

Long-chain fatty acid esters of sucrose have been manufactured in Japan since 1960 by a process originally developed under the auspices of the Sugar Research Foundation, by the Ryoto Co. Ltd., of Tokyo. Since 1975 they have operated a new process with a production capacity of 3000 tons/year in which ester exchange between sucrose and the methyl ester is effected without the use of the toxic dimethylformamide solvent.[30]

If the ester exchange reaction is from a triglyceride to sucrose, then at equilibrium a mixture of sucrose mono- and di-esters and mono- and di-glycerides is obtained. The mixture, known as a sucroglyceride, is suitable for use as a surface active agent without separation, especially in food applications. In the original process developed by the Italian Company, Ledoga, a solvent was used, but is not necessary for reaction.

In the solventless process, sucrose is heated at around 125°C in a stirred reactor with a natural triglyceride, such as tallow, palm oil or coconut oil, in the presence of potassium carbonate catalyst.[31] The product is suitable for direct use in many applications without further processing. If required, the sucrose esters may be separated by extraction with ethyl acetate.

Sucroglycerides are mainly used in compounding animal feeds to give improved digestion of fats. In the formulation of calf milk feeds from skimmed milk and added vegetable fats, the sucroglyceride disperses the fat and assists subsequent spray-drying of the product.

Sucroglycerides are also used as dispersing and emulsifying agents in the production of margarine, mayonnaise, ice-cream, artificial cream and dessert toppings, for example. Sucrose ester surfactants are also of use in chocolate manufacture to improve surface appearance, in spray-dried and freeze-dried beverages to improve dispersion, and in reconstitutable dried foods and mixes to aid wetting.

In baking, sucrose mono-esters are particularly effective as a dough strength improver, and to increase loaf volume and improve crumb texture and antistaling properties of bread.[32] The incorporation of non-wheat protein into bread is possible without loss of loaf quality, with the addition of sucrose esters. Up to 0.4% of sucrose monopalmitate, stearate or laurate would normally be used, or approximately 1.4 g/lb. Sucrose esters also improve ingredient mixing and product texture in cake, pastry and biscuit manufacture.

Sucrose esters are used in the preparation of cosmetics, having been found to be highly compatible with the skin and totally non-irritant. They appear to have the particular advantage, when used as a surface-active component of cleansing agents or shampoos, of maintaining the natural oily protective mantle of the skin, which is removed by ionic detergents.[33]

Chlorosucrose Derivatives. The extraordinary versatility of sucrose derivatives is seen in the range of basic flavour properties: several chlorosucroses are intensely sweet, the anticariogenic 1',4,6'-trichlorogalactosucrose being 650 times sweeter than sucrose itself.[34] It has a pure sweet taste, indistinguishable from that of sucrose, with no side-flavours or after-taste. It is completely non-toxic, non-metabolised and virtually non-absorbed. It is stable to enzymic hydrolysis and 60 times more stable than sucrose to acid hydrolysis. At the other extreme, 1',2,6,6'-tetrachlorotetradeoxymannosucrose is intensely bitter, having twice the bitterness of the octaacetate.

6,6'-Dichloro-6,6'-dideoxysucrose is effective as a male antifertility agent in rats and primates, allowing complete and reversible control of male reproductive fertility.[35] It is non-toxic, but depends specifically for its activity on hydrolysis under physiological conditions and absorption as 6-chlorodeoxyglucose and 6-chlorodeoxyfructose, which are interconvertible in the normal carbohydrate metabolic cycle.

Synthetic Resins from Sucrose. An important potentially large market for sucrose and sucrose derivatives is in synthetic resins. Sucrose itself has been evaluated as a filler or substitute for phenol in melamine and novalak-type resins,[36] though there is no evidence that unmodified sucrose will participate in the cross-linking reaction with formaldehyde. The expected hemiacetal link is presumably insufficiently stable at normal curing temperatures to provide any significant cross-linking to sucrose in the final polymer.

Octa-allylsucrose ether was investigated in some depth as a component of air-drying resin films, but they proved to be too brittle for practical use.[37] Sucrose polyesters of drying oil acids, derived from linseed oil, tung oil or soya oil, such as sucrose heptalinoleate, were found to be superior to linseed oil as an air-drying paint vehicle, in terms of film adhesion and alkali resistance. An ester with a lower degree of esterification could more readily be prepared, those with an average 4-5 ester groups being available by ester interchange between sucrose and methyl linoleate in dimethylsulphoxide solvent.[38]

The sucrose pentaester typically would be esterified with phthalic anhydride and further reacted with a suitable diepoxide, such as bisphenol A diglycidyl ether, to give a linear polymer. The varnish is prepared by disssolving the resin in the appropriate solvent (e.g. white spirit/diacetone alcohol) to which is added cobalt naphthenate drying catalyst and pigment to give an air-drying paint. Alternatively, the resin monomer can be polymerised with a diisocyanate to give a polyurethane-type resin. This product is less costly than the epoxy resin, and gives high-performance surface coatings. Generally the inclusion of sucrose did not give sufficiently marked advantages over conventional alkyd-type surface coatings to offset the higher cost, and further development of sucrose-based paint resins has not taken place.

Sucrose and its derivatives have been widely explored as components of polyurethane resins, in particular rigid polyurethane foams.[39] Sucrose itself can be used, but tends to lead to brittle products. The polyhydroxypropyl ether of sucrose is the preferred polyol in practice, having better miscibility with the diisocyanate and fluorocarbon blowing agent used in the preparation of rigid foams. In order to introduce flame retardant properties into the resin, halogen or phosphorus derivatives of sucrose are frequently included in the reaction.

A rigid polyurethane foam which is essentially non-flammable can be produced from a sucrose long-chain fatty acid ester, with starch as the cross-linking polyol. On exposure to heat the foamed resin tends to char without melting, and the flame does not propagate.

Future Developments

Despite the very large range of sucrose derivatives, both known and potential, very few have so far found industrial application. The largest non-food outlet for sucrose derivatives is in polyurethane foam resins, particularly rigid insulating foams.

The world-wide production capacity for sucrose mono-esters of fatty acids is relatively small being around 12,000 tons/year, though the demand could increase rapidly as its use in foods becomes more widely accepted.

The potential of sucrose as an industrial raw material obviously has yet to be exploited. Major factors in determining how quickly applications of the new derivatives of sucrose will be developed are raw material cost relative to oil-based chemical feedstocks and the cost of the derivatisation process as a commercial operation.

High-value sucrose derivatives with specialised or unique properties will develop a market in their own right, the cost of the sucrose raw material being a negligble factor. For example, in the production of a non-caloric high-intensity sweetener, such as trichlorogalactosucrose, the cost of sucrose would be insignificant, compared with the cost of the synthetic process, and would not directly influence the competitive position of the product in the artificial sweetener market.

The sucrochemical industry may thus be expected to develop along two lines: the utilisation of sucrose as a fixed carbon feedstock for the chemical industry and the exploitation of the unique properties of entirely new derivatives of sucrose.

References

[1] B.A.Rahmer, "Fuels for the Future", *Petroleum Economist*, London, 1980.

[2] J.Coombs, *Outlook on Agriculture*, 1980, 10, 235.

[3] D.O.Hall, *Fuel*, 1978, 57, 322.

[4] M.N.Sarbolouki and J.Moacanin, *Solar Energy*, 1980, 25, 303.

[5] J.G. da Silva, G.E.Serra, J.R.Moreira, J.C.Conçalves and J.Goldemberg, Science, 1978, 201, 903.

[6] E.S.Pankhurst, Biomass, 1983, 3, 1.

[7] F.Shafizadeh and P.P.S.Chinn, in "Wood Technology: Technical Aspects", I.S.Goldstein, Ed., American Chemical Society, 1977, Chapter 5, p.78.

[8] J.A.Russell, R.K.Miller and P.M.Molton, Biomass, 1983, 3, 43.

[9] W.P.M. van Swaaij, in "Energy from Biomass", W.Palz, P.Chartier and D.O.Hall, Eds., Applied Science Publishers, London, 1981, p.485.

[10] Anon., Chem.Eng. News, 1978 (Jan.30), 26.

[11] D.A.Stafford, in "Energy Conservation and Use of Renewable Energies in the Bioindustries", F.Vogt, Ed., Pergamon Press, Oxford, 1982, p.626.

[12] M.Calvin, E.K.Nemethy, K.Redenbaugh and J.W.Otvos, Petroculture J., 1981, 2, 26.

[13] R.A.Buchanan, F.H.Otey, C.R.Russell and I.M.Cull, J. Amer. Oil Chem. Soc., 1978, 55, 657.

[14] J.D.Ferchak and E.K.Pye, Solar Energy, 1981, 26, 17.

[15] J.D.Ferchak and E.K.Pye, Solar Energy, 1981, 26, 9.

[16] V.Yand and S.C.Trindade, Chem. Eng. Progr., 1979, 79, 11.

[17] N.K.Kochar, R.Merims and A.S.Padia, Chem. Eng. Progr., 1981, 81, 66.

[18] K.D.Sharma, UNIDO Workshop on Fermentation Alcohol for use as Fuel and Chemical Feedstock in Developing Countries, Vienna, Austria, 26-30 Mar., 1979, 293/14.

[19] C.J.Lawson and I.W.Sutherland, in "Primary Products of Metabolism", A.H.Rose, Ed., Academic Press, 1978, Chapter 9.

[20] L.F.Wiggins, Adv. Carbohydr. Chem., 1950, 4, 306.

[21] L.Rigal and A.Gaset, Biomass, 1983, 3, 151.

[22] W.N.Haworth, H.Gregory and L.F.Wiggins, J. Soc. Chem. Ind.(London), 1946, 65, 95.

[23] M.J.Kort, Adv. Carbohydr. Chem., 1970, 25, 311.

[24] O.K.Kononenko, Final Report of Project 155 to the Sugar Research Foundation, Herstein Laboratories, Inc., New York, 1967.

[25] S.D.Deshpande and S.N.Vyas, Ind. Eng. Chem. Prod. Res. Dev., 1979, 18,(1), 69.

[26] M.R.Jenner, in "Developments in Food Carbohydrates - 2", C.K.Lee, Ed., Applied Science Publishers, London, 1980, Chapter 2.

[27] C.H.Coney, in "Sucrochemistry", J.L.Hickson, Ed., American Chemical Society, Washington, D.C., 1977, Chapter 15.

[28] E.P.Lira and R.F.Anderson, in "Sucrochemistry", J.L.Hickson, Ed., American Chemical Society, Washington, D.C., 1977, Chapter 16.

[29] L.Bobichon, in "Sucrochemistry", J.L.Hickson, Ed., American Chemical Society, Washington, D.C., 1977, Chapter 8.

[30] T.Kosaka and T.Yamada, in "Sucrochemistry", J.L.Hickson, Ed., American Chemical Society, Washington, D.C., 1977, Chapter 6.

[31] K.J.Parker, K.James and J.Hurford, in "Sucrochemistry", J.L.Hickson, Ed., American Chemical Society, Washington, D.C., 1977, Chapter 7.

[32] P.A.Seib, W.J.Hoover and C.C.Tsen, in "Sucrochemistry", J.L.Hickson, Ed., American Chemical Society, Washington, D.C., 1977, Chapter 9.

[33] L.Noble, P.Rovesti and M.B.Svampa, Amer. Perf. and Cosm., 1964, 79, 19.

[34] L.Hough and R.Khan, T.I.B.S., 1978, 3, 61.

[35] W.C.L.Ford and G.M.H.Waites, Intern. J. Andrology, Suppl., 1978, 2, 541.

[36] W.Flavell and G.L.Redfearn, in "Sugar", J.Yudkin, J.Edelman and L.Hough, Eds. Butterworths, London, 1971, Chapter 7.

[37] M.Zief and E.Yanovsky, Ind. Eng. Chem., 1949, 41, 1697.

[38] R.N.Faulkner, in "Sucrochemistry", J.L.Hickson, Ed., American Chemical Society, Washington, D.C., 1977, Chapter 13.

[39] K.C.Fusch and J.E.Kresta, in "Sucrochemistry", J.L.Hickson, Ed., American Chemical Society, Washington, D.C., 1977, Chapter 17.

Gas Separations by Manganese(II) Phosphine Complexes

By C. A. McAuliffe
DEPARTMENT OF CHEMISTRY, UNIVERSITY OF MANCHESTER INSTITUTE OF SCIENCE AND TECHNOLOGY, MANCHESTER M60 1QD, U.K.

Introduction.

The manganese(II) phosphine complexes MnX_2(phosphine) (X = Cl, Br, I, NCS; phosphine = PPh_3, PPh_2R, $PPhR_2$, PR_3, R=alkyl), form an extremely unusual series, in as much as they have the ability to bind reversibly a number of small molecules (e.g. dioxygen, carbon monoxide, nitric oxide, ethylene) in a 1:1 ratio

$$MnX_2(\text{phosphine}) \underset{-L}{\overset{+L}{\rightleftharpoons}} MnX_2(\text{phosphine})(L)$$

$$(L = O_2, CO, NO, C_2H_4)$$

and can irreversibly bind sulphur dioxide to form adducts of most unusual stoicheiometry
e.g. $3MnI_2(PPhMe_2) + 2SO_2 \longrightarrow \{MnI_2(PPhMe_2)\}_3 (SO_2)_2$

Whilst we have spent most of our efforts on investigating the reversible co-ordination of molecular oxygen by these complexes[1,2], we have recently begun to address ourselves to the interactions of the manganese compounds with the other small molecules mentioned above. Much to our satisfaction we can clearly see trends emerging which make it possible for us to designate certain of the compounds useful for the binding of specific gaseous molecules. A brief mention of these is useful:

Dioxygen. We find that reversible co-ordination is both halogen (or pseudohalogen) and phosphine dependent. For example the $Mn(NCS)_2(PR_3)$ complexes form deeply red-coloured $Mn(NCS)_2(PR_3)(O_2)$ adducts, whereas the $Mn(NCS)_2$ complexes of PPh_3, PPh_2R and $PPhR_2$ do not bind dioxygen[2]. As regards the nature of the phospine there are two extremes: the $MnX_2(PMe_3)$ complexes are rapidly oxidised by dioxygen to manganese(III) complexes[3,4], whereas

$MnX_2(PPh_3)$ do not react with dioxygen even at pO_2=20 atmos. In between these two extremes the binding constant KO_2 can be related to the cone-angle, θ, of the phosphine[5] and a phosphine electronic perameter, ν, as defined by Strohmeier[6]. Designating the x, y and z axes to be θ, ν, and KO_2, respectively, Fig. 1, we find a graph in which overall activity appears to fall inside a narrow steric and electronic parameter range.[7]

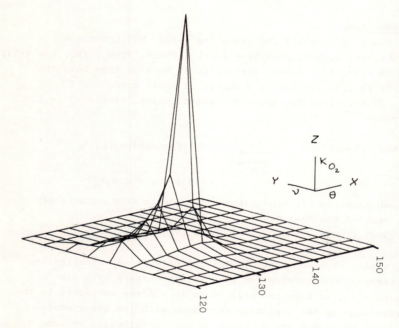

Figure 1 The 3D diagram of K_{O_2} vs. θ vs. ν for $MnCl_2(PR_3)$ complexes in toluene solution at 20°C

<u>Carbon Monoxide</u>. We have identified complexes which bind carbon monoxide reversibly in the solid state, <u>e.g.</u> $MnCl_2(PPhMe_2)$ and $MnCl_2(PPhEt_2)$, but which do not bind dioxygen[8].

<u>Ethylene</u>. Here we have observed a strong dependence of the nature of the halogen on the ability to co-ordinate ethylene reversibly. Thus all $MnCl_2(PR_3)$ complexes bind ethylene at 0°C in tetrahydrofuran, whereas no iodo-complexes, $MnI_2(PR_3)$, are

observed to bind ethylene at any temperature in this solvent. The bromo-complexes form an interesting intermediate series; thus, whilst $MnBr_2(PBu^n_3)$ binds ethylene at $0°C$ to form $MnBr_2(PBu^n_3)(C_2H_4)$ in tetrahydrofuran, $MnBr_2(PPr^n_3)$ does not. However, $MnBr_2(PPr^n_3)$ does not bind ethylene in this solvent at $-43°C$. We thus see a clear temperature effect for the bromo-complexes.

Sulphur Dioxide. Although we do not observe reversible co-ordination of sulphur dioxide by the manganese complexes, we nonetheless again see halogen dependence on the binding. Thus, in contrast to the ethylene co-ordination, all iodo-complexes (even that of PPh_3) react with sulphur dioxide, but no chloro-complexes do. Again, the bromo-complexes are intermediate and, in general, the $MnBr_2(PPh_2R)$ are not[9].

The potential importance of a wide-ranging and flexible gas separation system based on these manganese(II) phosphine complexes is obvious. The current pressure swing absorption process (PSA) for separating dioxygen from air employing molecular sieves generally yields dioxygen contaminated with noble gases (note, however, that allowing such a gas stream to interact with manganese compounds would leave a highly enriched noble gas mixture), but the manganese compounds can yield essentially 100% pure dioxygen by a method similar to that described below.

Other important gas separations are possible. With the present interest in visible light induced water photolysis to yield dioxygen and dihydrogen[10], then our ability to separate these two gases cleanly is clearly of interest. Similarly, we have demonstrated the clean separation of CO/H_2 and C_2H_4/H_2 mixtures, phenomena of interest to the petrochemicals industry.

Apparatus and Method.

The apparatus employed to demonstrate gas separation via the $MnX_2(PR_3)$ complexes is shown in Fig. 2, which demonstrates H_2/O_2 separation, but the system can be used to demonstrate other separations mentioned. In the process quantitative uptake of dioxygen in the mixture is followed via a gas burette; however the gas burette and tubing to the reaction flask contain only dihydrogen.

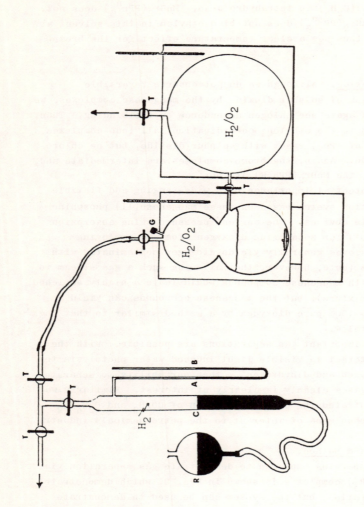

Figure 2   Apparatus for separation of gas mixtures: H_2/O_2 separation shown.

The system works on the principle that if the "dead-space" above the solution is a H_2/O_2 mixture and the gas burette filled with H_2/O_2 then as the uptake of O_2 progresses an O_2 concentration gradient would develop across the intervening tubing, the remaining O_2 uptake would then become diffusion controlled. To avoid this all the O_2 to be absorbed is initially in the dead space above the solution. As the uptake of O_2 proceeds the apparatus is maintained at atmospheric pressure _via_ additions of H_2 from the gas burette. This in turn yields the quantitative uptake data.

To begin an uptake the dead space volume above the solution is evacuated since the solution is normally stored under argon. This space is then filled with the pre-cooled gas mixture from the gas mixture resovoir. The whole apparatus is brought to atmospheric pressure _via_ the gas burette using H_2. The initial volume of the gas burette is noted and the initial gas mixture sampled by mass spectroscopy or glc _via_ the septum cap G employing a gas-tight syringe. This is repeated at regular time intervals during the experiment.

The complex chosen for this study was $MnI_2(PBu^n_3)$ in THF solution. There were a number of reasons for the choice of this particular complex; firstly, the complex is active only below $-40°C$ and, hence, is highly reversible when subjected to temperature and/or pressure swing. Secondly, the complex is, relatively, quite soluble and moderate uptakes of O_2 can therefore be accomplished on the uptake cycle. Thirdly, this complex is very robust to oxygenation/deoxygenation cycles and has been widely studied; its isotherm is shown in Figure 3. Finally, as can be seen in Figure 3, the complex absorbs dioxygen at low partial pressures; hence, the final sample of gas remaining will contain very small amounts of dioxygen and be essentially pure dihydrogen.

The experimental details are as follows:
200 ml of a 1.59×10^{-2} molar solution of $Mn(PBu^n_3)I_2$ was introduced into the gas separation apparatus, Figure 2. Since the overall volume of the reaction flask was found to be 660 ml, we therefore have 460 ml of the gas mixture above the solution. The partial pressure of dioxygen in this gas sample was found to be 128 torr. Hence we have 77.5 ml of O_2 in this gas sample.

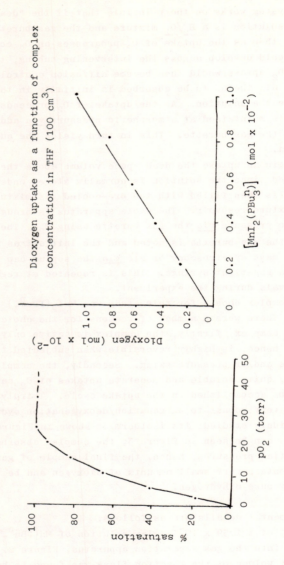

Figure 3 Solution isotherm for $MnI_2(PBu_3^n)$ in THF.

Now 200ml of a 1.59×10^{-2} molar solution of $Mn(PBu^n_3)I_2$ in THF at $-69°C$ will absorb 113.2 ml of O_2 (blank for 200 ml THF= 42ml O_2); thus the pO_2 of the gas sample should fall to ca. 0. The procedure followed was given above and the results are given in Table 1 and Figures 4 and 5.

Table 1: Percentage Composition of H_2 and O_2 During Separation

Time (mins)	pO_2 (Torr)	Vol O_2 absorbed (cm^3)	Composition %O_2	%H_2
0	128	0	16.8	83.1
1	115	7.9	15.1	84.9
2	100	17.0	13.2	86.8
3	98	18.2	12.9	87.1
4	90	23.0	11.8	88.2
5	85	26.0	11.2	88.8
6	71	34.5	9.3	90.7
72	32.5	57.8	4.3	95.7
18	0	77.5	0	100
24	0	77.5	0	100

The results clearly show that, within the levels of detection, all dioxygen is removed from the H_2/O_2 gas sample, yielding very pure H_2 above the solution and a solution of $MnI_2(PBu^n_3)(O_2)$ in THF. The pure H_2 can now be collected.

The dioxygenated solution releases dioxygen on warming to $0°C$; this has been fully described elsewhere[1].

In an exactly analogous manner the separation of CO/H_2 mixtures has been investigated, and the results are presented in Figure 6 and Table 2.

Syntheses and Manipulations.

The syntheses of these simple $MnX_2(PR_3)$ complexes have been thoroughly described[2,11,12]. It is extremely important to employ absolutely anhydrous manganese (II) salts and solvents, and to carry our preparations in an inert atmosphere. Green, Mingos and co-workers[13] have been unable to synthesise these simple molecules although Wilkinson[14] and Hill[4] and their co-workers have recently reported studies on MnX_2(phosphine) systems.

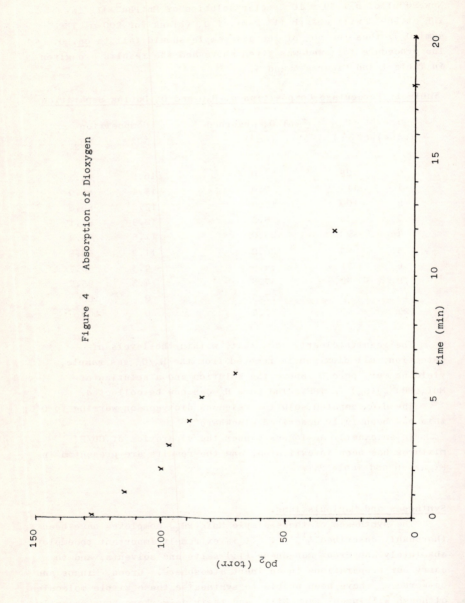

Figure 4 Absorption of Dioxygen

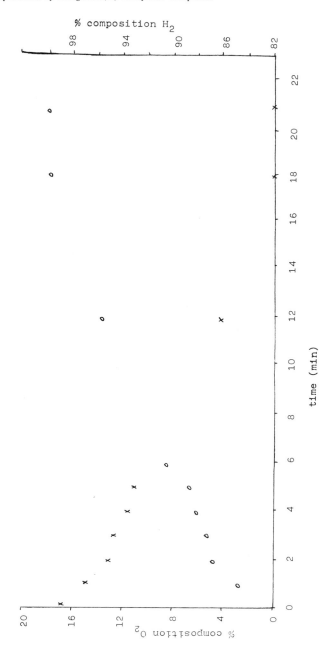

Figure 5 Change in % composition H_2 and O_2

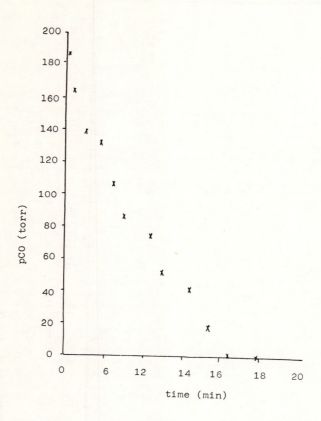

Figure 6   pCO during the gas separation experiment

Table 2: Percentage Composition of H_2 and CO during separation

Time (mins)	pCO (torr)	Vol CO absorbed (cm^3)	% CO	% H_2
0	187.0	0	24.6	75.4
1	165.0	13.3	21.7	78.3
3	140.0	28.5	18.4	81.6
5	133.0	32.7	17.5	82.5
7	108.0	52.7	13.2	86.8
9	87.0	60.6	11.4	88.6
13	75.0	67.8	9.9	90.1
15	52.0	81.7	6.8	93.2
19	42.0	87.8	5.5	94.5
22	18.0	102.3	2.4	97.6
25	0	113.2	0	100
30	0	113.2	0	100

Acknowledgements

We are grateful to The British Technology Group, The British Oxygen Company, Searle Life Support Systems Incorporated, and The Central Electricity Generating Board for support and interest in our work.

References.

1. M.G. Little, C.A. McAuliffe and J.B. Raynor, J. Chem. Soc., Chem. Comm., 1982, 68.
2. C.A. McAuliffe, H.F. Al-Khateeb, D.S. Barratt, J.C. Briggs, A. Challita, A. Hosseing, M.G. Little, A.G. Mackie and K. Minten, J. Chem. Soc., Dalton. Trans., Paper 2/1956, in press.
3. B. Beagley, D.W.J. Cruickshank, K. Minten, C.A. McAuliffe and R. Pritchard, manuscript in preparation.
4. H.D. Burkett, V.F. Newberry, W.E. Hill and S.D. Worley, J. Am. Chem. Soc., 1983, 105, 4097.
5. C.A. Tolman, Chem. Rev., 1977, 77, 313.

6. W. Strohmeier and F.J. Mueller, Chem. Ber., 1967, 100, 2812.
7. C.A. McAuliffe and K. Minten, J. Chem. Soc. Dalton, submitted for publication (August 1983).
8. C.A. McAuliffe, D.S. Barratt, C.G. Benson, A. Hoissing, M.G. Little and K. Minten, J. Organomet. Chem., Paper 83/27 in press.
9. C.A. McAuliffe, D.S. Barratt, C.G. Benson and S.P. Tanner, J. Chem., Soc., Dalton Trans., Paper 2/2097, submitted for publication.
10. J. Kiwi, K. Kalyanasundarampud, M. Gratzel, Struct. and Bonding 1981, 49, 37.
11. A. Hosseing, A.G. Mackie, C.A. McAuliffe and K. Minten, Inorg. Chim. Acta, 1981, 49, 99.
12. C.A. McAuliffe, J.Organomet. Chem.,1982, 228, 255.
13. R.M.Brown, R.E. Bull, M.L.H. Green, P.D. Grevenik, J.J. Martin-Polo and D.M.P. Mingos, J. Organomet. Chem., 1980, 201, 437.
14. J.I. Davies, C.G. Howard, A.C. Skapski and G. Wilkinson, J. Chem. Soc., Chem. Comm., 1982, 1077.

Engines and Future Liquid Fuels

By C. C. J. French
RICARDO CONSULTING ENGINEERS PLC, BRIDGE WORKS, SHOREHAM-BY-SEA, WEST SUSSEX BN4 5FG, U.K.

Introduction

While waterwheels go back to at least the first Century AD, wind mills date from the end of the 12th Century, and water power was and remains important in some areas, the Industrial Revolution was made possible by the invention by Thomas Newcomen of the atmospheric steam engine. Newcomen's first steam engine was built in 1712, and by the last quarter of the eighteenth century many were in use for pumping, blowing, etc., and Watt and others had invented the rotative engine.

From this time for a hundred and fifty years or more, coal was the dominant source of energy. Over the past forty years however petroleum has become more and more important, until today it provides 42% of World energy consumption.

There has also been a change in the use of petroleum fuels over the years. As may be seen from Figure 1, taken from reference [1], the proportion used by road transport has gradually increased and is expected to increase to the point that by 1990 half will be used for road transport and aircraft.

Figure 1 also shows another important factor, namely that even today the proportion of petroleum crude used as fuel oil and bitumen is less than the naturally occurring residue from light crudes and much less than the residue from heavy crudes. The situation is gradually expected to get worse with a consequent need for more and more secondary processing to produce the necessary fraction of lighter products to enable a balance between supply and demand to be achieved. This secondary processing will have an important impact on engine operation since there will be changes in the properties of both the light and heavy distillate fractions.

Before we discuss these implications, it is useful to consider the future usage of petroleum products. For some years,

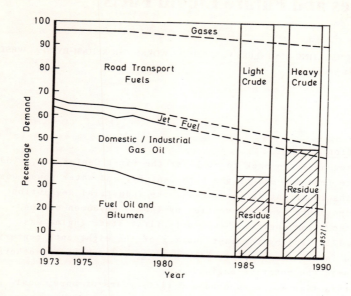

Figure 1 - Petroleum Fuels Demand - The Changing Barrel

especially since the first petroleum step price rise in 1973, warnings have been sounded as to the future availability of petroleum crude.

The present position is as follows. At current rates of consumption, the proved reserves will last approximately thirty years, and from previous experience it is certain that other reserves will be discovered to extend this life. In so far as consumption is concerned, the World consumption had until the 1970s been increasing at a substantial rate. With energy conservation and replacement of oil by coal, gas, and nuclear power, however, there was only a 7% increase in European petroleum consumption between 1970 and 1980, and it is expected to decrease by 15% between 1980 and 1990. With the high price of oil today it would seem unlikely that there will now be a substantial increase in the rate of worldwide consumption.

While there is concern about possible political effects on the security of supplies, since there have already been interruptions, it is clear that overall there are adequate reserves of petroleum products for some decades to come.

With gradually increasing fuel costs there will however be a growing emphasis on fuel economy and in some parts of the World, due to balance of payments problems or to particular concerns as to continuity of supply, alternative fuels have already come into widespread use, even though this would be difficult to justify on normal economic grounds.

Even without considering alcohols from plant products, vegetable oils, and biomass which, with a rapidly increasing World population, might be in conflict for growing space with food, there are very substantial reserves of tar sands, oil shales, and coal, which could be exploited for liquid fuels.

Over the whole period of engine development to date, the oil industry has collaborated with the engine builders in producing fuels having the properties which the engines required, without a need to maximise distillate fuel production or an over-riding concern for overall efficiency. This is most unlikely to continue, and it will become more and more important to optimise engine design and fuel properties for the greatest efficiency in the usage of petroleum products in their declining years - no matter how long these may be.

In this paper we are particularly concerned with the engines for and the fuels used in road transport. We shall almost certainly have to continue to use light liquid fuels from wherever they are derived. These are clean and easy to handle, have few problems of toxicity, and are compatible with normal engine and vehicle constructional materials.

There is also much to be said for continuing to use conventional spark ignition and diesel reciprocating engines, since the capital investment for their manufacture is very large and has already been made. There would have to be either very substantial advantages in favour of an alternative powerplant, over-riding considerations with regard to non-availability of a suitable fuel, or the discovery of additional environmental hazards to justify a change of powerplant type on any but a very gradual scale.

Properties of Fuels
    1. Petroleum Based
Diesel Fuels. The most important single property in so far as the use in a diesel engine is concerned is cetane number. This is a measure of the time delay between the commencement of injection into the cylinder and the start of combustion. With reduction in

cetane number a point will be reached at which combustion fails to occur. Before this point is reached, however, the quantity of fuel which is vaporised within the cylinder by the start of combustion may be so large that the initial rates of combustion and hence of pressure rise give unacceptable levels of engine noise. There will also be problems with starting, especially under cold ambient conditions, resulting in failure to start or to misfire with high exhaust hydrocarbon concentrations.

The effects of secondary processing of the crude oil may be seen from Table 1, taken from reference[1]. With catalytic cracking, it will be seen that there is a very substantial increase in the aromatic content of the diesel range. Since there is a correlation between increase in aromatic content and reduction in cetane number - Figure 2 - the effect of a wider use of catalytic cracking is to reduce cetane number.

	Gasoline Range - %		Diesel Range - %	
	Aromatics	Olefins	Aromatics	Olefins
Light catalytic	15	31	70	<5
Heavy catalytic	47	15	70	
Steam cracked	58	18	-	
Visbroken	9	39	25	<5
Straight run	4	1	20	<5

Table 1 - Effect of Secondary Processing of Residuals

The trend to a wider use of catalytic cracking to maximise gasoline production has already proceeded a long way in the United States, and as a result average cetane numbers are of the order of 45 as against 50-51 in Europe. The spread in the USA is however much wider than in Europe, depending on the source of the fuel, and cetane numbers lower than 40 can be encountered.

As cetane levels approach 40, it is desirable that ignition-improving additives are employed so that cetane numbers do not drop below 40-42. A number of chemical compounds have been employed over the years, and some are in fact used commercially. Amyl nitrate was an early material which was studied, and more recently a substantial amount of work has been carried out with nitrated glycols.

The future diesel fuels are also likely to have a higher boiling point and to be denser. Since diesel fuels are

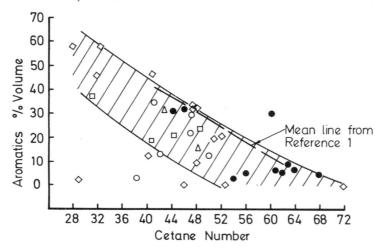

Figure 2 - Aromatic Content versus Cetane Number for a Wide Range of Diesel Fuels - Reference[2]

metered volumetrically and since it is the gravimetric calorific value which tends to be constant, the use of denser fuels brings the operating point closer to stoichiometric for a fixed fuel setting, and this will result in more exhaust smoke unless the fuel pump is re-set, which would not be easy to do if the fuel density varied over the geographical area of use of a vehicle.

There may also be problems due to increases in cloud and pour point of the fuel and hence to fuel filter blockage under cold conditions. Engine deposits may be helped by detergent additives, and cloud and pour point improvers are available, and the result is likely therefore to be a much wider use of fuel additives in future diesel fuels.

Gasoline. Until recently one would not have forecast any rapid change in gasoline quality. It is true however that lead-free gasoline has had to be available in Japan and the USA for some years past to allow for the use of catalytic converters in engine exhaust systems so that carbon monoxide and hydrocarbon levels could be reduced to meet very low legislative limits. The use of three-way catalysts has also enabled oxide of nitrogen levels to be substantially reduced.

In Europe there has been much less emphasis on reduction of

exhaust emissions and it appeared unlikely that very low levels would be legislated. There was however an interest in the removal of lead as a toxic substance, and legislation had been introduced in Germany to reduce lead levels in gasoline, with similar proposals in the U.K. Medical evidence showed however that gasoline was a minority source of lead in the environment, and as recently as 1980 the Lawther Committee set up by the British Government recommended only gradual reduction of gasoline lead levels - reference[3].

Very recently however the British and West German Governments have announced their intention of requiring that lead-free gasoline be made available for new cars, and the European Commission will be considering legislation next year.

Lead is added to gasoline in the form of tetraethyl- and tetramethyl-lead in order to increase the octane number of the fuel, which is the measure of detonation resistance. If lead is no longer added, there are three alternative approaches.

Firstly, gasoline can be supplied at a lower octane number, say 91-92 octane, as in the United States and Japan. Engines to burn lower octane fuel must have a lower compression ratio with a consequent loss in efficiency, although steps can be taken to improve the 'mechanical' octane requirement of the engine to minimise the loss.

Secondly, the octane rating could be improved in the refinery, although as shown in Figure 3, this entails a substantial energy loss which negates any engine efficiency gain - reference[4].

Thirdly an alternative anti-detonant may be used. It is likely that other metallic-based compounds such as manganese would be unacceptable. The current front runners are oxygenates such as methanol, TBA (tertiary butanol), or MTBE (methyl tertiary butyl ether). The methanol would be manufacturerd from natural gas or coal, MTBE is made at present by reacting refinery gas with methanol derived from natural gas, and limited supplies of refinery gas may lead to difficulties in the production of substantial quantities. The supply of increased quantities of TBA is also difficult since it is currently produced as a co-product of propylene oxide, and the price and availability depend on the demand for propylene oxide - reference[4].

With the wider use of secondary processing, gasoline characteristics will also change. With catalytic cracking higher aromatic levels will occur, with lower front-end volatility, lower stability, and the presence of heavy hydrocarbon components. Catalytic cracked fuels also lead to a more sensitive fuel, i.e. for a given research octane number, the motor octane number is lower, and engines will be more prone to high speed knock, and it will not of course be possible to offset this by using tetra-methyl-lead.

As a result of these changes, there is likely to be a worsening of engine fuel economy, the introduction of a number of engine operating problems, and an increase in environmental impact.

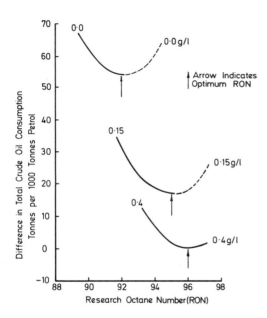

Figure 3 - Effect of Lead Content on Crude Oil Consumption and Octane Number - Base Case 1000 Tonnes Gasoline, 96 RON, Lead Content 0.4 g/ℓ

2. Alternative Fuels

Alcohols. Ethanol - produced by fermentation of sugar cane - is widely used in Brazil. It can also be produced by the fermentation of other agricultural products such as manioc, grain, and grapes, and it is in many ways an ideal fuel for the spark ignition engine, having a high octane rating. The most serious problem arises from the low vapour pressure which will lead to a lack of vaporisation and a failure to start under cold conditions as shown in Figure 4, taken from reference[5]. Furthermore, due to the high latent heat requirement of alcohol fuels, engine problems can exist over an extended warm-up period after a cold start, and it may be desirable to start and warm up on gasoline.

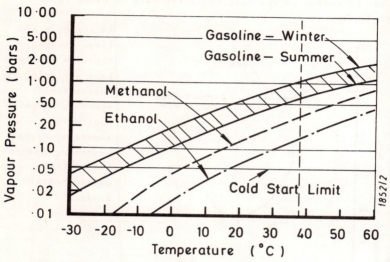

Figure 4 - Vapour Pressure of Ethanol, Methanol, and Gasoline *versus* Ambient Temperature

If alcohols are used in blends with gasoline, problems can arise from phase separation in the presence of water, and it is desirable to eliminate water if such blends are employed.

Methanol can be produced from a number of raw materials, including forest products, natural gas, and coal. In general its use in engines is very similar to that of ethanol, but it is much less compatible with many normal fuel system and engine materials such as aluminium and copper, zinc-based die castings, and many normally fuel-resistant polymer materials. Methanol is also highly toxic.

Due to their high octane ratings, methanol and ethanol have very low cetane numbers, and it is difficult to obtain good performance with them in diesel engines. Dual injection systems with a pilot charge of diesel fuel for ignition purposes can be employed, but such systems are complicated and expensive. Work has also been carried out with highly active ignition promoters such as nitrated glycols, but due to the very large lift in cetane number which is required the difficulties are much larger than in trimming the cetane number of secondary processed middle distillate petroleum fuels.

Vegetable Oils. Vegetable oils such as palm oil, soya bean oil, sunflower oil, and rape seed oil are of potential interest as diesel engine fuels since some at least have cetane ratings which make them suitable for direct use in diesel engines. The major problem which arises from direct use is nozzle coking and the formation of combustion chamber deposits which occur after a very short running time - reference[6]. These are so severe in fact that pretreatment of the fuel is essential. The most convenient treatment is transesterification with alcohol, converting the tri-ester vegetable oils into mono-esters which have lower viscosities and boiling points. This greatly reduces combustion chamber deposits, e.g. reference[7].

With unsaturated mono-esters however, such as those obtained from transesterised soya bean oil, solution of the ester in the lubricating oil will deplete the antioxidant, and problems can then arise with rapid thickening of the lubricating oil.

Gaseous Fuels. LPG and natural gas, either as compressed natural gas or as liquefied natural gas, are also ideal fuels for spark ignition engines although LPG is now in shorter supply than at one time seemed likely, and there are problems with the distribution of natural gas, which might therefore be better suited to fleet use.

Fuels from Oil Shales, Tar Sands, and Coal. The World's reserves of these raw materials are many times the reserves of petroleum crude, and light distillates can be produced from all three.

Canada is currently producing 10% of its crude production from tar sands, reference[8], but the resulting fuels have rather different properties from conventional petroleum-based fuels - Tables 2,3, and 4.

The Naphtha is highly paraffinic with a low octane rating, and hence it is a poor spark ignition engine fuel without catalytic reforming in the refinery.

The Kerosene has a higher aromatic content - higher than is acceptable for a jet-engine fuel - which could lead to smoky combustion unless blended to a relatively small percentage into conventional fuels.

	Synthetic	Conventional
Boiling Range	C5-150°C	
Density, kg/dm^3	0.697	0.718
Sulphur, ppm	1	30
GC Hydrocarbon Analysis, Vol %		
Paraffins	69	43
Naphthenes	25	43
Aromatics	6	11
Benzene	1.1	1.4
Toluene	2.7	4.3
Xylene	2.0	0.5
Research Octane No.	58	63
Motor Octane No.	57	63

Table 2 - Properties of Naphtha from Oil Shale - Synthetic and Conventional Crudes

	Synthetic	Conventional	Jet A-1 Specs.
Density, kg/dm^3	0.830	0.803	0.839 max
Viscosity, cSt @ 40°C	1.3	1.2	-
Freeze pt, °C	<-60	-50	-47 max
Sulphur, ppm	38	100	2000 max
Nitrogen, ppm	4	2	--
Hydrogen, Wt.% (D3701)*	12.9	13.9	-
Aromatics, LV% (D1319)*	32	19	22 max
Naphthalenes, Wt.%(D1840)*	0.6	2.2	3 max
Smoke Point (D1322)*	13	22	20 min
Luminometer No. (D1740)*	29.3	47.3	45 min

* ASTM Test Methods

Table 3 - Properties of Kerosene from Oil Shale - Synthetic and Conventional Crudes

The middle distillate fraction is highly aromatic with a low cetane number, which also limits the amount which can be blended in with petroleum-based fuels.

The properties of distillate fuels produced from coal depend strongly on the production process. The gasification/synthesis route starts with the complete gasification of the coal by

	Synthetic	Conventional
Density, kg/dm^3	0.874	0.833
Viscosity, cSt @ 40°C	3.3	2.9
Cloud Point, °C	-27	-10
Pour Point. °C	<-50	-18
Sulphur, Wt.%	0.03	0.30
Nitrogen, ppm	25	15
Aromatics, Vol.%(D1319)*	44	32
Cetane No. (D613)*	34	48
Cetane Index (D976)*	42	51
Distillation (D68)*		
IBP °C	157	158
10%	192	226
50%	274	271
90%	329	311
FBP	334	320

* ASTM Test Methods

Table 4 - Properties of Diesel Fuel from Oil Shale - Synthetic and Conventional Crudes

reaction at high temperatures with steam/oxygen mixtures to produce a carbon monoxide/hydrogen syngas mixture which is then built up into hydrocarbons, employing the Fischer-Tropsch process. This method which was used in Germany during the war is used in South Africa, producing straight run naphtha compounds which are highly paraffinic and hence have a low octane rating, as shown in Table 5, taken from reference[9]. The syngas can as an alternative be used as the basis for the production of methanol, which could be used directly as a fuel.

If a solvent extraction route is taken for the production of liquid fuels from coal, Table 5 also shows that the naphtha fractions are high in aromatic content and low in paraffins and hence have a high octane rating, which is an attractive starting point for the production of a spark ignition engine fuel.

The middle distillates produced by the syngas route are also paraffinic, leading to a high natural cetane number, whereas those from solvent extraction are aromatic with a low cetane rating. This can however be increased to an acceptable level by secondary hydrotreatment.

The choice between the various processes for producing liquid fuels from coal will obviously ultimately be made on technical and economic grounds, and the author is not qualified to make this choice. There is evidence however - for example from Table 6, taken from reference[9] - that the solvent extraction process considered is almost twice as efficient in terms of returns of the energy in the feed coal as is the Fischer-Tropsch process used at the Sasol II plant in South Africa. Whether these figures are typical for the two types of processes is not clear.

Process	Nominal Boiling Range	Specific Gravity	Heteroatom Content ppm			Paraffins (P)% Naphthenes (N) Aromatics (A) Olefins (O)				Research Octane No. (RON)
			S	N	O	P	O	N	A	
Petroleum*	76-185	0.742	400	<5	<5	66		23	11	65
Sasol	65-200	0.7275	4	4	23000	70	30	-	-	40
H-Coal	56-310	0.8076	1289	1930	5944	16	5	55.5	18.6	80
Kohleol	65-200	0.863	146	2400	35000					
NCB-LSE	85-180	0.801	50	50	<0.3%	5		66	29	80

* Light Arabian

Table 5 - Characteristics of Straight Run Naphthas

When considering the whole range of synthetic fuels from tar sands, oil shales, and coal it is clear that due to the much lower hydrogen/carbon ratio than that of petroleum crude, the natural tendency without substantial hydrogen addition would be to produce highly aromatic middle distillates, which would be unsatisfactory as diesel fuels due to low cetane ratings, and unsatisfactory as gas turbine engine fuels due to high combustion chamber liner temperatures and high exhaust smoke.

Product	Process Yields (wt.% daf coal)*	
	Sasol II	NCB-LSE
C_1-C_4 gases	nil	6
Gasoline	13	17
Jet and diesel	11	28
Overall thermal efficiency	36%	66%

*Total coal energy - self-sufficient plant, dry ash free coal

Table 6 - Yields from some Coal Liquefaction Processes

With substantial hydro-treatment and the use of conventional refinery techniques, however, it is possible to produce synthetic fuels which will cover the whole range of properties of current petroleum-based fuels used in internal combustion engines.

It would appear therefore that these fuels are fundamentally expensive and that there could be advantages in developing new engines to use for, example, middle distillates with a low cetane number to eliminate the secondary hydro-treatment which is necessary with all synthetic fuels, apart from those manufactured by the Fischer-Tropsch process.

Synthetic fuels can therefore be produced which can be used in today's engines and which might initially be employed as blends with petroleum-based fuels but which could ultimately be used on their own. In terms of energy efficiency, however, and in the interest of conservation of the primary energy source - be it tar sands, coal, or oil shales - this may not be the best approach. It could be preferable to use synthetic fuels as a 'broad cut' with the minimum necessary refinery treatment and with low cetane and octane ratings in engines which have been optimised for these types of fuels.

Future Engines for Petroleum Based Fuels

The diesel engine is the most efficient engine available today and as such it is the preferred powerplant for ships, locomotives, trucks, buses, farm tractors, and taxis. It is likely to continue to be used so long as fuels with a suitable cetane rating are available and, as has already been mentioned, additives are available to give some trimming of cetane rating as future cetane numbers decline.

Due to their high thermal efficiency, direct injection combustion systems are universally employed for all but vehicles requiring an engine with a wide speed range - passenger cars, taxis, and light delivery vans. Here indirect injection has been and still is being used since it considerably eases the technical requirements for the fuel injection equipment, gives a quieter engine, and has inherently low exhaust emissions.

With a potential further improvement in fuel economy of 8-15% or so, direct injection has obvious advantages which will become even more pressing if fuel prices continue to rise, and there is considerable interest therefore in this application.

The classic direct injection engine, as employed by large engines with limited speed ranges, uses multi-hole type nozzles, which places severe pressure requirements on the fuel injection equipment at high engine speeds, and this has been a major limiting factor on the use of direct injection for light duty use. Conventional distributor type of fuel injection equipment has however been developed in recent years so that it can be used to give higher injection pressures, and the direct injection delivery van running at 3500 to 4000 rev/min is now a possibility. Figure 5 shows comparative full load performance data for a single-cylinder version of such an engine and an IDI equivalent.

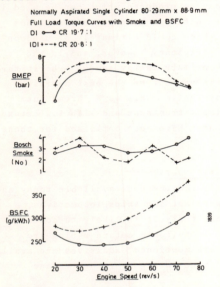

Figure 5 - Comparative Performance of Light Duty Diesel Combustion Systems

For a wider speed range there are three possibilities. The first would require the development of conventional fuel injection pumps capable of operating at still higher pressures, which may yet prove to be a practical proposition. The second involves the use of unit pump injectors, one for each cylinder. With the unit injector it is possible to run to very high injection pressures, and furthermore the absence of a substantial volume of compressible fuel between the pump and the nozzle gives a better control of the injection process.

The use of unit injectors, almost certainly with electronic control of timing to give the variation of timing with load and speed which is required for low exhaust emissions and good fuel economy, is a possibility for the future, although such engines may have to be sold at a further premium in price to cover the extra cost of the fuel injection equipment. The engine is also likely to suffer a penalty of additional engine height.

Multiple hole nozzle types of direct injection engines, with or without the use of unit injectors, tend to be noisier than their indirect injection counterparts. They also have higher maximum cylinder pressures which may give problems with bearing loads, particularly when one is considering the diesel conversion of a Vee engine, whose bearing area tends to be limited due to cylinder spacing considerations.

The third alternative is to use a high swirl, wall-wetting type of combustion chamber, as for example that developed by MAN and called the 'M' system, as shown in Figure 6. The original 'M' system, as developed by Professor Meurer some 20 years or so ago, was quite widely used in truck engines, both in Europe and the United States. While such engines displayed good starting characteristics and were quiet in operation, there was a hydrocarbon emissions problem resulting in excessive exhaust odour, and most of the engines were gradually replaced by conventional direct injection engines.

MAN have however in recent years developed the engine further, and it now employs a pintle nozzle and a novel injector, containing two springs on a central plunger which give a controlled needle lift at light load and idle to assist in giving low noise and low hydrocarbon emissions. The system is known as the MAN CDI - controlled direct injection - system.

With the use of a pintle nozzle, such an engine employs similar peak fuel injection pressures as does

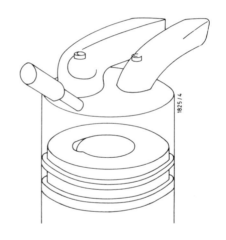

Figure 6 - Wall Wetting Type of Combustion Chamber

an indirect injection engine, and hence the system is particularly suitable for wide speed range light duty vehicles.

Typical comparative vehicle performance data for cars fitted with IDI and MAN CDI engines are given in Table 7. The improvement in fuel consumption and the acceptable exhaust emission levels of the MAN CDI vehicle may be clearly seen.

<u>Reduced Heat Loss Diesel Engines</u>. With the drive for improved economy has come a growing interest in the reduction of engine

			CDI		IDI	CDI % Difference
			T.5,8,9	T.40		
	Fuel economy	mile/US gal	40.4	42.5	38.1	+6, +11.5
US FTP Urban Cycle	Emissions:					
	HC	g/mile	0.33	0.36	0.28	+23
	NOx	g/mile	1.58	1.55	0.98	+60
	CO	g/mile	1.77	1.64	0.88	+94
	Particulates		0.26	0.22	0.33	-20
	Performance:					
	0-100km/h	s	20.1	(18.2)*	17.6	+14 (+3)
	4th Gear Flexibility km/h:					
	40-80		18.7		15.1	+24
	40-100	s	28.5		25.6	+11
	40-120		46.1		40.7	+13

Table 7 - Comparison of CDI Golf with 1.6 Litre IDI Golf

losses - thermal, fluid-dynamic, and frictional. The reduction of heat losses can also have some interesting side effects in so far as fuel requirements are concerned.

Heat loss reductions are effected by the use of insulating materials, such as ceramics, or by the use of insulating air gaps in the design of the combustion chamber components, as can be seen in Figure 7. Either method increases the gas side surface temperature of the combustion chamber components and hence reduces the temperature step between the gas and the surface, reducing the heat flow. Due to the heat transfer between the wall and the incoming air charge, the temperature of the air into which the fuel is injected is increased, giving a reduction in the delay period between injection and the start of combustion. This results of course in a reduction in the cetane number requirement of the engine once the cylinder walls have been warmed up. Auxiliary aids such as heater plugs or spark plugs would be necessary for starting, and it might also be necessary to use these aids for idling and low load operation.

Due to the air charge heating, the volumetric efficiency of the engine would be considerably reduced, and an increased turbocharging boost pressure is required to avoid substantial derating.

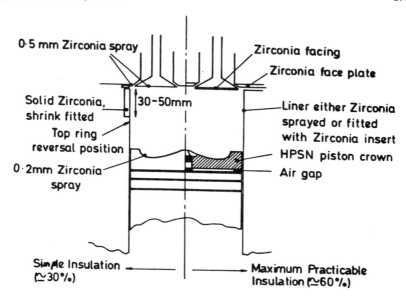

Figure 7 - Combustion Chamber Insulation Schemes

<u>Spark Ignition Engines</u>. Despite the likelihood of a much wider use of diesel engines in light duty vehicles, the majority of such vehicles must continue to burn volatile fuels with low cetane numbers which are not suitable for use in a diesel engine. Gasoline is likely to remain the primary fuel, although as has been emphasised earlier, with a growing environmental pressure to remove lead alkyls from gasoline, we are likely to see a drop in octane number. The elimination of lead will require the use of hardened valve seats or valve seat inserts, but such technology is well established.

When high octane gasolines are available - 97 to 98 - there has been a growing interest in recent years in the use of high compression ratio, compact charge engines, with the combustion chamber either in the piston or in the cylinder head, and in the use of four valve engines. The results of Ricardo work with these combustion systems have been published and are summarised in reference[10]. Both systems give scope for an improvement in vehicle fuel economy of 15 to 20%, the four valve slightly less than the HRCC, coupled with significant increases in engine power and torque.

Fast-burn, open-chamber combustion systems, as exemplified by the Nissan NAPS-Z engine operating at lower compression ratios, such as 8.5:1, and giving high burn rates due to a combination of dual spark plugs and high inlet air swirl can, however, from Ricardo tests, offer potentially important advantages in part load, low speed fuel economy when burning lower octane fuel, as may be seen from Figure 8. This shows a wide range of test data obtained on Ricardo Hydra single cylinder diesel engines, and the band of data covers combustion chambers of Bathtub, HRCC, Disc, and Four-Valve form. The single point outside the band employs a Nissan NAPS-Z cylinder head, which with 92 octane fuel offers a 6% improvement in fuel economy over the best of the other chambers.

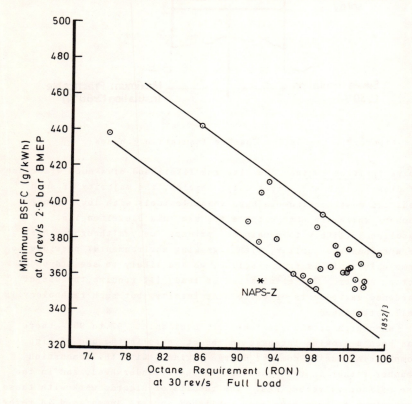

Figure 8 - Fuel Economy - Octane Requirement Trade-Off All Results from Ricardo Hydra Single Cylinder Engine.

The fuel consumption at 40 rev/s and 2.5 bar bmep has been chosen as being typical at one representative point of the duty cycle for a light-duty vehicle, and similar arguments might not necessarily apply when considering a high-output engine driven over a more stringent operating cycle.

Future Engines for Non-Petroleum Based Fuels

Volatile fuels, such as alcohols, with an octane number of 90 and upwards - and even somewhat lower - are best burnt in a spark ignition engine. With alcohols, as has already been mentioned, it may be desirable to start and warm up on gasoline. Alternatively, it may be possible to heat the fuel electrically for start up, or fuel additives may be employed. One possibility is isopentane, but substantial quantities are required for very cold conditions, as may be seen from Figure 9, taken from reference[11].

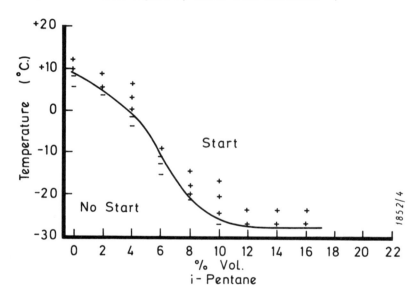

Figure 9 - Necessary Quantity of Additive for Cold Start
i-Pentane with Methanol Fuel

With alcohol fuels, due to the low volumetric calorific value, it is necessary to provide a fuel tank approximately twice as large as for gasoline to obtain the same range. Methanol and ethanol both however have a high octane rating, of the order of 110, and alcohol engines can utilise a higher compression ratio. Due to this and an increase in the mol constituents of the combustion products, engines burning alcohol have a higher thermal efficiency, as may be seen for example from Figure 10, also taken from reference[11], which compares the specific energy consumption of gasoline and methanol fuelled spark ignition engines. Due to the high latent heat of vaporisation and the consequent charge cooling, alcohol fuelled engines also give a higher power than their gasoline fuelled equivalents - Figure 11, taken from reference[5] - and this effect will be accentuated by increases in compression ratio.

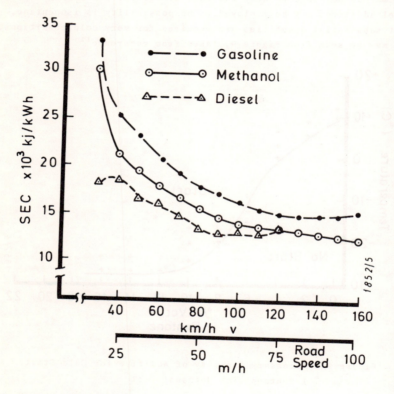

Figure 10 - Specific Energy Consumption of a Gasoline, Methanol, and Diesel Powered Car of the same Type at Road Speed

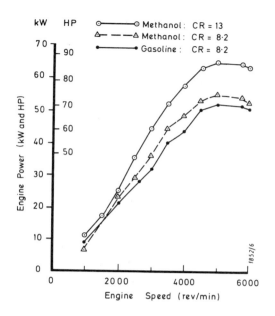

Figure 11 - Wide-Open-Throttle (WOT) Power vs. Engine Speed of 4-Cylinder Engine Operating on Methanol or Gasoline

With ethanol as a fuel - and it is of course widely used in Brazil - the major operating problems reported have been deposits in the inlet system and corrosion of carburetter float chambers - reference[12].

There are clearly more fundamental problems with methanol due to its corrosive nature. The problems described in reference[12] are deposits in the inlet system, severe bore wear, and corrosion of bearings, camshaft and crankshaft with, under some condition's, sludge deposits. Improved lubrication oils helped to reduce inlet system deposits, and both bore wear and corrosion were substantially reduced by an increase in coolant temperature - Figures 12 and 13 taken from reference[12]. More work is clearly desirable if methanol is to be used as a fuel for spark ignition engines.

Vegetable oils, in the light of current knowledge, are probably best modified to monoesters and burnt in diesel engines. Once again however further work is desirable to ensure an adequate engine life.

With regard to synthetic fuels from shale oils, tar sands, and coal, we have as has already been explained a choice. Fuels can certainly be produced which could be used in conventional spark ignition and diesel engines - see for example diesel performance data in Figures 14 and 15, taken from reference[13] - or we can consider engines which have neither octane nor cetane requirements.

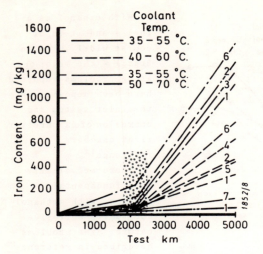

Figure 12 - Effect of Coolant Temperature with Various Oil Formulations on Iron Content of Used Oil

Four alternatives present themselves, two with intermittent combustion and two with continuous combustion. The two with intermittent combustion are late injection stratified charge and catalytic combustion, and the two with continuous combustion are gas turbines and Stirling engines.

Late injection stratified charge may be exemplified by the Texaco system and the MAN-FM. The start of fuel injection immediately precedes combustion which is initiated by a spark. The principle is old, and Hesselmann engines were sold around the time of the last war. There were formerly problems in producing an engine obtaining satisfactory performance over a wide range of loads and speeds, but these problems have been overcome by using high-energy ignition systems, giving a very long duration spark.

Having operated five prototypes with satisfactory results,

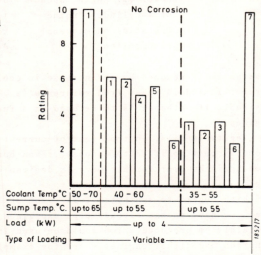

Figure 13 - Effect of Coolant Temperature & Oil Formulation on Corrosion

United Parcel Services in the United States are about to procure 500 engines to run on a wide cut type of fuel. While this type of engine is smoke limited like a diesel engine and hence is substantially derated as compared with the equivalent carburetter or fuel injection gasoline engine, the fuel economy can be equivalent to that of a diesel engine and hence is particularly good in stop/start city driving service. Similarly good fuel economy has been recorded by MAN and by Ricardo employing the MAN-FM system.

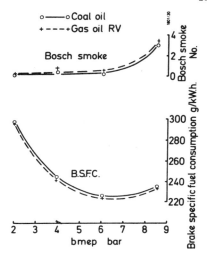

Figure 14 - Engine Performance with Coal Oil Fuel at 20 rev/s

The most serious problem which remains to be solved with stratified charge engines is the tendency to high exhaust hydrocarbons, although this may not be as serious with the lower vapour pressure hydrocarbons as with gasoline. The use of lean mixtures with a resulting reduction in exhaust temperature also makes it more difficult to employ catalytic afterburners to reduce these hydrocarbon levels.

It was as a result of a study of this problem that Ricardo suggested the use of catalytic combustion, as described in references[14,15]. Figure 16 shows the 'indirect injection' layout which has been found so far to be the best. The fuel consumption is comparable over a limited

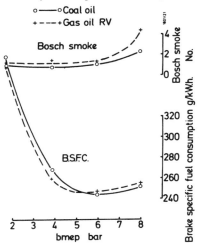

Figure 15 - Engine Performance with Coal Oil Fuel at 35 rev/s

part of the operating range to that of the IDI diesel and stratified charge engines, as may be seen from Figure 17, and Figure 18 shows that in fact very low hydrocarbon levels are experienced.

Much work remains to be done however to extend the good fuel economy zone to both low and high load areas of operation, and it is likely that very substantial problems remain before an adequate catalyst life is obtained.

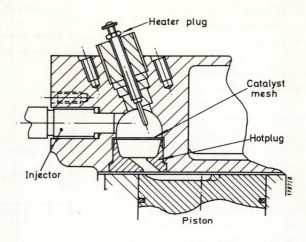

Figure 16 - Arrangement of Catalytic Comet Combustion Chamber

Gas turbines and Stirling engines owe their theoretical fuel tolerance to the use of continuous combustion in specially developed combustion chambers. The problems which arise with smoky combustion with highly aromatic fuels, which have already been mentioned, show however that combustion problems can occur with some fuels. Furthermore, considerable development problems must be overcome with both engines before they can be widely used in service.

Gas turbines require the use of higher peak cycle temperatures before an adequately good fuel economy can be obtained. This may prove to be possible by the use of ceramic components, if high-temperature engineering ceramics prove ultimately to be successful. Currently however there are also problems of first cost and of obtaining a good fuel economy as engine size is reduced, and this question will be accentuated by the increased specific output which could arise from the use of higher peak cycle temperatures.

Stirling engines have their own particular difficulties. While fuel economy is fundamentally good, the engines tend to be bulky

Engines and Future Liquid Fuels

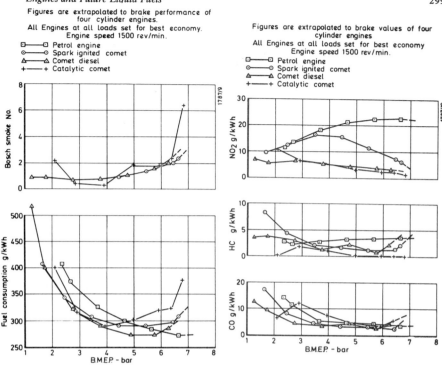

Figure 17 - Comparison of Performance from Various Engines of Similar Cylinder Size

Figure 18 - Comparison of Exhaust Emissions from Various Engines of Similar Cylinder Size

and expensive. They tend to have problems with regard to the control of load, and there are also difficulties in retaining the high-pressure working fluid, be it hydrogen or helium, within the system.

Electric Vehicles and Hydrogen Fuelled Vehicles. It may be wondered why these alternatives have not been mentioned in a paper on transport engines for use in the long term.

Electric vehicles require a breakthrough in battery capacity before they can be considered for anything but a 'cars-for-cities' type of operation. Lead-acid batteries are very bulky and heavy for wide application, and despite substantial research, alternatives still appear to be a long way away. With electric vehicles there still remains the fundamental problem of the primary energy source. Coal-fired power stations would be possible, or nuclear power, but the latter is likely to involve considerable objections from the anti-nuclear lobby.

Similar difficulties arise from the use of hydrogen. While hydrogen can be burnt in spark ignition engines, hydrogen is not a primary fuel but it has to be produced by the use of some other fuel, such as by the electrolysis of water, using electricity from coal-fired or nuclear power stations. As such it is less attractive than the use of the alternative fuels which have been considered.

The author would like to thank the Directors of Ricardo Consulting Engineers for permission to publish the paper. He is also indebted to Mr. James Izard and Dr. Calvert Stinton, and others of his colleagues and friends, who have given advice on sections of the paper.

References

[1] J.M. Tims, "Trends in Transportation Fuels", NATO Advanced Research Workshop 'Polycyclic Organic Matter from Exhaust Gases', Liege, August 30- September 2, 1982.

[2] W.R. Wade and C.M. Jones, "Current and Future Light Duty Diesel Engines and Their Fuels", National Petroleum Refiners Association Annual Meeting, San Fracisco, March 20-22, 1983.

[3] "Lead and Health", Report of a DHSS Working Party on Lead in The Environment, Her Majesty's Stationery Office, 1980.

[4] "Lead in the Environment", Ninth Report of the Royal Commission on Environmental Pollution, CMND 8852, April 1983.

[5] "Status of Alcohol Fuels Utilization Technology for Highway Transportation",Vol.1, Spark Ignition Engines, U.S. Deptartment of Energy DOE/CE/56051-7, May 1982.

[6] D. Collins and M.L. Monaghan, "Operating Passenger Car and Truck Diesel on a Varied Diet", Fifth International Alcohol Fuel Technology Symposium, Auckland, New Zealand, May 13-18, 1982.

[7] L.M. Ventura, A.C. Nascimento, and W. Bandel, "First Results with Mercedes-Benz DI Diesel Engine Running on Monoesters of Vegetable Oils", Presented at Vegetable Oil Fuel Conference, ASAE, Fargo, N.Dakota, 1982.

[8] D.E. Steere and C.B. Rupar, "Quality of Synthetic Fuels Derived from the Canadian Tar Sands", Paper D19, CIMAC Conference, Helsinki 1981.

[9] G.O. Davies, "The Preparation and Combustion Characteristics of Coal Derived Transport Fuels", Institute of Mechanical Engineers, Paper C85/83, 1983.

[10] D. Downs, "The Passenger Car Power Plant: Future Perspectives", Paper 82010, presented at the XIX FISITA Conference, Melbourne, Australia, 8-12 November 1982.

[11] H. Menrad, W. Lee, and W. Bernhardt, "Development of a Pure Methanol Fuel Car", Alcohols and Motor Fuels, published by the Society of Automotive Engineers as 'Progress in Technology' No.19.

[12] A.A. Reglitzky, H.-J. Halter, H. Krumm, "Lubrication of Alcohol-fuelled Engines", Presented at 4th International Seminar on Developments in Fuels, Lubricants, Additives, and Energy Conservation, Cairo, Egypt, March 7-10, 1983.

[13] G.O. Davies and R.G. Freese, "The Preparation and Performance of Coal Derived Diesel Fuel", Paper No. D74, CIMAC Conference, Helsinki 1981.

[14] "A Study of Stratified Charge for Light Duty Power Plants", U.S. Environmental Protection Agency, Contract No.68-03-0375 EPA-460-374-011A, October 1975.

[15] C.C.J. French, "Fuel Efficient Engines for Light Duty Vehicles", presented at FISITA Conference, Hamburg, 1980.

Concluding Remarks

By C. F. Cullis
DEPARTMENT OF CHEMISTRY, THE CITY UNIVERSITY, NORTHAMPTON SQUARE,
LONDON EC1V OHB, U.K.

The fact that oxygen constitutes almost one half of the weight of the Earth's crust is of course a manifestation of the high chemical reactivity of this element which, even as gaseous oxygen, combines directly with nearly all other elements. As was pointed out in the opening lecture, by far the largest part of the oxygen in the atmosphere is used up in respiration and combustion, but oxygen, mainly as air, has been widely used in chemical processing for many years.

Traditionally syngas has been produced from hydrocarbon feedstocks - derived mostly from low-value, high-sulphur crudes - and from natural gas, the process consisting essentially of catalytic steam reforming which does not of course involve oxygen per se. However the enormous increase in the price of oil, especially since the mid-1970's, has led to the development of processes for the production of syngas from coal gasification in the presence of oxygen as in the entrained-bed gasifier process. The ratio of carbon monoxide to hydrogen in syngas is a function of both the initial feedstock and the process used and in the gasification of coal by oxygen the proportion of carbon monoxide is maximised. Large amounts of carbon monoxide are ideal for the production of oxygenated organic compounds such as methanol, acetic anhydride and many carbonyl-containing high-tonnage materials. In this way oxygen is now being used in industrial processes which are competing in the open market for the manufacture of many chemicals. It is interesting to observe this return to coal as a feedstock since by the mid-1950's the direct oxidation of coal to chemicals had been more or less completely abandoned as uneconomic, being replaced by the use of syngas originating from oil. We have heard in this Conference, about various ways in which this so-called "traditional" method of producing syngas can be catalysed and also about how syngas can be purified, the need for and the extent of which depends on the end-use involved.

We have also had interesting descriptions of several heterogeneous and homogeneous catalytic oxidation processes and of particular interest has been the use of precious metals, metal oxides, copper chloride, transition metal peroxides and alumina, silica and zeolites as catalysts. As was emphasised in the opening lecture, many of the compounds produced using oxygen as one of the initial reactants do not in fact contain oxygen.

As regards the future, it appears that liquid fuels will be produced to an increasing extent from unconventional starting materials, such as tar sands and shales, and even from coal; and it will obviously make more economic sense to turn to coal as the source of the feedstocks for chemical industry. This source may however be complemented by the recycling of waste (particularly polymeric materials) and by the use of biomass. We have heard how vitally important it is to obtain process improvements before biomass can compete successfully with other feedstock sources. Nevertheless, the use of novel catalysts, particularly zeolites, is having a great impact.

The use of oxygen as an oxidising agent for chemical industry's feedstocks therefore seems likely to grow. It is of course almost certain that the feedstocks in use in even 10 or 20 years' time will differ substantially from those of the 1960's and 1970's but perhaps it is possible to hazard the guess that at least some of these feedstocks will be ones which would have been familiar to Joseph Priestley.

Priestley Lecture: The World's First Chemical Explosive — in China and the West

By Dr. Joseph Needham, FRS, FBA
EAST ASIAN HISTORY OF SCIENCE LIBRARY, 16 BROOKLANDS AVENUE, CAMBRIDGE
CB2 2BB, U.K.

This lecture is going to be devoted to man's first and oldest chemical explosive, namely gunpowder, particularly its peaceful uses, in civil engineering - and, as is much less well known, as a power-source in the mechanisms of the external, and internal, combustion engines. First developed by Chinese alchemists looking for elixirs of immortality in the middle of +9th century, it then underwent a rapid development for weapons of war, coming to Europe, probably in three stages, between +1260 and +1315. I will expand this presently; here it is only necessary to mention that we are not far from the subject of oxygen, because that oxidising gas was built in to the composition of the mixture in the form of nitrate, ready to combust with explosive violence the sulphur and the carbon also present. But first we must say something about Joseph Priestley himself, that great man whom we are honouring today.

Priestley's life spanned the second half of the +18th century, for he was born in +1733 and died in 1804, the very same year that Napoleon was proclaimed emperor in France. So much has been written about his work that we can be very brief, but still it is necessary to say that all his connections were with that 'dissenting' world which gave a home to the scientific traditions of the previous century before the schools and universities of the establishment were open to them. Priestley was always devoted to three things, scientific research, protestant unitarian theology, and progressive social movements such as universal suffrage. Today he might be called a pioneer of Christian socialism, and veneration is therefore due to his memory.

Although a northerner by birth, he got his first job at Needham Market in Suffolk where he was assistant minister of a dissenting congregation from 1755 to 1758. In 1762 he married, and then, from 1765 onwards, his scientific interests blossomed

forth, for, going to London, he met John Canton, and later other
Fellows of the Royal Society such as Richard Kirwan, together with
Richard Price, Wm. Watson, Henry Cavendish, and last but not least
Benjamin Franklin. Afterwards he belonged to the Lunar Society
at Birmingham, so called because it was convenient for the members
to meet together at each other's country houses on nights when
the moon was full; and there he was familiarly acquainted with
Mathew Boulton, James Watt, John Wilkinson, William Withering
(of digitalis fame), and Erasmus Darwin. As the eighties drew to
a close, however, politics outbalanced science and theology, and
with the taking of the Bastille in 1789 the French Revolution
got under way. Priestley joined those who expressed much sympathy
with it. And so it happened that in July 1791, a Birmingham mob,
intoxicated by 'church-and-king-ism', as well as by other more
chemical substances, wrecked his home, his laboratory, and library,
and burnt down the chapel where he was accustomed to preach. Two
of the ringleaders were afterwards executed, but that could not
restore Priestley's books, apparatus, and manuscripts; nor could
a monetary compensation do so. After staying for a while at
Hackney, he decided therefore to emigrate to America, a land which
was then the home of liberty, and the abode of all progressive
ideals. His last decade was spent at Northumberland, Pennsylvania,
where his house still stands as a permanent memorial to him and a
museum of his activities. Only last March a pilgrimage honouring
him there was organised by the American Chemical Society.

The flavour of Joseph Priestley's enlightenment can be gained
from one or two quotations. In 1790 he wrote:

'The expence of the late American war only, would have converted
all the waste grounds of this country into gardens. What canals
and bridges, what noble roads, etc. would it not have made for
us? If the pride of nations must be gratified, let it be spent
in things like these...'

If Priestley had been living today, what would he have said about
nuclear armaments, essentially evil, for which every family in
this country has to pay some £18 a week? In the same year Priestley
wrote as follows:

'The very idea of distant possessions will one day even be rid-
iculed... Only those divisions of men, and of territory, will take
place, which the common convenience requires, not such as the mad
and insatiable ambition of princes demands. No part of America,
Africa, or Asia, will be held in subjection to any part of Europe,
and all the intercourse that will be kept up among them, will be
for their mutual advantage.'

Prophetic words for the end of the eighteenth century – though I do not know what he would have said of the Falklands dispute. Another feature of Priestley's character was his remarkable catholicity (with a small c); there was little sectarianism in him. 'If liberality of sentiment [he wrote on one occasion] be the result of general and various acquaintance, few men now living have had a better opportunity of acquiring it than myself. This has arisen from the great variety of my pursuits, which has naturally brought me to be acquainted with persons of all principles and characters. One day, I remember, I dined in company with an eminent Romanist priest; the evening I spent with philosophers who were determined unbelievers; the next morning I breakfasted, at his own request, with a most zealously orthodox clergyman, Mr. Toplady, and the rest of the day I spent with Dr. Jebb, Mr. Lindsay and some others, men in all respects after my own heart. I have since enriched my acquaintance with that of some very intelligent Jews; and my opponents, who consider me already as half a Mahometan, will not suppose that I can have any objection to the society of persons of that religion.'

Now we all know Joseph Priestley as one of the greatest pioneers of pneumatic chemistry, for he isolated and studied the properties of upwards of twenty gases. Some of the apparatus which he devised for doing this was exhibited at the Royal Society conversazione last June. The discursive diary style in which he wrote up his results at length is to this day very engaging. The only drawbacks were that he would never adopt the convenient term 'gas' which van Helmont had introduced in the seventeenth century, preferring to call them all different kinds of 'air', and secondly that he would never give up the theory of 'phlogiston' which J.J. Becher had first introduced. As we all know, phlogiston was supposed to be the principle of inflammability, or 'food of fire', having a negative weight. The calcination of a metal was therefore thought of as a loss of phlogiston, while we should describe it as a combination with oxygen to form the oxide; similarly reduction, which we think of as a loss of oxygen, was visualized as a gain of phlogiston. The whole theory is now considered almost a type-specimen of an erroneous way of looking at things, which rose and fell within little more than a century. Thus for Priestley, oxygen was 'dephlogisticated air', hydrogen was 'inflammable air', and carbon dioxide 'fixed' or 'mephitic air'. Among the other gases which he studied were ammonia (as 'alkaline air'), hydrogen sulphide (as 'sulphuretted inflammable air'), nitric oxide (as 'nitrous air'), and sulphur dioxide (as 'vitriolic acid air').

Now what about the connection of Priestley with gunpowder? It might be divided into two heads, the chemical and the rhetorical.

As for the former, he got oxygen from saltpetre in 1772, and was fond of using iron gun-barrels as part of his apparatus; this was how he got hydrogen from the decomposition of steam. As John McEvoy has said, after giving his explanation of the long-standing problem of the 'detonation' of gunpowder and nitre in terms of dephlogisticated air, Priestley also expressed his confidence that many other phenomena in chemistry would 'admit of the greatest illustration of discovery'. He wrote:

'The discovery of dephlogisticated air throws great light on many very important facts in chemistry, but upon none more than that very difficult and striking one of the detonation of nitre, concerning which the most improbable conjectures have been advanced by the most eminent philosophers and chemists. This detonation is the sudden inflammation produced by contact of various substances containing phlogiston and nitre, when either of them is red-hot. The hypothesis that has been thought the most satisfactory is that of Mr. Macquer, who supposes that, in these circumstances, an union is formed between the pure nitrous acid and phlogiston, similar to that which is formed between vitriolic acid and phlogiston in the composition of sulphur. He therefore supposes that, in this case, a nitrous sulphur is formed, and that the substance is of so inflammable a nature, that it cannot exist a moment without actual ignition... This nitrous sulphur is capable of the most violent inflammation in the closest vessels, where there is no access of air; and it is well known that compositions of gunpowder are made to burn even under water...'

It was no wonder that gunpowder would explode even more strongly in dephlogisticated air, though Priestley would have been interested to know that some of this air was already contained, as it were, within the explosive mixture itself.

The great force of the explosions of inflammable air, and gunpowder in dephlogisticated air, says John McEvoy again, suggested its use in 'blasting instruments' in excavation and mining. Similarly, a blast of dephlogisticated air could be used to 'augment the force of fire to a prodigious degree', thereby facilitating new chemical operations such as the melting of platinum. Priestley wrote:

'I rashly conjectured, that inflammable air [hydrogen] would explode with more violence, and a louder report, by the help of dephlogisticated air [oxygen] than of common air; but the effect far exceeded my expectations, and it has never failed to surprise every person before whom I have made the experiment.
Inflammable air requires about two-thirds of common air to make it explode to the greatest advantage; and if a phial, containing about an ounce-measure and a half, be used for the experiment, the explosion with common air will be so small, as not to be heard further than, perhaps, 50 or 60 yards; but with little more than one-third of highly dephlogisticated air, and the rest inflammable air, in the same phial, the report will be almost as

loud as that of a small pistol; being, to judge by the ear, not less than 40 or 50 times as loud as with common air... The repercussion is very considerable, and the heat produced by the explosion very sensible to the hand that holds it...'

And he goes on:

'It may be inferred, from the very great explosions made in dephlogisticated air, that, were it possible to fire gunpowder in it, less than a tenth part of the charge, in all cases, would suffice; the force of the explosion, in this kind of air, far exceeding what might have been expected from the purity of it, as shewn in other kinds of trial. But I do not see how it is possible to make this application of it. I should not, however, think it difficult to confine gunpowder in bladders, with the interstices of the grains filled with this, instead of common air; and such bladders of gunpowder might, perhaps, be used in mines, or for blowing up rocks, or digging for metals, etc...'

This brings us very near to the subject of our lecture.

But Joseph Priestley also used the explosion of gunpowder as a rhetorical device, and this got him into great trouble. In his 'Importance and Extent of Free Enquiry' of 1785 he urged that we should think far beyond the limits reached by Luther and Calvin. So of course we all have, but Priestley's oratory ran away with him. In a celebrated passage he said:

'Let us not therefore be discouraged, though for the present we should see no great number of churches professedly unitarian ... We are, as it were, laying gunpowder, grain by grain, under the old building of error and superstition, which a single spark may hereafter inflame, so as to produce an instantaneous explosion; in consequence of which that edifice, the erection of which has been the work of ages, may be overturned in a moment, and so effectually as that the same foundation can never be built upon again... And till all things are properly ripe for such a revolution, it would be absurd to expect it, and in vain to attempt it.'

This passage, with its inflammatory metaphor (which William Wedgewood had strongly advised him against), was destined to be, as much as anything else, the fuse of the troubles that befell him six years later. Edmund Burke called Priestley 'Gunpowder Joe', and it was inevitable that the Guy Fawkes episode of 'gunpowder, treason and plot' should be resurrected. But no one can be always on guard against injudicious remarks, and today we must greatly honour the memory of one who was at the same time scientist, theological philosopher, and outstanding democrat.

But what, and where, was the origin of that unique mixture which constituted the first of all chemical explosives known to mankind? This is the question to which we must now turn.

The development of gunpowder was certainly one of the greatest achievements of the mediaeval Chinese world. One finds the beginning of it towards the end of the Thang, in the ninth century A.D., when the first reference to the mixing of saltpetre (i.e. potassium nitrate), sulphur, and carbonaceous material is found. This occurs in a Taoist book which strongly recommends alchemists not to mix these substances, especially with the addition of arsenic, because some of those who have done so have had the mixture deflagrate, singe their beards, and burn down the building in which they were working.

The beginnings of the gunpowder story take us back to those ancient practices of religion, liturgy, and public health which involved 'smoking out' of undesirable things in general. The burning of incense was only part of a much wider complex in Chinese custom, fumigation as such (hsün). That this procedure, carried on for hygienic and insecticidal reasons, was much older than the Han appears at once from a locus classicus in the Shih Ching (Book of Odes), where the annual purification of dwellings is referred to in an ancient song, datable to the -7th century or somewhat earlier. It is perhaps the oldest mention of the universal later custom of 'changing the fire' (kuan huo, huan huo), a 'new fire' ceremony annually carried out in every home. The medical fumigation of houses, after sealing all the apertures, with Catalpa wood, is referred to in the Kuan Tzu book not many centuries later, and the Chou Li, archaising in character even if a Foomer Han compilation, has several descriptions of officials superintending fumigation with the insecticidal principles of the plants Illicium and Pyrethrum. From later literature we know that among Chinese scholars it was long the custom to fumigate their libraries to minimise the damage caused by bookworms, a great pest, especially in the centre and south.

As an extension of techniques like these, we find that the uses of scalding steam in medical sterilisation were appreciated as early as the +10th century. In his Ko Wu Tshu Than (Simple Discourses on the Investigation of Things) about +980, Lu Tsan-Ning wrote: 'When there is an epidemic of febrile disease, let the clothes of the sick persons be collected as soon as possible after the onset of the malady and thoroughly steamed; in this way the rest of the family will escape infection.' How general this practice was it would be hard to say, but it probably formed part of traditional hygienic practices from the Thang onwards.

Not only in peace, moreover, but also in war, the ancient Chinese were great smoke-producers. Toxic smokes and smoke-screens generated by pumps and furnaces for siege warfare occur in the military sections of the Mo Tzu book (-4th century), especially as part of the techniques of sapping and mining; for this purpose mustard and other dried vegetable material containing irritant volatile oils was used. There may not be sources much earlier than this, but there are certainly abundant sources later, for all through the centuries these strangely modern, if reprehensible techniques were elaborated ad infinitum. For example, another device of the same kind, the toxic smoke-bombs (huo chhiu) of the +15th century recall the numerous detailed formuale given in the Wu Ching Tsung Yao of +1044. The sea-battles of the +12th century between the Sung and the Chin Tartars, as well as the civil wars and rebellions of the time, show many further examples of the use of toxic smokes containing lime and arsenic. Indeed, the earth-shaking invention of gunpowder itself, some time in the +9th century, was closely related to these, for it was at once seen to be connected with incendiary preparations, and its earliest formulae sometimes contained arsenic.

The whole story from beginning to end illustrates a cardinal feature of Chinese technology and science, the belief in action at a distance. In the history of naval warfare, for instance, one can show that the projectile mentality dominated over ramming or boarding, with its close-contact combat. Smokes, perfumes, hallucinogens, incendiaries, flames, and ultimately the use of the propellant force of gunpowder itself, form part of one consistent tendency discernible throughout Chinese culture from the earliest times to the transmission of the bombard, gun, and cannon to the rest of the world about +1300. And indeed we believe that the following sub-sections will demonstrate beyond doubt that the entire development, from the first discovery of the gunpowder formula to the perfection of the metal-barrel gun emitting a projectile of dimensions closely fitting the bore, took place in China before other peoples knew of the inventions at all.

Before going any further it must be emphasised that the invention of gunpowder had implications far transcending military history. The viewpoint of the civil engineer is not to be ignored. His attitude towards explosives is very different from that of the soldier, for he thinks of them as rock-blasting and earth-moving facilities, means for carving out the formations for roads, water-ways, railways, pipe-lines, and all the multifarious veins and arteries of civilised intercourse; nor could the achievements of modern mining and quarrying be thinkable without the use of explosives. These things we shall take a look at later on as we see them growing out of the very ancient technique of 'fire-setting'. Other civil uses of gunpowder and the more sophisticated explosives that derived from it can be found in religious, ceremonial, and meteorological rockets, whether exploratory or weather-modifying. But the mechanical engineer is also in the picture. Later on we shall have something to say about the efforts to make gunpowder-engines before steam-engines came into their own, and indeed it was the former that led directly to the latter. As everyone knows, the steam-engine had its day, and it was a great one, not yet quite over; but when men's thoughts returned to internal combustion a fuel was needed to explode obediently in the cylinder, and what was it? Nothing other than the antecedent of gunpowder, namely the distilled petroleum that had constituted Greek Fire. And so these substances, the effects of which have been so terrible in warfare, turn out to be most intimately related to the development of the heat-engine, on which all modern civilisation has depended.

Clearly this entire subject is concerned with power, might and power placed in the hands of man as social evolution has gone on, power and might which form a couple of chapters only in the line of development which in the end has now given him mastery over the sub-atomic processes of suns, sources of inextinguishable energy, a mastery which has outstripped (it may be greatly feared) his ethical and moral maturity. Yet mastery over Nature remains the second grandest of ideals, as Robert Boyle wrote long ago, in 1664. He is well worth listening to.
'And though it be very true [said he] that man is but the Minister of Nature, and can but duely apply Agents to Patients (the rest of the Work being done by the applyed Bodies themselves) yet by his skill in making those Applications, he is able to perform such things as do not only give him a Power to Master Creatures otherwise much stronger than himselfe; but may enable one man to do such wonders as another man shall think he cannot sufficiently admire. As the poor Indians lookt upon the Spaniards as more than Men, because the knowledg they had of the Properties of Nitre, Sulphur and Charcoale duely mixt, enabled them to Thunder and

Lighten so fatally, when they pleas'd.

And this Empire of Man, as a Naturalist, over the Creatures, may perchance be, to a Philosophical Soul preserved by reason untainted with Vulgar Opinions, of a much more satisfactory kind of Power or Soveraignty than that for which ambitious Mortals are wont so bloodily to contend. For oftentimes this Latter, being commonly but the Gift of Nature, or Present of Fortune, and but too often the Acquist of Crimes, does no more argue any true worth or noble superiority in the possessor of it, than it argues one Brasse Counter to be of a better Metal than its Fellows, in that it is chosen to stand in the Account for many Thousand Pounds more than any of them. Whereas the Dominion that Physiologie gives the Prosperous Studier of it (besides that it is wont to be innocently acquired, by being the Effect of his knowledge), is a Power that becomes Man as Man. And to an ingenious spirit, the Wonders he performes bring perchance a higher satisfaction, as they are Proofes of his Knowledge, than as they are Productions of his Power, or even bring Accessions to his Store.'

Here at the outset it would not be inappropriate to say something, for the benefit of those less familiar with the Chinese literary tradition than others, on what we might call the 'philological network'. Chinese historical writing cannot just be dismissed as unreliable, for no civilisation has had a greater historical tradition than China, and the accounts of what really happened in all the ages have been the work of thousands of scholars. All that historians can do, they did, and archaeological finds have proved then right again and again, sometimes spectacularly. No other civilisation produced a body of work like the twenty-four dynastic histories (erh shih ssu shih), and these were supplemented by a vast body of unofficial historical writings; besides which there were encyclopaedists with high scholarly values in all ages, as well as biographers and authors of memorabilia. Modern philology has had a great part to play in the evaluation of all this, for the authenticity of texts can be cross-checked in many ways - who quotes whom, and is quoted by whom, who was a contemporary of whom, and what do we know about their life and times. Occasional false attributions and anticipatory ascriptions of course there are, but a whole literature of historical criticism and elucidation is available in Chinese, whereby the texts of erroneous, composite or doubtful date (wei shu) can be distinguished from the majority which have impeccable authenticity. As we noted at an earlier point, the study of the history of science and technology in China is in fact aided by the very circumstance that these pursuits were not highly regarded by the Confucian literati, so it would not have occurred to anyone that credit could be gained by falsifying matters so as to ascribe a given discovery or invention to a date earlier than that at which it actually happened. The same circumstances prevented dealers from forging non-artistic things such as scientific equipment or military weapons so as to give an erroneous appearance of antiquity. No one wanted to collect such things; there was no profit in it. The Confucian bureaucrats always had a supercilious attitude towards the soldiers, whose commanders were invariably lower than the corresponding civilians in official rank. From the texts of the military compendis one gets the impression that they were in deadly earnest, lacking the allusions and literary graces which other books possessed. Interpolations in them are very rare indeed. All in all, we believe that what the Chinese historians and military writers say is almost always credible. Such is our view of the reliability of what we shall be telling in the following sub-sections.

It is well to be clear from the beginning that broadly speaking the term 'fire-chemical' or 'fire-drug' (huo yao) never means anything other than that mixture of saltpetre, sulphur, and charcoal which we call gunpowder. To this there is, so far as we know, but a single exception - a recondite one - and that lies in the field of physiological alchemy, or the making of the 'inner elixir' (nei tan), where the juices and fluids of the human body, wrought upon by divers techniques and exercises, were believed to generate an enchymoma or macrobiotic drug which would confer material immortality upon the adept. Here, in order to make manifest the intimate relations of these entities with the Five Elements, it was necessary to coin special adjectives, and to translate chin i as 'metallous juice' (not as potable gold), or mu yao as 'lignic medicine'. Accordingly, encountering huo yao in nei tan texts of late date, it has to be translated as 'pyrial salve', i.e. the salivary Yang descending, in contrast with the 'aquose salve', the seminal Yin ascending; essential components of the enchymoma to be formed at the centre of the body. But the lore of these two pro-enchymomas had an extremely limited readership, and we can be sure that very few Chinese scholars throughout history ever understood huo yao in any sense other than gunpowder.

Out of the remote depths of history come the incendiary substances, needing ignition, and burning, sometimes quite fiercely, in air. Attached to arrows they cross the stage and must have lasted down well into the Sung time or even later. One of these incendiaries was naphtha, derived from natural petroleum seepages; but a great step forward was made in +7th century Byzantium, when Callinicus successfully distilled it to give low boiling-point fractions something like our petrol, which could be projected at the enemy by pumps which constituted flame-throwers. We think we can identify naphtha under the name <u>shih yu</u>, and 'Greek Fire', as it was called, under that of <u>mêng huo yu</u>. The 'siphon' or force-pump was of particular importance because it was the site of the first use of gunpowder in war; this was the appearance of a slow-match impregnated with the material in the ignition chamber (<u>huo lou</u>) of the machine - and the date was +919. This was a century which saw great commerce in these petrol fractions; they often came through from the Arab trade, but so much of the spirit was circulating among the rulers of the Five Dynasties period that the Chinese must surely have been distilling it themselves.

Without doubt it was in the previous century, around +850, that the early alchemical experiments with the constituents of gunpowder, with its self-contained oxygen, reached their climax with the appearance of the mixture itself. We need not harp upon the irony that the Thang alchemists were essentially looking for elixirs of life and material immortality. But it is only reasonable to recognise that once their elaboratories had jars containing (among many other things) all the constituents (more or less purified) of the deflagrative and explosive substance on their shelves, and once the alchemists started mixing them in all possible combinations, gunpowder was sure to be found out one day. If its formulae did not appear in print until +1044, that was a full two hundred years before the first mention of the mixture in the Western world, and even then no information was given there about the proportions necessary.

By about +1000 the practice of using gunpowder in simple bombs and grenades was coming into use, especially thrown or lobbed over from trebuchets (<u>huo phao</u>). Here the progression was from bombs with weak casings (<u>phi li phao</u> or 'thunderclap bombs') to those with strong ones (<u>chen thien lei</u> or 'thunder-crash bombs'). This paralleled a slow but steady rise of the percentage of saltpetre (potassium nitrate) in the composition, so that by the +13th cnetury brisant explosions became possible. In the meantime there was also a development of devices for mines, both on land and in the water. As long as the nitrate content remained low, there was a tendency to use gunpowder just as an incendiary better than those before available, but this did not outlast the +12th century.

So far all the containers had been in principle spherical, but the way to the true barrel gun - and to the piston of all engines too - lay through the cylindrical container. In China people had a natural cylinder ready to hand, the bamboo stem, once cleared of its septa, and any contents of the internode removed. This transition occurred first in the middle of the +10th century, as we know from a silk banner belonging to one of the Buddhist cave-temples at Tunhuang in Kansu. The scene depicts the temptation of a Buddha by the hosts of Mara the Tempter, many of whose demons are in military uniforms and carry weapons, all aiming to distract him from his meditation. One of them, wearing a head-dress of three serpents, is directing a fire-lance (huo chhiang) at the seated figure, holding it with both hands and watching the flames shoot out horizontally. This is the earliest representation we have of a weapon which had enormous repercussions between +950 and +1650; it played a very prominent part for example in the wars between the Sung and the Jurchen Chin Tartars from +1100 onwards. It was then for the first time described, about +1130, in the Shou Chhêng Lu of Chhen Kuei, relating the defence of a certain city north of Hankow. Essentially the fire-lance was a tube filled with rocket composition, relatively low in nitrate, but not allowed to fly loose, held instead upon the end of a spear. An adequate supply of these five-minute flame-throwers, passed from hand to hand, must have been effective discouragement to enemy troops from storming one's city wall.

The development of the fire-lance from the petrol flame-thrower pump must have been an easy and logical process. It turned that flame-projector into a portable hand-weapon for spouting fire, and since gunpowder, even though very low in saltpetre, had been used in the projector as a slow-match igniter, the new development was not far to seek. Also it was in a way a more effective method of using the incendiary properties of gunpowder, which must have been apparent even before the +10th century had begun. But the basic point was that the cylinder had been born. Most probably it originated with the natural gift of the bamboo tube, but as time went on all kinds of materials were employed for it, even paper (another Chinese invention), a substance which by appropriate treatment can be made so hard that it was actually used for armour.

What is important to note is that as the fire-lance period went on, through the +10th and +12th centuries, metal, both bronze and cast iron, perhaps also brass, was used to make the tube. This was one outstanding precursor aspect of the true metal-barrel gun or cannon, but the other was the addition of projectiles which issued forth along with the flames.

Here in this phase we have been obliged to coin two technical terms. The projectiles which were spurted forth in this way needed a special name, so we call them 'co-viative', distinguishing them thus from the true bullet or cannon-ball, which, in order to use the maximum propellant force of the gunpowder charge, must fill the bore of the barrel. The fire-lance projectiles could be anything offensive, such as bits of scrap metal or broken porcelain, but they could also be arrows. None of them would have issued with great velocity, but they could have been effective enough against unarmoured attackers, especially if the arrows were poisoned, as the texts often say they were. Secondly, when the fire-lances grew large, they were mounted on specially designed frames or carriages, almost like field-guns, and these we call 'eruptors'. These in their turn emitted miscellaneous co-viative projectiles, including arrows and containers of poisonous smokes, containers which in some cases may have been explosive, and therefore merit the name of shells or proto-shells. We often have to utilise these ambiguous prefixes; for example a gunpowder which contains carbonaceous material rather than charcoal may be usefully called proto-gunpowder. Similarly, we cannot always be sure whether a projectile fitted the bore of a gun or not, in which case it is convenient to call the weapon a quasi-gun or a proto-gun.

Thus the fully developed firearm had three basic features; (1) its barrel was of metal; (2) the gunpowder used in it was rather high in nitrate; and (3) the projectile totally occluded the muzzle so that the powder charge could exert its full propellant effect. This device may be called the 'true' gun, hand-gun, or bombard, and if it appeared in late Sung or early Yuan times, about +1280, as we believe it did, its development had taken just about three and a half centuries since the first cylindrical barrels of the fire-lance flame-throwers. This was not bad going for the Middle Ages, and it is important to realise that none of these early tentative phases had existed in Islam or Europe at all. The bombard appears quite suddenly full-fledged in the famous illustration of Walter de Milamete's Bodleian MS of +1327. Give or take a few decades, the bombard cannot have come to Europe much before +1310.

There, however, great sociological changes were about to happen - the Renaissance, the Reformation, the growth of capitalism, and the scientific revolution. Hence the speed of change in Europe began to outstrip the slow and steady rate of advance dictated by Chinese bureaucratic feudalism. The merchant-adventurers and the bourgeois entrepreneurs were to the fore once the +15th century had begun; the patricians of the mercantile city-States, the ironmasters, the mining proprietors and the factory builders, all these took charge as European aristocratic military feudalism died. Hence the way in which the gunpowder weapons first worked out by the Chinese began to come back to them in improved form. The serpentine lever, which applied the smouldering match to the touch-hole of guns, may have been invented in China, and the Turks may have improved it into the matchlock musket; certain it is that this superior weapon reached China either direct through Central Asia by +1520, or at the latest via the Portuguese and Japanese by +1548. Similarly, the Portuguese breech-loading culverin or small cannon came up from Malaya by +1510 or so, and its replaceable chambers were greatly appreciated by the Chinese gunners. And later the flintlock musket appeared, and later still the rifle. In the +17th century the Jesuits were 'drafted', so that John Adam Schall von Bell could be seen superintending the Western-style cannon foundry of the last Ming emperor in the +1642-43, while Ferdinand Verbiest had to undertake the same duty for the Chhing court in +1675. Thus did the inventiveness of the Chinese reverberate and recoil across the length of the Old World. Some eastern nations in modern times have been accused of being able only to copy and improve; but of no one was this more true in the +15th and +16th centuries than the Westerners. To be sure, with ballistics and dynamics they soon became 'airborne', but that was quite a time after the first knowledge of the first of all chemical explosives reached Europe.

It may seem surprising that until now nothing has been said about the rocket. In this day and age, when men and vehicles have been landed on the moon, and when the exploration of outer space by means of rocket-propelled craft is opening before mankind, it is hardly necessary to expatiate upon what the Chinese engineers started when they first made rockets fly. After all, it was only necessary to attach the tube of the fire-lance to an arrow, with its orifice pointing in the opposite direction, and let it soar away free, in order to obtain the rocket effect. Exactly at what date this 'great reversal' happened has been a debatable question. Twenty years ago, when our contribution to the 'Legacy of China' was written, we thought that rocket-arrows were developed by about +1000, in time for the Wu Ching Tsung Yao. That depended on one's interpretation of the 'gunpowder whip-arrows' (huo yao pien chien) described therein, but we now believe that these were not rockets, nor yet the huo chien either, which it also mentions and illustrates. All these were still incendiary arrows, desigend to set on fire from a distance the enemy's camps or city buildings; but in later times this same phrase was universally used to mean rockets. Here was another example of terminological confusion, when the thing fundamentally changed, while the name did not.

There would be a very good case for a linguistic analysis of such problems over the whole range of science and technology, and Hollister-Short has made a valuable contribution to it. How, he enquires, is a technical vocabulary generated in order to denote some new machine or technique? Language has often failed to keep up with technical change. Already we have come across the difficulties of precise nomenclature with regard to water-raising machinery, and vertical or horizontal wheels. We had to define our terms. Hime long ago encountered the same problem in relation to the subject of the present Section. He wrote:

'Take for example a word W, which has always been the name of a thing M. It is then applied to some new thing, N, which has been devised for the same use as M and answers the purpose better. W thus represents both M and N for an indefinite time, until M eventually drops into disuse, and W comes to mean N, and N only. The confusion necessarily arising from the equivocal meaning of W during this indefinite period is entirely due, of course, to (the failure of people) to coin new names for new things. If a new name had been given to N from the first, no difficulty would have ensued... But as matters have fallen out, not only have we to determine whether W means M or N when it is used during the transition period, but we have to meet the arguments of those ... who insist that because W finally meant N it must have meant N at some bygone time when history and probability alike show that it meant M, and M only.'

This is exactly the case with the fire-arrow and the rocket. We can recall a similar situation in China when the invention of the escapement for mechanical clocks was made, yet no one could think

of a new name to distinguish such horological machines from clepsydras. As for Hollister-Short, he took for his study the term Stangenkunst (rod-engine), which had two entirely different meanings, (1) a water-wheel placed above a mine-shaft, with rods descending from its cranks to actuate tiers of suction-pumps, and (2) the transmission of power across country from a water-wheel by means of horizontal rocking pantograph-like 'field-rods'.
It took all of two and a half centuries to clarify this. Fifty years ago I drew attention to the development of technical terms as a prime limiting factor in the history of science.

So when then did the rocket really start on its prestigious career? It is clear now that the fire-lance long preceded it; the Tunhuang banner of about +950 settled that question. We have to search for rocket beginnings in a rather different direction, and a couple of centuries later. During the second half of the +12th century we find the appearance of two kinds of fireworks, the one called 'ground-rats' (ti lao shu), and the other 'meteors' (liu hsing). Probably the former was the older, just a tube, probably of bamboo, filled with gunpowder and having a small orifice through which the gases could escape; then when lit, it shot about in all directions on the floor at firework displays. Alternatively, if attached to a stick, it flew off into the air, as at the night-time celebrations on the West Lake at Hangchow. That the two things were closely connected appears from late appellations such as 'flying rat' (fei shu) and 'meteoric ground-rat' (liu hsing ti lao shu). Ground-rats are contained in many specifications for bombs, where they are often equipped with hooks, and they must have been quite effective, especially when used against cavalry. As a firework they were certainly capable of frightening people, as we know from the story of a Sung empress who was 'not amused' by them.

Such civilian uses would have reminded the soldiers of the recoil effect of fire-lances which they must always have had to withstand, whereupon someone in the last decades of the century, perhaps about +1180, tried a fire-lance fitted backwards on a pike or arrow, with the result that it whizzed away into the air towards a target. Thenceforward, rockets were very commonplace, both in peace and war, through the Southern Sung, the Yuan and Ming, indeed down to the late Chhing, when they appeared in action against foreign invaders in the Opium Wars. Many developments of great interest occurred during this long period. First, there were several types of multiple rocket-arrow launchers, designed so that a single fuse would ignite and despatch more than fifty projectiles. Later on these were mounted on wheelbarrows, so that whole batteries could be trundled into action positions like regular artillery in modern times. But even more interesting were the rockets provided with wings, and carrying a bomb with a bird-like shape, early attempts to give some aerodynamic stability to the missile's flight, prefiguring the fins and wings of modern rocket vehicles. And just as the Chinese had invented the rocket itself, so it was natural that they should be the first to construct large two-stage rockets; propulsion motors ignited in successive stages, and releasing automatically towards the end of the trajectory a swarm of rocket-arrows to harrass the enemy's troop concentrations. This was a cardinal invention, foreshadowing the Apollo space-craft, and the exploration of the extra-terrestrial universe.

A word may be said here of how the path led from the ground-rat to the space-rocket. We shall see how for a time it was the Indians who excelled in the use of rocket missiles, a circumstance which led to a great development of war-head rockets in the first half of the nineteenth century in Europe. But this was a phase which came and went, for high explosive and incendiary shells could be fired from more advanced artillery with much greater accuracy of aim; so that the rocket batteries of the West died out after about 1850, and little use was made of rockets during the First World War. Meanwhile, however, another fundamental step forward had been made, to join the cluster of inventions which had happened in China in the first place; this was the study and development of liquid fuels, rather than the deflagrative gunpowder with which it had all begun. And this development was not inspired by war, rather by the science fiction writers, some of whom had appreciated the crucial fact that the rocket is the only vehicle known to man which can overcome earth's gravity, leave earth's atmosphere, and voyage among the planets and the stars. Truly, 'meteoric' was no bad name that the Chinese of the +12th century had coined for their 'flying rats'.

We have now passed in review the whole procession of inventions, with all their implications so fateful for the human race, between the earliest experiments with the gunpowder mixture in the +9th century and the appearance of the multi-stage rocket in the +14th. This had occupied some five centuries or so, with the transmission to the Western world coming right at the end of the period. And so, as we view the wheelbarrow rocket-launcher batteries passing off behind the curtains on the right of the stage, we must feel bound to salute those ingenious men of the Chinese Middle Ages 'that were Authours of such great Benefits to the universal World'.

With this our introduction may be ended, but before throwing open the vast museum of historical detail which justifies the statements that have been made, there are a very few concluding considerations we ought not to omit.

For example, there is a classical notion, a cliché perhaps, an idée reçue, a vulgarism, a false impression, which still circulates in the wide world - namely that though the Chinese discovered gunpowder, they never used it for military weapons, but only for fireworks. This is often said with a patronising undertone, suggesting that the Chinese were just simple-minded; yet it has an aspect of admiration too, stemming from the Chinoiserie period of the eighteenth century, when European thinkers had the impression that China was ruled by a 'benevolent despotism' of sages. And indeed it was quite true that the military were always (at least theoretically) kept subservient in China to the civil officials. Like scientists in England during the Second World War, the soldiers and their commanders were supposed to be 'on tap, but not on top'. No other civilisation in the world succeeded as well as China in keeping the military under tight control for all of two millennia, in spite of massive and extended foreign invasions, as well as peasant rebellions ever renewed. So the cliché could have been justified, but, as we shall abundantly see, it never was.

Then the Chinese invention of the first chemical explosive known to man should not be regarded as a purely technological achievement. Gunpowder was not the invention of artisans, farmers, or master-masons; it arose from the systematic, if obscure, investigations of Taoist alchemists. We say systematic most advisedly, for although in the +6th and +8th centuries they had no theories of modern type to work with, that does not mean that they worked with no theories at all. On the contrary, we have shown that the theoretical structure of mediaeval Chinese alchemy was both complex and sophisticated. An elaborate doctrine of categories, foreshadowing the study of chemical affinity, had grown up by the Thang, reminiscent in some ways of the sympathies and antipathies of the Alexandrian proto-chemists, but more developed and less animistic. Thus it remains to be seen what elements in this thought-complex were dominant when the fateful mixture was for the first time made. To sum the matter up, its first compounding arose in the course of century-long systematic exploration of the chemical and pharmaceutical properties of a great variety of substances, inspired by the hope of attaining longevity or material immortality. The Taoists got something else, but in its devious ways also an immense benefit to humanity.

Robert Boyle had something to say on this subject in +1664.

'Those great Transactions [he wrote] which make such a Noise in the World, and establish Monarchies or ruine Empires, reach not so many persons with their influence, as do the Theories of Physiology.

To manifest this Truth, we need but consider what changes in the Face of things have been made by two Discoveries, trivial enough, the one being but of the inclination of the Needle, touched by the Load-stone, to point toward the Pole; the other being but a casual Discovery of the supposed Antipathy between Salt-Petre, and Brimstone. For without the knowledge of the former, those vast Regions of <u>America</u>, and all the Treasures of Gold, Silver, and precious Stones, and much more precious Simples they send us, would have probably continued undetected; And the latter giving an occasional rise to the invention of Gunpowder, hath quite altered the condition of Martial Affairs over the World, both by Sea and Land. And certainly, true Natural Philosophy is so far from being a barren Speculative Knowledg, that Physick, Husbandry, and very many Trades (as those of Tanners, Dyers, Brewers, Founders, &c.) are but Corollaries or Applications of some few Theorems of it.'

Thirdly, in the gunpowder epic we have another case of the socially devastating discovery which China could somehow take in her stride, but which had revolutionary effects in Europe. For decades, indeed for centuries, from Shakespeare's time onwards, European historians have recognised in the first salvoes of the +14th century bombards the death-knell of the castle, and hence of Western military aristocratic feudalism. It would be tedious to enlarge on this here. In one single year (+1449) the artillery train of the King of France, making a tour of the castles still held by the English in Normandy, battered them down, one after another, at the rate of five a month. Nor were the effects of gunpowder confined to the land: they had profound influence also at sea, for in due time they gave the death-blow to the multi-oared war galley of the Mediterranean, which was unable to provide sufficient space for the numerous heavy guns carried on the full-rigged ships of the North Sea and the Atlantic. Chinese influence on Europe even preceded gunpowder by a century or so, because the counter-weighted trebuchet, an Arabic improvement on the projectile-hurling device most characteristic of China (the phao), was also most dangerous for even the stoutest castle walls.

Here the contrast with China is particularly noteworthy. The basic characteristics of bureaucratic feudalism remained after five centuries of gunpowder weapons just about the same as they had been before the invention had developed. The birth of this form of chemical warfare had occurred before the end of the Thang, but it did not find wide military use before the Wu Tai and Sung, and its real proving-grounds were the wars between the Sung empire, the Chin Tartars, and the Mongols, from the +11th to the +13th centuries. There are plenty of examples of its use by the forces of agrarian rebellions, and it was employed at sea as well as on land, in the siege of cities no less than in the field. But since there was no heavily armoured knightly cavalry in China, nor any aristocratic or manorial feudal castles either, the new weapon simply supplemented those which had been in use before, and produced no perceptible effect upon the age-old civil and military bureaucratic apparatus, which each new foreign conqueror had to take over and use in his turn, if he could. If he could not, his dynasty would not last very long, and always the Confucian bureaucracy, with their more or less obedient military inferiors, were ready to sweep back and run the country as it had been run from the very beginning of the Empire.

Finally, the sting in the tail, which shows once again how unstable Western mediaeval society was in comparison with that of China, is the foot-(or boot-) stirrup (teng). The conclusion now is that it was a Chinese invention, for tomb-figures of about +300 clearly show it, and the first textual descriptions come from the following century (+477), about which time there are numerous representations, Korean as well as Chinese. Foot-stirrups did not appear in the West (or Byzantium) till the +8th century, but their sociological influence there was quite extraordinary. The foot-stirrup welded the horseman and the horse together, and applied animal-power to shock combat. Such riders, equipped with the spear or the heavy lance, and more and more enveloped in metal armour, came in fact to constitute the familiar feudal chivalry of nearly ten European mediaeval centuries; that same body of knights which the Mongolian archers had overcome on the field of Liegnitz. There is no need to stress all that the equipment of the knights had meant for the institution of mediaeval military aristocratic feudalism. Thus one can conclude that just as Chinese gunpowder helped to shatter this form of society at the end of the period, so Chinese stirrups had originally helped to set it up. But the mandarinate went on its way century after century unperturbed, and even at this very day the ideal of government by a non-heritary, non-acquisitive, non-aristocratic élite holds sway among the thousand million people of the Chinese culture-area.

The social effects of gunpowder have of course often been meditated. A great Victorian writer, H.T. Buckle, saw its chief effect in 1857 as the professionalisation of warfare. Gunpowder technology was complicated and difficult to handle, and therefore there inevitably arose a separate military profession and ultimately standing armies; no longer was every man potentially a soldier. Hence there occurred a reduction in the proportion of the population entirely devoted to war, with the result that more people were shunted into peaceful arts, techniques and employment, hence also a 'diminution of the warlike spirit, by diminishing the number of persons for whom the practice of war was habitual'. Gunpowder technology was also expensive, more so than any individuals could afford, so only wealthy republics, or kings backed by merchants and endowed with rich estates, could manufacture, own, and operate musketry and artillery. Hence the rise of what Buckle called the 'middle intellectual class', so that 'the European mind, instead of being, as heretofore, solely occupied with either war or theology, now struck out into a middle path, and created those great branches of knowledge to which modern civilisation owes its origin'.

As a description of one aspect of the rise of the bourgeoisie this was all well siad, but Victorian optimism erred only in the belief that the situation would last. It might have been better to note what Robert Boyle had said in his 'Usefulnesse of Experimental Natural Philosophy' (+1664). Speaking of 'Engines so contriv'd, as to be capable of great Alterations from slight Causes', he wrote:
'The faint motion of a mans little finger upon a small piece of Iron that were no part of an Engine, would produce no considerable

Effect; but when a Musket is ready to be shot off, then such a Motion being applied to the Trigger by virtue of the contrivance of the Engin, the spring is immediately let loos, the Cock fals down, and knocks the Flint against the Steel, opens the Pan, strikes the fire upon the Powder in it, which by the Touch-hole fires the Powder in the Barrel, and that with great noise throws out the ponderous Leaden bullet with violence enough to kill a Man at seven or eight hundred foot distance.'

Thus a single touch could already mean life or death. And the touch would in time be open to everyone. It might have been wiser to foresee that science and technology would, as time went on, and by the very impetus of the industrial revolution itself, which Buckle so much admired, immensely improve, and enormously cheapen, the production of these lethal weapons, not only on the mechanical side but also on the chemical, producing a vast variety of explosives which would come within the reach of almost every man, whether dubbed 'terrorist' or 'freedom-fighter'. History has passed through a complete cycle, and alas, once again, 'every man is potentially a soldier'. This is our plight today, and nothing but universal social and international justice will relieve it.

All the peaceful applications of gunpowder as in meteorology and civil engineering tend to pale into insignificance by comparison with the part it played in the genesis of the steam-engine; and by the same token its predecessor, Greek Fire, the oil that we call petrol or gasoline, was destined to fuel, when the time was ripe, the internal-combustion engine.

For half a dozen decades past the idea has been hovering among the minds of historians that the cylinder and the cannon-barrel are essentially analogous, and that the piston and piston-rod may be considered a tethered cannon-ball. The piston and cylinder of course long preceded the metal-barrel gun and bombard, going back to the Alexandrian mechanicians and the Roman force-pumps, as also to the Chinese piston-bellows; but the military engineers of China can have had little idea of what they were starting when they first used metal to make their fire-lance tubes, and then later their metal-barrel hand-guns and bombards. One can now see the significance of our definition of true guns as opposed to co-viative projectiles and proto-guns, for only when the projectile exactly fitted the bore did the analogy with cylinder and piston make itself manifest. It all began with incendiary and hurtful fire as such, but when it ended with propellant explosion, then the door was open for all piston engines. The djinn was now well and truly in the bottle, and it was the Chinese military inventors who put it there in the first place.

There was really a convergence here of two strains of cylindrical structures; in the pumps and bellows the force was applied from the exterior to the contents, but in the cannon the force, and a very great one, was applied from the inside outwards, doing work. We already long ago came across this antithesis when we found that the morphology of the rotary steam-engine of the early nineteenth-century, with its classical solution of the problem of inter-conversion of longitudinal and rotary motion, (piston-rod, connecting-rod, and eccentric crank) had been anticipated by that of the water-powered reciprocating blowing-engines of China. But the physiology was exactly the inverse, for the water-power bellows applied the force to the piston from outside, while the steam-engine applied it from inside. As for the dating we used to say that the reciprocating furnace-bellows and flour-sifters were in general use by about +1300, but we now know that the whole assembly developed much earlier. First it was pushed back to the +10th century by Cheng Wei, who studied a painting of +965; and then Jenner, translating ch.3 of the <u>Loyang Chhieh-Lan Chi</u> (Description of the Buddhist Temples and Monasteries of Loyang), found unmistakable terminological evidence of it - a bolting- or sifting-machine about +530. The book says that at Ching Ming Ssu, south of the city:

'there were roller-mills and mills for grinding, trip-hammers for pounding, and bolting-machines for sifting and shaking, all driven by water-power (<u>yu nien wei chhung pho</u>, <u>chieh yung chui kung</u>). Of all the marvels of the monasteries these were considered the most remarkable (<u>chhieh lan chih miao</u>, <u>tsui wei chhêng shou</u>).'

This water-powered shaker or shifter assuredly worked by the mechanism which we find depicted later on. The reciprocating conversion design thus preceded the rotary steam-engine by no less than thirteen centuries. And the steam-engine, and later the internal-combustion engine, were in the trust sense children of the cannon. But now the work they did was beneficent work.

Perhaps Bernal was the first to formulate the analogy when he remarked that

'the steam-engine has a very mixed origin; its material parents might be said to be the cannon and the pump. Awareness of the latent energy of gunpowder persistently suggested that uses other than warfare might be found for it, and when it proved intractable there was a natural tendency to use the less violent agents of fire and steam.'

And he also wrote:

'A new and important connection between science and war appeared at the breakdown of the Middle Ages with the introduction... of gunpowder, itself a product of the half-technical half-scientific study of salt mixtures... In their physical aspect the phenomena of explosion led to the study of the expansion of gases, and thus to the steam-engine; and this was suggested even more directly by the idea of harnessing the terrific force that was seen to drive the ball out of the cannon to the less violent function of doing useful civil work.'

Seven years later, Vacca was speaking at an Italian symposium on the origins of specifically modern science in Europe rather than in China.

'A further advance was made [he said] by the invention of firearms, and of machines to use the expansive force of steam. Gradual familiarisation with the mechanics of explosions as they occur in firearms led to an almost ceaseless series of attempts to harness their power, from Papin's rudimentary efforts down to the modern internal-combustion engine.'

In 1948 Bernal discussed the connection again.

'Ultimately [he wrote] it was the effects of gunpowder on science rather than on warfare which were to have the greatest influence in bringing about the Machine Age. Gunpowder and the cannon not only blew up the mediaeval world economically and politically; they were major forces in destroying its system of ideas. As John Mayow put it: "Nitre, that admirable salt, hath made as much noise in philosophy as it hath in war, all the world being filled with its thunder". The force of the explosion itself, and the expulsion of the ball from the barrel of the cannon was a powerful indication of the possibility of making practical use of natural forces, particularly of fire, and was the inspiration behind the development of the steam-engine.'

But it was left for Lynn White in 1962 to express the matter even more clearly.

'The cannon [he wrote] was not only important in itself as a power-machine applied to warfare; it is a one-cylinder internal combustion engine, and all of our more modern motors of this type are descended from it. The first effort to substitute a piston for a cannon-ball, that of Leonardo da Vinci, used gunpowder for fuel, as did Samuel Morland's patent of +1661, Huygens' experimental piston-engine of +1673, and a Parisian air-pump of +1674. Indeed, the conscious derivation of such devices from the cannon continued to handicap the development [of internal-combustion engines] until the nineteenth century, when liquid fuels were substituted for powdered.'

And in 1977 he returned to the same theme, saying that

'Francis Bacon had more reason to be excited about a cannon than perhaps he himself realised. The cannon constitutes a one-cylinder internal-combustion engine, the first of its genus .... Lamentably, inventors along this line of technological growth fell into the very trap against which Bacon had warned them; focus on

tradition rather than on the qualities of Nature itself. They were so conscious of the cannon and gunpowder as precedents for their efforts that it was not until the mid-nineteenth century that they finally realised that the Chinese chemical mixture .... was inherently too awkward to give power to continuously operating engines. Only then did they turn - and with immense technical success - to the lighter distillates of petroleum which during the Middle Ages had been developed by the alchemists of Byzantium, Islam [and China] primarily for use as "Greek Fire". Two of the more conspicuous results were the automobile and the piston-engined aeroplane.'
But in the meantime the gunpowder engine's failure had led directly, as we shall see, to the steam-engine's success.

So far we have been dealing in generalities. By +1500 it was becoming clear that the force of gunpowder ought to be made to do something useful, instead of just propelling projectiles. It was the merit of Hollister-Short that he recognised the first appearance of such a use in the 'gunpowder triers', devices introduced by gunners to test the quality of their powder by making it perform some effect, some measurable work, and that he then saw the close relation of these to the gunpowder-engines which appeared rather later. Broadly speaking, one may say that the heyday of the triers or testers occupied the century +1550 to +1650, while that of the engines followed in the half-century ending about +1700. The latter were thus directly proemial to the development of the steam-engine, and intimately connected with it.

In +1540 Biringuccio was still taking a piece of paper and burning a small amount of gunpowder on it to see whether it would go off in a puff without burning the paper or no. But soon afterwards designs for more sophisticated mechanical triers were beginning to be pondered. The oldest which has come down to us is that of William Bourne in his book of +1578, 'Inventions or Devises ...'. He had a cylindrical metal box within which the powder was set off, and according to its strength the explosion pushed up the hinged metal lid so that it caught on one or other tooth of a quadrant ratchet, giving thereby a crude quantitative measurement. This device was again described by John Bate in his 'Mysteries of Nature and Art' of +1634: 'So, by firing the same quantity of divers kindes of powders at severall times, you may know which is the strongest.' The hinged cover appears again in a fine plate of John Babington's 'Pyrotechnia' (+1635); now the lid when blown off upwards rotates a graduated discoidal plate which must have been braked in some way so that the strength of the gunpowder could be empirically measured. Finally, the hinged cover reached its apogee in the trier which Robert Hooke demonstrated to the Royal Society, a much more work-manlike machine than any that had gone before. On 9 September 1663 'Mr Hooke brought in a scheme of the instrument for determining the force of gunpowder by weight, together with an explication thereof; which was ordered to be registered as follows...'The explosion cylinder had a hinged lid with a touch-hole closed by a strong spring, and at the other end the lid narrowed to a tooth which engaged with a cam or wheel-ratchet; this was on the same axle as a beam or arm which could be loaded with a variable weight. During the following months several tests were made but all failed, yet when notices of gunpowder experiments resume in +1667 the deficiencies of construction had been overcome, and the emphasis had shifted to making the engine do some other kinds of useful work. In January of that year an experiment was ordered for the applying of the strength of gunpowder to the bending of springs, thus storing energy, and this was successfully accomplished. Hooke was also asked to see if weights could not be raised by gunpowder. Robert Boyle suggested that the force of gunpowder might be tried by making it raise a weight of water (which it would expel out of a vessel). How exactly the springs were wound up, or the weights raised, the Journal Books of the Royal Society do not say, and on subsequent experiments they are silent too, but the whole sequence is of the greatest interest for it shows a gunpowder trier in the very act of turning into a gunpowder engine.

The blowing off of lids continued to the end of the century and beyond, as can be seen in the book of Surirey de St. Remy published in +1697. Though often mounted like pistols, they were quite similiar to Babington's device, for, upon firing, the cap of the explosion chamber rotated a graduated wheel, which came to rest upon a ratchet tooth and so assessed the force of the charge.

Boyle's suggestion reminds us that another of Babington's triers involved precisely the expulsion of water from one vessel to another. A given weight of a gunpowder mixture when exploded sent its gases into one vessel and expelled a measurable quantity of water into another. For comparing powders, this, said Babington, was 'the certainest way, although the most troublesome'. But now he was measuring the volume of gases formed rather than the mechanical force of the explosion, and the result was directly proportional to the percentage of nitrate in the composition, since that was the oxygen-provider, giving mostly CO_2 with smaller amounts of the oxides of sulphur and nitrogen. The displacement of water by air or steam was an ancient principle, going back to the Alexandrians, and therefore very familiar, but in +1635 the properties of the vacuum had still not been explored, so that Babington's device was only obliquely a predecessor of the water-raising systems of de Hautefeuille and Savory.

The other two peices of apparatus in Babington's plate derive from experiments of Joseph Furtenberg published in +1627. The former was not very practical, driving up a cover-plate with two holes along a vertical graduated scale marked on one of the columns, but the latter was a useful and workable device. Here the cap of the explosion chamber was blown up vertically guided by two wires, tripping as it went a series of 20 hinged ratchet-arms or 'keys', upon one or other of which it eventually came to rest, thus giving a measure of the gunpowder's strength. This system has a descendant among the pieces of apparatus used for determining explosive force at the present day; this is the 'whirling height éprouvette'. The upper conical rifled opening of a combustion chamber is closed by the tapered end of a 10 kg weight, and this is whirled upwards by the explosion between a cage of slide-bars, clicking in at the culimination point by means of a catch.

It should be mentioned here, however, that the most widely used contemporary device for measuring explosive force is the 'ballistic pendulum' developed and used at Fort Halstead. This employs a principle quite different from those used in any of the old triers, namely retro-active rocket propulsion. A mass of steel 150 kg in weight is suspended from a rigid framework by wire, just over 2 m long, and within this mass there is a steel tube of 25 mm bore taking a charge of about 10 g and with an unconfined orifice. When detonated electrically the heavy weight is propelled forwards in an arc by the energy release and its swing recorded by a stylus; then the excursion is expressed in percentage terms of a standard charge of picric acid. This has quite superseded the rather qualitative Trauzl test, which assessed the deformation produced by explosions set off within a block of lead.

Hollister-Short remarks that Furtenberg's flying-cap trier could have been a precursor of, or at least a stimulus for, Huygens' gunpowder engine of +1673, sonce the guide-wires directed the cap just as the cylinder-walls directed the piston. We should not lightly dismiss this idea, which after all is no more far-fetched than the comparison of the piston and cylinder with the cannon-ball and cannon, an analogy accepted on all hands as justified.

But Huygens was not the first to make a gunpowder-engine; he had been anticipated by Leaonardo da Vinci (as so often happened with that great Renaissance inventive genius) and Leonardo had a cylinder and piston. We must compare what he wrote with the diagram which he drew. His words of +1508 were these:

'To lift a weight with fire, like a horn or a cupping-glass. The vessel [i.e. the cylinder] should be 1 braccio wide and 10 in length, and it should be strong. It should be lit from below like a bombard, and the touch-hole rapidly closed, and then [all] immediately closed at the top. [You will see] the bottom [i.e. the piston], which has a very strong leather [ring] like a [pump-] bellows, rise; and this is the way to lift any heavy weight.'

As his diagram shows, a weight was suspended from the downward-pointing piston-rod, and gunpowder ignited above the leather-packed piston; then as soon as the gases had rushed out (expelling most of the air) all openings were closed, and, as the remaining gases cooled and contracted, a partial vacuum was generated, thus sucking up the piston and raising the weight. Here Leonardo came nearer to the ultimate gateway of success, the vacuum, than any before him, anticipating in a sense the +17th century physicists by 150 years or so; but without a deeper analysis of the phenomenon of the cupping-glass he could go no further.

What was this thing? It was one of mankind's most ancient medical instruments, a cup-shaped vessel placed on the skin at a suitable site and emptied of air by the burning of a small piece of wool or other combustible material inside it. The vacuum so formed sucks up the skin and flesh so as to bring about cutaneous vaso-dilation, and if the place has first been scarified it encourages transudation and bleeding. Historians of medicine regard the procedure as prehistoric in origin, and describe it from all the civilisations. The oldest vessel used was probably a hollow buffalo horn, which accounts for one of its Chinese names, chio fa, but later short tubes of bamboo were used, and these are still called huo kuan, tow or paper being burnt in them. There are many early literary mentions, and cupping was part of one of the seven departments (kho) of the Thang Medical Administration. Now Leonardo certainly did not have the concept of the vacuum and its uses, subsequently so clear, but it was a fine thing to recognise that certain procedures would make vessels suck other things in, and burning gunpowder could be even more effective than the small combustibles used by the physicians. All he had to go on was the Aristotelian truism that 'nature abhors a vacuum', but it was enough.

Perhaps the most extraordinary aspect of the situation was that Leonardo also conceived what one might call the standard experimental set-up afterwards used by Huygens and others, namely a cylinder and piston, the piston-rod of which was attached to a cord passing over two pulleys and then suspending a counter-balance weight. But he did not use this for weight-raising by gunpowder; he set it up about +1505 in order to see how much steam coming off from heated water would expand. All this was connected, no doubt, with his steam cannon, the 'Architronito', in which a jet of high-pressure steam was suddenly admitted behind a ball to shoot it forth through a long barrel. Here again was a striking link in the connections we are unravelling between the cannon barrel and the steam-engine cylinder.

Yet still the expansive force of steam was not the clue or key which would open the gate into the future; that key was nothing at all, the absence even of air, just emptiness. Ctesibius had been responsible, about -230, for a simple and fundamental machine, the piston air-pump, known from the descriptions of later mechanicians. This simplest of pumps entered upon a new incarnation in the +17th century, when the virtuosi began to explore with excitement the properties of vacuuous spaces, for what had been invented originally as a bellows for pumping air into something now found fresh employment as the 'air-pump' for getting as much air as possible out of it. The closer scrutiny of the alleged horror vacui began with Galileo himself in +1638 and was continued in +1643, by his desciple Evangelista Torricelli (+1608 to +1647) whose mercury barometer was the start of many experiments showing that the air weighs down on everything with a pressure of some 14 lbs per square inch. This opened men's minds to the recognition of the fact that air has weight, and that the vacuum was a physical reality. Then came the long-continued work of Otto von Guericke (+1602 to +1686) who invented the evacuating air-pump by about +1650, and then four years later performed the sensational experiment of the 'Magdeburg hemispheres'. He also demonstrated

the weight required to tear them apart, the crumpling of
evacuated copper globes, and the pistons which raised men or
weights into the air when sucked down by the vacuum. In +1659
there followed the improved air-pump of Robert Boyle (+1627 to
+1691), and with assistance from Robert Hooke it had attained its
final form by +1667.

Thus was established that all these effects were due to an
omni-present force - 'the spring and weight of the air', from
which man might draw infinite profit if he could get it to work.
As Cardwell has written:

'To the newly discovered agent, so powerful that it could overcome
the strongest horses, and rival the largest water-wheels and wind-
mills, science and common sense could set no obvious limit. The
wind might blow where, and when, it listed, but the atmosphere
<u>always</u> exerted a pressure of 14-15 lbs per square inch <u>everywhere</u>
on the face of the earth. The point was, could one utilise this
immense force, this 'head' of power? Could one, in effect, invent
an atmospheric water-wheel or windmill, driven by the dead-weight
pressure of the atmosphere rather than by the pressure of moving
air or water against sails or blades? To the speculative and
enterprising of the +17th century the possibilities may have
seemed as revolutionary as those of nuclear power in our generation,
probably with greater reason, certainly with fewer moral
reservations...'

By +1670 an acutely interesting question had arisen. What
could one do to create a vacuum underneath a piston (and so make
it do useful work), otherwise than by the previous use of another
piston in an exhaustive air-pump according to the Magdeburgian art?
When young Denis Papin from Blois took up his post as assistant
curator of experiments under the great Christiaan Huygens at the
Academie Royale des Sciences in Paris in +1671 this must certainly
have been one of the things they discussed, and the explosion of
gunpowder was certainly one of the methods proposed. The example
of the powder triers would assuredly have been in mind. Huygens
set to work in earnest towards the end of +1672 and by 10 February
in the following year described the engine he had constructed for
obtaining 'a new motive power by means of gunpowder and the pressure
of the air'. It consisted of a cylinder the piston-rod of which
was attached to a cord running over two pulleys and suspending a
weight. At the movement of the explosion the gases formed swept
out most of the air through valves which were them immediately
shut, so that as everything cooled a partial vacuum was produced,
and the piston was sucked down with great force - but not the
whole way, for, as Huygens noted, about one-sixth of the air and

gas remained. The down-stroke was therefore incomplete. Nevertheless he found the effect striking enough.

'The force of cannon powder has served hitherto [he wrote] only for very violent effects such as mining, and blasting of rocks, and although people have long hoped that one could moderate this great speed and impetuosity to apply it to other uses, no one, so far as I know, has succeeded in this, or at any rate no notice of such an invention has appeared.'

Huygens went on to prophesy that this motive power could be used for raising water or weights, working mills, or even for driving vehicles on land or water.

'I think [he wrote] that a flexible wooden tail at the stern of a boat, as I have visualised it, and as I believe they make use of in China, would be a good application for this, if moved by the force generated in this cylinder.'

Here was unquestionably a reference to the 'yuloh' scull (yao lu) or self-feathering propulsion-oar propeller, characteristic of small Chinese craft from Han times onwards. Huygens also sketched a form of ballista operated by linkwork as the piston descended.

So matters remained until Denis Papin, now occupying a chair at Marburg, returned to the problem of improving the gunpowder-engine. In +1688 he published a new version, the chief difference in which was that the piston was now furnished with a spring valve closed by atmospheric pressure when the gases had left, after which it was allowed to make a powerful down-stroke. But the fifth or sixth part of the air and gases always remained. Ruminating on this, Papin made a pregnant statement in his paper of +1690:

'Since it is a property of water that a small quantity of it, turned into vapour by heat, has an elastic force like that of air, but upon cold supervening is again resolved into water, so that no trace of the said elastic force remains, I readily concluded that machines could be constructed wherein water, by the help of no very intense heat, and at little cost, could produce that perfect vacuum which could by no means be obtained by the aid of gunpowder...'

Thus was born the first of all steam-engines. It looked just like the earlier gunpowder-engines, but a spring catch fitted into a notch on the piston-rod so that the down-stroke could be delayed until it was as powerful as possible, and then it went down right to the bottom of the cylinder. Here the boiler, engine-cylinder, and condenser were all in one; it was given to Thomas Newcomen to separate the boiler from the cylinder, and to James Watt to introduce a separate condenser - otherwise all the essential parts were present. Here at last was an effective cycle, the removal of air and the condensation of steam, so that the way was open to the 'atmospheric, or vacuum, steam-engine' (+1712). Though Denis Papin never harnessed his piston-rod to anything, his historical position in the transition from gunpowder to steam is a central one, and Thomas Newcomen himslef would surely never grudge him his statue among the flower-sellers and vegetable-stalls that overlook the Loire on the great flight of steps at Blois.

Steam had been on Papin's mind for quite a long time. Nine or ten years before, he had produced his steam pressure-cooker or 'digester'. In China steam had traditionally been used for many things, especially in cooking, and bread was (and is) generally steamed there rather than baked. In his 'Travels in China' (1804) John Barrow wrote:

'In like manner they (the Chinese) are well acquainted with the effect of steam upon certain bodies that are immersed in it; that its heat is much greater than that of boiling water. Yet although for ages they have been in the habit of confining it in close vessels, something like Papin's digester, for the purpose of softening horn, from which their thin transparent and capacious lanterns are made, they seem not to have discovered its extraordinary force when thus pent up; at least, they have never thought of applying that power to purposes which animal strength has not been adequate to effect.'

What Barrow perhaps failed to appreciate was that the way to the steam-engine historically lay not directly through high-pressure cookers, but indirectly through evacuated vessels, and the understanding of the vacuum was a characteristic result of the methods of modern science born in the Scientific Revolution. In other words the way to high-pressure steam, and all that it could do, lay dialectically through its precise opposite; and the whole historical process was an extraordinary justification of the classical idea of Taoist philosophy that emptiness would be the gateway to all power.

Papin could never have guessed that seventeen centuries earlier in China some experimentalist stubled upon the creation of a vacuum by condensing steam, and then proceeded no further. Among the procedures in the <u>Huai Nan Wan Pi Shu</u> (Ten Thousand Infallible Arts of the Prince of Huai-Nan), probably of the -2nd century, there is one which runs as follows:

'To make a sound like thunder in a copper vessel (<u>thung wêng</u>). Put boiling water into such a vessel [which must be closed extremely tightly], and then sink it in a well. It will make a noise which can be heard several dozen <u>li</u> away.'

If the vessel was full of steam when it was let down into the cold water, condensation would have created a vacuum, and if the copper was thin an implosion would have followed, echoing far beyond the well. Perhaps it was characteristic of the place and time that the invention served only military or thaumaturgical purposes, with no attempt to use the strong force that was evidently present.

Here there was no piston, but nor was there any in a collateral development which also preceded the steam-engine, and also arose from the properties of gunpowder, namely the vacuum displacement systems for water-raising. As early as +1661 Samuel Morland got a patent or warrant for pumping water from mines or pits more effectively 'by the force of Aire and Powder conjointly', but it was never finalised. Then in +1678 Jean de Hautefeuille published his tract entitled 'Pendule Perpetuelle' which included 'a way of elevating water by gunpowder'. Actually there were two ways. In the first, a rising pipe from the water 30 ft below delivered into a vessel that was partially evacuated by exploding a charge of gunpowder in it, and the water so sucked up was drawn off by a tap into a reservoir; this in turn could act as a second-stage sump for a further 30 ft lift arranged in the same way. Such cisterns in pairs, with gunpowder successively let off, would give a continuous discharge. But this was doing no better than a set of suction-pumps, so a second system was described, for use where force-pumps were necessary. Here a horizontal pipe was set under the water-surface down below, with an inlet-valve at its central point. At one end of this pipe there rose above the water level a short vertical tube leading to a gunpowder combustion chamber. At the other end a much longer tube rose up having a succession of non-return valves. As one charge after another was ignited, the water was driven up the rising main as high as the materials would stand. In this second system there was no dependence on the partial vacuum, and the gunpowder could be supplied in culasses like breech-loading cannon. All in all, these devices would obviate the expense of great numbers of men and horses in mines, drainage-schemes, and the like.

But the vacuum came back with a bang in Captain Thomas Savery's 'water-commanding engine'. In this machine, so often described, 'water was sucked up some 30 ft into a vessel made vacuous by condensation of steam, then forced higher still by a second admission of steam, suitable cocks being turned by hand at the several phases of the system. A continuous discharge was gained by having two vessels in parallel, one being filled by suction while the other was emptied upwards by pressure. This was ready by +1698, but ran into many difficulties largely because of the inferior strength of the materials available. William Blakey improved it nearly a hundred years later, but by then Thomas Newcomen's atmospheric steam-engine working a rocking beam, the ancestor of all later steam-engines, had been set up in many places since +1712, and the need for displacement systems was no more felt. Still, they are justifiably numbered among the predecessors of steam power.

From all that has now been said it will be evident that the explosive force of gunpowder played a fundamentally important part in the development of the steam-engine. But there is a second chapter yet to relate, that of the internal-combustion engines. Of these the very first was the hand-gun and cannon or bombard itself, which we have traced back to China about +1285; and the gunpowder-engines of Huygens and Papin were of the same category since they exploded the mixture within the cylinder itself and not in any separate vessel. The water of Papin's first steam cylinder was heated directly in it, so his experimental engine was an 'internal' one although there was no 'combustion', but as soon as Thomas Newcomen decided to have a separate boiler, as he did in the early years of the +18th century, the line of descent of the steam-engine separated off from all true internal-combustion engines. Still, for a century and more thereafter men's minds continued to be haunted by the idea of having an explosion right in the cylinder, and somehow taming its violence to give useful power. But the purpose of the explosion was now quite different from that of Huygens; it was no longer to drive out air and gases with a view to forming a vacuum so that the piston would get sucked in (at least some way), it was rather to effect a working stroke more closely similar to that of the cannon itself - though the piston was not free to depart from the machine.

It is interesting that the evolution of the steam-engine was just about complete by the time that engineering inventors began producing designs for internal-combustion engines. The separate condenser had been evolved by James Watt between +1765 and +1766, the double-acting principle came in about the same time as reciprocating rotary motion, ca. +1783, and high-pressure steam was introduced by Richard Trevithick from 1811 onwards. In the light of this it is quite interesting that gas-engines date from about 1826, and all the oil-engines (among which one must include those running on Diesel oil and petrol) from about 1841.

The first way of getting an explosion in the cylinder was to make a mixture of air and coal gas, and then to ignite it on each stroke. This was accomplished more or less by Samuel Brown from 1823 onwards, but his engine was not a success. Ignition at reliable intervals was always the problem, and William Barnett used coupled gas flames in 1838. Others turned to different gases, such as hydrogen and air, or pure methane and air, as in the work of Eugenio Barsanti and Felice Mateucci between 1843 and 1854, and it was these inventors who were the first to introduce that electrical ignition to which the future belonged. But the Newcomen of gas engines was J.J.E. Lenoir (1822 to 1900) who in 1859 made the first practical types, resembling horizontal double-acting steam-engines, with flywheels, slide-valves, and water cooling. The next greatest step forward came, however, when Alphonse Beau de Rochas (1815 to 1891) described in 1862 the four-stroke cycle basic to the successful operation of all internal-combustion engines. The first outward stroke of the piston draws the explosive mixture into the cylinder and the first inward stroke compresses it; ignition then takes place at or about the dead-centre position and the explosion drives the piston on its second outward stroke, after which its second inward stroke expels the burnt gases from the cylinder. Now at last the engineers had got their explosions under control, so to that extent the cannon was by 1860 firmly mastered. The real dénouement from our present point of view was, however, yet to come. A few gas-engines are still running, though most of them exist today only in museums; naturally they could never go far from gas supplies, though of course there is a sense in which all internal-combustion engines are gas-engines since the combustible material enters the cylinder as a spray mixed with air.

There followed an entr'acte or deviation somewhat analogous to the steam-vacuum displacement water-raising systems in the history of the steam-engine - namely that of the hot-air engine. John Stirling in 1826 and Eric Ericsson in 1849 had the thought of substituting for steam some new motor fluid more economical and easy to deal with. They therefore fell back on air itself, noting that its volume increases by a third between 0 and $100^{O}C$, doubles by $272^{O}C$, and triples by $544^{O}C$. Most of the older generation have memories of seeing part of an engine heated by a blow-torch, after which a swing of the flywheel would set the machine going; but although a number of engineers sought to perfect it there were many disadvantages, such as the fire danger, and deformations of the working parts, with the result that like the gas-engine it now survives only in museums, and on a small scale for toys and working models. The hot-air engine lies on a siding because no explosion, no internal combustion, was involved, only the expansion of heated air; but some source of heat remained imperative, so a heat-engine it certainly was. But the motive power of the future it was not.

The dénouement of the whole story came in 1836, when Luigi de Cristoforis (+1798 to 1862) began to think of making an internal combustion engine run on naphtha, a project which he perfected by 1841. Now at last those light fractions of distilled petroleum, orginally as Greek Fire so hurtful an incendiary weapon, burning men as well as things, was to become a beneficent power-source for daily use. What the Byzantine +7th century had begun and the Chinese +10th century had continued, now, after a thousand years, found its ideal place within the cylinders of internal-combustion engines. Perhaps we should pause here an instant to consider all the oily substances of this kind which can be used as combustibles, for oil-engines, Diesel engines, and petrol engines form a single family. We can tabulate the boiling points of those hydrocarbon fuels as follows:

	b.p. $^{\circ}$C	
petroleum ether or petrol	40-70	
gasoline	70-90	
ligroin or light petroleum	80-120	
benzene and toluene (from coal-tar)	82-110	
cleaning oil (turpentine-substitute)	120-150	
naphtha (from coal tar)	140-170	mostly xylene, pseudocumene, mesitylene
kerosene (paraffin oil)	150-250	
Diesel fuel oil	250-300	
carbolic oil (from coal-tar)	170-230	mostly naphthalene and carbolic acid
creosote oil (from coal-tar)	230-270	
anthracene oil (from coal-tar)	270-	
lubricating oil	300-	

Many of the lighter fractions of these oils have been used in internal combustion engines at one time or another, but eventually engineers settled for the lower b.p. oils in what we universally know nowadays as the automobile and aero engine. Higher hydrocarbons are commonly 'cracked' to give the lower, lighter ones.

The history of these power-sources can be briefly told. In 1873 J. Hock made an engine work with kerosene, and two years later Siegfried Marcus introduced petrol much like that of today; both worked in Austria. At the same time another petrol engine was improved by Enrico Bernardi of Verona, and in the following decade Gottlieb Daimler and Karl Benz (1883-85) brought it almost to its present form, attaining 800 r.p.m. In a parallel development many types of oil-engine appeared, but the greatest advance was made by Rudolph Diesel (1858 to 1913), who in a certain sense married the hot-air engine to the oil or petrol engine by compressing air violently to a temperature of $800^{\circ}C$, sufficient to ignite spontaneously a quite heavy oil injected into the cylinder. As everyone knows, there has been a vast expansion in the use of Diesel engines, especially for railway locomotives. Meanwhile, by 1895 the internal combustion petrol-burning high-speed automobile engine had reached essentially modern design in the hands of the Count de Dion and M. Bouton.

So now, reflecting on what we have found, one can see that the inventions of Greek Fire in the +7th century and of gunpowder in the +9th were not the unmitigated disasters that many people, even Shakespeare, speaking through his characters, have thought. Without them we would have had neither the steam-engine nor the internal combustion engine. And the moral is the same as that which we saw in the case of the rocket - all depends on what you do with it. Like fire itself, which can be used either for cooking food and warming people, or alternatively for torturing and killing people, the uses of every invention depend upon human ethical judgments; a problem for mankind as a whole, and common to all the civilisations.

Priestley and the Dissenting Academies

By J. W. Ashley Smith
5 SLACK, HEPTONSTALL, HEBDEN BRIDGE, WEST YORKSHIRE HX7 7EX, U.K.

Priestley would turn in his grave. He published some fifty theological works, and regarded these as his principal activity. He produced also a score of educational works. His numerous scientific papers and some books are difficult to compare for total bulk, but he himself regarded them as a sideline. His scientific research was done for fun, as a change from the important things of life.[1]

How did Priestley come to have these scientific interests? It would be pleasant if I could show you in detail how the interests arose from his education - that he was encouraged by some schoolmaster or tutor to think at an early stage about chemical research or about gases. But this is not so. His education did introduce him to scientific ideas and to the importance of experiment, but not to the fields which he made his own.

That protestantism went with an interest in experimental science seems to be established - certainly Christopher Hill and Bronowski[2] think so, though there are dissentients. Exceptions can be found to a proposition that those who wished to reform the church were also progressives in science. In the fifteenth century, for instance, it was complained, by Buonaventura and his English translator Nicholas Love, that the Lollards were addicted to Aristotle.

But the pioneer anti-Aristotelian was Ramus (Pierre de la Ramée), whose protestantism led to his death in the St. Bartholomew massacre of 1579. The title of Ramus' thesis at the University of Paris was 'Quaecumque ab Aristotele dicta essent, commentitia esse', which seems to mean (I'm not a Latin expert) 'All that Aristotle said was bunk'. It is to the credit of the very aristotelian University of Paris that Ramus was awarded his degree for that thesis.

What are the reasons for the linkage between extreme protestantism and experimental science? First is the likelihood that those who are overthrowing established authorities in one field will do the same thing in another. Second is the calvinistic emphasis upon the sovereignty and omnipotence of God - exerted directly, as compared with the medievalist idea of various saints who might seem to be acting (for example in acceding to prayer-requests) more or less independently of each other, thus destroying possibilities of causation. Third, the doctrine of transubstantiation, by which the bread and wine are converted (as to their substance but not their accidents, to use the philosophical terms) into something else by the use of certain words by the priest, may seem to tell against a scientific approach to matter.

Such reasons were likely to single out as pioneers of experimental science the more extreme protestants. One can distinguish to some extent between the attitudes of congregationalists and presbyterians. The congregationalists - including Ramus - believed that human sinfulness is so complete that we are unable to discern any general laws - for instance of behaviour, but the same applies in science. Perhaps, indeed, God hasn't made any general laws. When a lot of us agree in finding some general law, we know that it is only provisional; we have not seen into God's mind. A most desirable attitude for scientists.

Presbyterians, on the contrary, believed that it is possible for us to understand from God his laws for the organization of the church and for behaviour, and so also in the realm of science. (Presbyterians were thus outraged by the refusal of congregationalists to conform to norms of proper behaviour - such as old Thomas Goodwin, who was discovered exceeding limits of sartorial modesty by wearing simultaneously three embroidered nightcaps.)

But it is time we got on to the nonconformist academies, involving both those varieties of churchmanship. In 1660 Charles II came to the throne promising toleration. But parliament - to whom he owed his throne - thought otherwise, and by 1662 had enacted laws which effectively excluded nonconformists from the universities, whether as dons or as students. So it came about that men who had, or would have, taught at Oxford or Cambridge began - in secret for a generation - teaching young men who would have gone to university. In 1689 nonconformists received toleration under William and Mary, but this did not extend to the universities. The academies therefore continued throughout the eighteenth century, and indeed additional ones were founded, becoming in some cases well-known institutions.

Now, going back to the beginning of the academies in the years following 1662: if we had been in that situation, what should we have done? We should, I think, have found out what books and methods were used at Oxbridge, and tried to follow as far as possible exactly the same course as the students would have found at university. The odd thing is that the academy tutors did no such thing. Any educational history book will ascribe to the academies various innovations. Amongst these is the introduction of experimental science.

An early example was Charles Morton at Newington Green, near London. (We who are used to efficient persecution by modern states find it difficult to imagine how he was able to escape!) His academy was described as:

> 'having annexed a fine Garden, Bowling-Green, Fish-pond, and within a Laboratory, and some not inconsiderable Rarities, with Air-Pumps, Thermometers, and all sorts of Mathematical Instruments ...'[3]

Two generations later Charles Owen, to take another example, wrote an Essay towards a Natural History of Serpents (1747) which includes an accurate account of the properties of asbestos and the statement that a bee has four feet.

Thus by the middle of the century, when Priestley entered the Daventry Academy, science was established in the curriculum. There is no evidence, however, of any chemistry at Daventry. Thus Priestley's earlier interests were in electricity and the vacuum pump. In his second pastorate, at Nantwich, he was able to assemble some apparatus, and to use it with his school pupils and also with an adult class. In 1761, joining the staff at Warrington, Priestley happened to be introduced by Josiah Wedgwood to a Liverpool physician, Dr. Turner, whom he persuaded to visit Warrington to lecture on chemistry. Accounts of Warrington imply that Priestley merely helped Turner in those lectures; but when, thirty years later, Priestley gave chemistry lectures at Hackney College, he said that he was repeating his Warrington lectures.[4]

Priestley's justification for science in the curriculum was utilitarian:

> 'It is by increasing our knowledge of <u>nature</u>, and by this alone, that we acquire the great art of commanding it, of availing ourselves of its powers, and applying them to our own purposes; true <u>science</u> being the only foundation of all those <u>arts</u> of life, whether relating to peace or war, which distinguish <u>civilized</u> nations from those which we term <u>barbarous</u>...'[5]

In this, Priestley departed from the normal academy tutor's reason for teaching science. This was not often mere utility. Sometimes the motive given first place was enjoyment. But the dominant motive was shown by the title of the wellnigh universal academy textbook. This had been written by John Ray, of Trinity College, Cambridge, the seventeenth-century precursor of Linnaeus. His book was <u>The Wisdom of God manifested in the Works of Creation</u> (1691, and later, enlarged, editions). This was the idea uppermost in the minds of tutors; it does, of course, connect with utility - the 'wisdom of God' will be displayed by the ways in which creation can be used for the benefit of the human race.

We saw earlier that attitudes to science can be expected to differ between congregationalists and presbyterians. Priestley was brought up a congregationalist, and destined initially for a strictly orthodox academy. When the time came, however, he preferred the more liberal atmosphere at Daventry. If it is true that his early congregationalism set Priestley on the path of scientific adventure and willingness to doubt existing interpretations of nature, can it perhaps be suggested that his later presbyterianism gave him a hope of arriving at some law of nature - or, equally, gave him confidence in the results of human investigations which caused him to cling so tenaciously to phlogiston?

<u>References</u>

[1] <u>E.g.</u> 'Experiments and Observations on Different Kinds of Air', 1774, Preface v.
[2] J.E.C. Hill, 'Intellectual Origins of the English Revolution', Oxford, 1965, p.20; J. Bronowski, 'Magic, Science and Civilization', Columbia, 1978.

[3] Charles Wesley, 'Letter from a Country Divine ... on the Education of Dissenters ...', 1704, p.6.

[4] J. Priestley, 'Heads of Lectures on a Course of Experimental Philosophy, particularly including Chemistry; delivered at New College 1794', subtitle.

[5] 'Experiments ....Air', op. cit., Preface xxxii ff.

Priestley in Caricature

By M. Fitzpatrick
DEPARTMENT OF HISTORY, HUGH OWEN BUILDING, THE UNIVERSITY COLLEGE OF
WALES, ABERYSTWYTH, DYFED, WALES SY23 3DY, U.K.

Dr. Johnson defined caricature as ' v. to hold to ridicule ... A drawing ... intended as humour, satire and comment'. His definition reflected contemporary usage; printsellers of the period described any print which was in some sense humorous as a caricature. Yet strictly speaking the word caricature refers to a specific tradition of humorous representation and not to humorous prints in general. Carracci, one of the founders of the caricatura tradition believed that the aim of the caricaturist was 'to grasp the perfect deformity, and thus to reveal the very essence of a personality'. Hogarth, who regarded the caricatura tradition as perverse, frivolous and foreign, softened and naturalized it while remaining true to its aim, namely to convey an essential truth about the inner personality by the portrayal of an external appearance. His famous engraving of John Wilkes casually holding a pole supporting a pileus or liberty cap was intended to show by realistic portrayal Wilkes's unfitness for the cause he purported to support. One has only to glance at the leering squinting Wilkes to realise that he was a profligate rake and an unworthy proponent of any higher cause. [1] A portrait of Joseph Priestley, published in July 1791 with the title 'Dr. Phlogiston', was very much in this tradition. [Pl. 1]* Priestley's coarse features, his long heavy nose, thick lips, elongated chin, sly shadowy eyes, and his overall sinister appearance alert the viewer to his base and untrustworthy character.

Caricature in the caricatura tradition needs few props, and if successful is the least ephemeral of the caricaturist arts. It thrives naturally in an age of mass visual communications, but in an earlier age caricaturists could not depend upon making their point, or selling their work, by such means. There were very few caricatures purely in this tradition in the late eighteenth century. Even the Wilkes portrait drew upon another satirical tradition, for it incorporated emblems, in particular the ancient emblem of the liberty cap, in order to point the

*Plates are between pp. 360 and 361.

meaning. In the Priestley portrait, the emblems used are the contemporary ones of his own works, or works supposedly his own, which provide ominous evidence of his sinister intent: his firebrands are his Political Sermon and Essay on Government; his pockets are stuffed with works inscribed Revolution of Toasts, Essay on Matlin (sic) Spirit and Gunpowder, and his left foot rests on a book entitled Bible Explained Away.

The emblematic tradition of satire developed during the Reformation crisis when artists were afforded ample opportunity of drawing upon Biblical symbolism to ridicule their Protestant or Catholic opponents. Such symbolism naturally lent itself to the hostile portrayal of Dissent. In the late eighteenth century, mitres in the dust, fallen crosses, crowns in the gutter, and demons abounding are the stock-in-trade of caricaturists of Dissent. They are used to provide an easy frame of reference and, in themselves, carry little potency. This is the case in the caricature entitled Puritanical Amusements revived! of 27 February 1790. [Pl. 2] It shows Priestley, Stanhope, the Dissenting ministers and their fiendish friends creating mayhem. St. Paul's collapses, bishops fly through the air, and the latter-day Puritans punish those who have offended against their own strict standards: a man for kissing his wife on the sabbath; a lady for 'stealing to the playhouse last night'; and a cock for copulating on the sabbath. It is all rather comic and absurd, trite rather than properly satirical, and, if intended seriously, was surely a failure. The demons and fiendish creatures hardly add a fearful demonic dimension to the activities of Priestley and his fellow travellers. The emblems cannot make up for the weakness of the design and content. There was, however, one ambitious attempt to attack the Rational Dissenters in a totally emblematic way, and it is a testimony to the difficulty of such an enterprise at this time that it required a page of printed explanation. Entitled The Unitarian Arms, published on 12 July 1792, it was 'Addressed to thosePeaceable Subjects of this Kingdom who prefer the Present Happy Constitution to that Anarchy & Bloodshed so zealously sought for by these restless advocates for Priestly & Paine's Sophistical Tenets'. [Pl. 3] On the left, religion provides the cloak under which hypocrisy (on the right) pursues her devilish schemes with the Unitarians hatched by the harpy. Although the message is the same as that of other caricatures - the seditious evil intent of the Rational Dissenters, who, despite their professions of good faith, aim to undermine the blessed constitution in Church and State with the assistance of the

Foxite Whigs - the means chosen to convey it were unusual. Most caricaturists preferred to be freer and wittier in their inventions than the emblematic tradition allows. Following the example of James Sayers (1748-1825), they learned the value of establishing easily identifiable and repeatable likenesses and of educating the public in the visual identity of the leading figures of the day. Yet, although the growing assurance of the caricaturists that the public would recognise their subjects gave them greater freedom in the treatment of the issues of the moment, very few caricaturists made their point solely through their art. Caricatures at this time have to be read literally as well as visually. They were poured over in the print shops such as Hollands and Fores, where permanent exhibitions were held, entrance one shilling, and caricaturists crowded their drawings with comment and witty asides, often to the detriment of the main point and the visual impact.

Given the fact that caricatures were a species of literature as well as of art, it is suprising that Joseph Priestley was so slow to come to the attention of the practitioners in this unique genre. Eminent in science, philosophy and theology, he was an unflagging pamphleteer and controversialist. His writings were littered with apocalyptic denunciations of the establishment in Church and State and his rhetoric lent itself to emblematic representation. In particular, he gained notoriety from a passage in the published version of a sermon preached on 5 November 1785, The Importance and Extent of Free Inquiry in Matters of Religion, in which he declared that the Rational Dissenters were by

> The present silent propagation of truth ... laying gunpowder, grain by grain, under the old building of error and superstition [the church establishment].

so that it required only a single metaphorical spark for the building to be overturned for good. Nevertheless, even when the Dissenters began to campaign for the repeal of the Test and Corporation Acts in 1787, and even when Priestley attacked William Pitt in forthright terms for his opposition to repeal, the caricaturists held their fire. A print of 1780 did, however, offer Priestley and the Dissenters a clue as to what was in store for them.[2] Entitled Opposition Defeated and published on 27 February 1780 at a time when the government was suffering from the growing pressure of parliamentary reformers and critics of its conduct of the war against the American Colonies, it shows John Bull ridden by Lord North, having charged down the foreign foe, France, Spain and America, kicking back at the Earl of Shelburne, who appears to be saying, 'I die, d(amnation) stares me in the face'. Behind him the devil exhorts Price and Priestley to support their master, which indeed they attempt while offering him comfort. Price with his back to the viewer, declares, 'Natural moral religious and civil liberty authorises the murder of a minister. Thou hast nought to fear'. Priestley, for his part, informs Shelburne, 'Damn(ation) is a rest, the soul is not immortal'. In the background on the right, Shelburne's ally, Charles James Fox carrying the Prince of Wales on his shoulders, declares that his hopes and those of the Jews were at an end. Thus sturdy, virtuous John Bull triumphs over his enemies as represented by an evil combination of heterodox, rebellious Dissent, opportunism, political ambition, and hostile foreign powers.

When Price and Priestley appeared again in satirical prints they would be represented in the company of a similar assemblage of evil forces, although, in keeping with the trends in caricature already noted, they would in future usually be identifiable by their profiles and not solely by their words as in <u>Opposition Defeated</u>. But that future was ten years hence, for during the 1780s caricaturists paid scant attention to Dissent. One can only conclude that Dissent was insufficiently interesting and their leaders, notably Price and Priestley, were not public enough figures for the caricaturists to pay attention to them. Indeed it required a unique combination of circumstances to bring them into the caricaturists' eye: the revival of the London Revolution Society with the commemoration of the anniversary of the Glorious Revolution, an event in which the leading Rational Dissenters were closely involved: the French Revolution applauded by Dr. Price in his <u>Discourse on the Love of Our Country</u>, preached at the anniversary meeting of that society a year later, on 4 November 1789; the final and strenuous efforts of the Dissenters to obtain the repeal of the Test and Corporation Acts; and the choice of Charles James Fox to propose the motion for repeal when it was put, unsuccessfully, on 2 March 1790. With these events, the Dissenters moved to the centre of the national stage, and the prejudices which existed against them came to be expressed in caricature after caricature. The first of the extant caricatures which appeared was a powerful and telling print by Sayers. It is worth spending a little time reading it.

The Repeal of the Test Act a Vision satirizes the sermon preached by Price to the Revolution Society at the Old Jewry on 4 November the previous year. [Pl. 4]   Price is the central figure in the pulpit and is intoning the words:

> And now let us fervently pray for the abolition of all unlimited and lim(ited) monarchy, for the Annihilation of all ecclesiastical Revenues and Endowments, for the Extinction of all Orders of Nobility and all rank and Subordination in civil Society and that Anarchy and Disorder may by our pious Endeavours prevail throughout the Universe.

He concludes by referring his hearers to his Sermon on the Anniversary of the Revolution.   With his left hand he grasps a message passed to him by a clerk which reads, 'The Prayers of the Congregation are desired for the Success of the patriot Members of the national assembly now sitting in France'.   This was a reference to the congratulatory address which Price moved at the formal proceedings of the society following his sermon, and which was later presented to the National Assembly by the duc de la Rochefoucauld.   Stanhope, who chaired the meeting of the Revolution Society and who signed the address, is also present at Sayer's concocted occasion. Standing to the right below the pulpit, his association with church reform as well as that of political reform is indicated by his tearing up the 'Act of Parliament for the Uniformity of Common Prayer and Service in ye Church and Administration of the Sacrament', and by his having ready a 'bill for the Abolition of Tithes and other ecclesiastical endowments'.

Besides Price in the pulpit are Priestley and Theophilus Lindsey.   The latter is tearing up the Thirty-nine Articles while Priestley is emitting smoky hot air labelled, 'Atheism, Deism, Socinianism and Arianism', which drives off an angel carrying a cross, much to the applause of Charles James Fox below.   Just above and behind the trio is an inverted triangle with their initials, 'P P L', inscribed within.   This is a puzzle, for in the Unitarian Arms, the triangle was a symbol of the trinity.   Is this triangle inverted to denote the trio's anti-trinitarianism?   If so, why is it still radiating light?   More likely, it links the three preachers with the forces of the enlightenment and republicanism; the symbol of the triangle was adopted by the enlightenment, by the American Revolution and by the French Revolution when liberty, equality and reason came to form a new trinity.   At any rate, the adherence of the preachers to the modern republican cause is also signified by the American flag hanging by the window. The fusion of transatlantic republicanism is suggested by the portrait of Oliver

see original Page 356.

be stopped in their tracks by refusing to repeal the Test and Corporation Acts, and he conveyed his message with a wealth of imagery and detail and carefully identified the main enemies. Published some eight months before Burke's Reflections on the Revolution in France, it anticipates, as Dorothy George has also noted, in a rather crude way his attack on Price.

There were at least three other prints published before the parliamentary debate on the repeal of the Test and Corporation Acts: one, the burlesque Puritanical Amusements revived! and two others, one attributed to Rowlandson, the other by Dent, both of which take up some of the themes of the Sayer's print. In the putative Rowlandson, entitled The Test (20 February), the attack on the church is portrayed in graphic detail. [Pl. 5] Priestley features in the background supervising a gang demolishing a church. He exhorts them, 'Make haste to pull down that old Whore and we'll build a new one in its place'. In the Dent, the Meeting of Dissenters Religious and Political 1790, the main theme is the cynical alliance of the Dissenters with Fox. Priestley here plays a minor role. [Pl. 6] Dorothy George has suggested that he is standing by the minister resembling Price who is declaiming against the Church though in a speech fit only for a particularly cynical Methodist. An alternative candidate, in terms of likeness and even in terms of symbolism, is the person, immediately behind Sheridan (on the left), who is holding the padlock marked 'motive' which secures the door marked 'Church and State'. The print, unsatisfactory in many ways, contains the usual symbols evoking the dangers of heterodox Dissent, and the crown and a mitre lie upon a heap of garments labelled 'Deism, Arianism, Presb(yterianism), Socinianism &c., Republicism &c.'

Dent, who was well known for his hostility to Fox, pursued the theme of his relationship with the Dissenters in a print entitled A Word of Comfort (22 March 1790). [Pl. 7] Fox in company which includes the Prince of Wales and Mrs. Fitzherbert, Sheridan and other friends including a motley collection of female hangers-on, listens with sly intent as Priestley offers an unequivocal 'no' to his question, 'Pray Doctor is there such a thing as a devil?'. But Priestley, though his works have crushed a crucifix and though he preaches from a barrel marked 'Fanaticism', is portrayed as naive and misguided rather than as the personification of evil: a grinning devil at his back comments, 'if you had eyes behind you'd know better my dear doctor'. This relatively innocuous print has, however, a nasty sequel.

The view that Priestley and the Rational Dissenters were misguided and were not really aware of what they were doing, was the theme of one of the most unusual caricatures which treated of the Dissenters' campaign for greater toleration. This was an anonymous print entitled Sedition & Atheism Defeated which was published in the wake of the defeat of the repeal motion of the Dissenters in March 1790. [3] Although the characters are quite well drawn and are recognizable, the visual impact of the caricature is limited. It has to be read, and one wonders how many persevered in that head-twisting task. Its particular interest lies in its sensitivity to the association of Priestley and the Rational Dissenters with the cause of toleration for the Roman Catholics. No other caricature noticed this association despite comment in the press upon it from the beginning of the campaign in 1787. The caricature shows on the top line the outlines of an infernal plot. The devil instructs his demons to take their parts. The writing is difficult to read, but it appears that one of the demons decides to possess Priestley with the imprecation, 'Blasphemy, Atheism &c. Priestley are Mine'. At the next stage of the plot, the Rational Dissenters meet with Father Joseph Berington, a leading liberal Catholic and a friend of Priestley. A toast is proposed to the Catholics, republican sentiments expressed, subversion countenanced, an alliance with Fox mooted, religious establishments denounced - by Berington, who offers an aside 'except Popery' - and Priestley, at the head of the table on the left, sums up, 'And thus we will lay Gunpowder grain by grain till we blow up the --------- Church'. At the end of the line, however, we see that the final stage of the intended plot was not disestablishment, as the Rational Dissenters hoped but a Catholic establishment, with the King and Queen observing Catholic devotions. But that was not to be. According to the caricature, though this detail was erroneous, Priestley's proposal was turned down by 'the good Body of worthy Dissenters' as shown in the third box on the top line. Moving to the bottom line, from 'intention' to 'fact', the print shows Fox, in the first frame on the left, informing the conspirators of the crushing defeat of the repeal motion. They express their disappointment in various ways. Priestley confesses that, 'the conversion of Silas Deane to atheism was but an Introduction to what I intended to do on ye Repeal'. The reference is explained in the next frame in which the

American Revolutionary, Silas Deane, is shown confessing on his deathbed that Priestley had taught him that there was no God. In the final scene, the plotters are led off by the nose by their respective demons. Priestley, seeing the prospect of the descent into Hell, recants his opinions, but the demons will have none of this: one tugs a string attached to his nose and declares, 'Come along most learned Doctor - I have always had you by the nose - so don't mind it now'; while another gives him a kick from behind and tells him to, 'get along Joey 'tis too late to recant now besides we love you too well to part with you'. As a final humourous touch, some demons express reservations about the impact of this 'parcel of divinity' upon the government of the nether regions.

As we have seen, the point that this caricature makes, that the Rational Dissenters were naïve though dangerous stool pigeons of evil forces beyond their comprehension (despite the reservations of one or two demons!), was made in other prints, but the identification of those forces with Catholicism, traditionally treasonous, immoral and on a par with atheism, was not. Yet this Catholic dimension appears thus briefly, for anti-Catholicism was eclipsed by the pressing and immediate fear of Jacobinism. It could yield no satirical mileage in the period when the French Revolution was at its height. Moreover, although the appeal to the good sense of orthodox Protestants, Anglicans and Dissenters alike, to mind the wiles of the Rational Dissenters was enduring, caricaturists came to take the view that good sense along was not enough to halt their wicked plans. During the repeal campaign of the Dissenters, they had allowed themselves to laugh at least a little bit at the schemes and machinations of the sectarians. The exception to this was Sayers whose print set the tone for the future. The doctrines and actions of the Rational Dissenters increasingly came to be portrayed as highly dangerous, furnishing the literal as well as the metaphorical gunpowder which would blow up the existing order. Following Sayers, attention was focused on those who preached the pernicious doctrines and who led the Rational Dissenters into dangerous friendships. Despite the continuing satire against Burke, caricaturists began to share his views on the danger of the French distemper crossing the channel. They repressed the memory of their earlier enthusiasm or even indifference and vented their spleen on those who had the temerity to continue to regard the French Revolution with some sympathy and/or detachment.

As long as Richard Price was alive, and indeed for a little while after his death, he provided an alternative or partner to Priestley as a figure for execration. They were almost interchangeable. In January 1791, they appeared in consecutive prints in a series entitled <u>Attic Miscellany</u>. Price was shown preaching from a tub marked 'Political Gunpowder'. [Pl. 1a] This was followed by the print of a particularly villainous Priestley brandishing his fiery works which we have already discussed. The images could so easily have been reversed. With the death of Richard Price on 19 April 1791, Priestley was left to epitomize all the evils of Rational Dissenting thought and action. The emblematic <u>Unitarian Arms</u> provides the only exception to this rule. Apart from that print, the Unitarians are attacked through Priestley as their great representative figure. Of their other leaders, only Dr. Joseph Towers features at all prominently and then primarily in a political role. Priestley alone encapsulated for the satirists all the vices of heterodoxy and dissent, and his appearances in caricature embraced all the themes which Sayers had incorporated in his crucial print. Although he was at all times an emblem of all that was obnoxious about Unitarians in these caricatures, they may, at least for the period up to his emigration to America in 1794, be treated in three separate categories: those which feature his (supposed) alliance with Charles James Fox; those which highlight him as an English Jacobin and associate of Thomas Paine; and those which focus on him as a Dissenting minister preaching an evil message and who is made to suffer for it in the Birmingham riots. The categories are partly for convenience, but they also offer three rather rough markings on the Richter scale of opprobium.

So long as Priestley appears in the company of the larger-than-life Charles James Fox, there is always the chance of a bit of a laugh or snigger. Since Fox was a much more public figure than Priestley, these caricatures predominate in the Priestley corpus. They also demonstrate Gillray's satirical genius and his ability to cast a plague on both houses. The first extant caricature drawing attention to the (supposed) connection between Priestley and Fox is a print by Dent of the Westminster election of 1790. [Pl. 8] Two years earlier, the Whigs had spent perhaps as much as £50,000 to defeat Lord Hood, but chose in 1790 to avoid such expense by concluding a deal with Hood whereby he and Fox agreed to share rather than contest the two seats. Dent

Cromwell in the <u>Sancta Sanctorum</u> and by the puritan hats, one worn by a demon above the sounding board and another replacing the crown on the royal coat of arms in the pew at the bottom right of the picture. Appropriately that archetypal English American, Thomas Paine, is one of the hearers. He can be identified as the exciseman guaging a communion cup. This is his first appearance in English political caricature. His <u>Rights of Man</u> had yet to be published, but scattered around the chapel are the subversive works of Price and Priestley and other emblems of treason and subversion, in which the caricature abounds. These include the headless lion on the coat of arms, the church and steeple being pulled down in the distance, the swooning woman below the pulpit holding a book marked, 'Margt. Nicholson her Book' (Margaret Nicholson was an insane woman who had attempted to assassinate George III in August 1786) and the despoiling of the Bishop's chest in the foreground. Below these pregnant symbols of revolutionary sectarianism Sayers inscribed a fervent prayer for the preservation of 'the church and throne' from 'such implacable Tormentors, Fanatics, Hypocrites, Dissenters, Cruel in Power, and restless out'.

The Repeal of the Test Act a Vision created quite a stir, and received special notice in the press. The <u>Public Advertiser</u> detected 'a trait, quite à la Hogarth' in what it interpreted as the Long Parliament soldiers pillaging the chest of the Church of Canterbury', but it was uneasy about the ridicule of 'Dr. Price and one or two other Dissenters' in libellous caricatures. The <u>St. James's Chronicle</u> had no such reservations. It took immediate notice of the print, and was happy to report that it contained ' the most forcible strokes of satire that, since the time of Hudibras, have been aimed at the cause of fanaticism'. Thus in different ways contemporaries perceived that this was, as Dorothy George has subsequently noted, a key print. Not only was it the most complete indictment of the Dissenters printed before the motion for the repeal of the Test and Corporation Acts was put on 2 March, but also it was the first sign (in caricature) of a reaction to the French Revolution. Sayers objected in principle to the Dissenting Ministers preaching politics and objected even more fiercely to what they preached: their favourable views of recent revolutions, their attacks upon the church articles, and their heterodox theology, especially that of Priestley, all of which constituted not merely a criticism of the <u>status quo</u>, but an evil threat to the established order in Church and State. He suggested that these atheistical regicidal republicans had to

shows Fox and Hood in an ecstatic embrace. Priestley is a minor figure
appearing on one of the emblems representing the stages of Fox's cynical careerism.
The second flag on the left designated 'Dissenting Interest' shows Fox in another
happy embrace, this time with a Dissenting minister resembling Priestley: Fox
clasps a book entitled 'New Faith', Priestley one entitled 'Hypocrisy'.

Within a year of the publication of Dent's
print, Priestley had become a major figure representing Fox's attachment to
subversive values. Sayers, who had already set the pattern, returned to the
political fray in only his second (extant) print since the Repeal of the Test Act a
Vision. Published on 12 May 1791 with the title Mr. Burke's Pair of Spectacles
for Short Sighted Politicians, it was another complex design, heavily symbolic and
also prophetic in that it foreshadowed the Whig split over two years later. [Pl. 9]
A hand is proferring a pair of Burke's spectacles to Fox and Sheridan who
desperately need to have their sense of perspective restored to them, for they are
attacking a tree symbolising those things which are the source of their own power
and authority: the crown, the church and the great Whig families. Two of the
escutcheons on the branches are inscribed hereditary nobility, and the other two
bear the arms of the Cavendish and Portland families. Little wonder that
Portland, sitting on a log views the scene with the utmost apprehension. The
Whig lumberjacks, however, ignore his dismay and are assisted in their iconoclasm
by Priestley. Mounted on an infernal charger, he tilts at religion and the church
establishment as represented by a mitre and chalice resting upon a Bible. Price,
who had recently died, rises bewigged from the grave, and rejoices in words which
he had used to welcome the French Revolution: 'Lord lettest now thy Servant depart
in Peace ... (for) mine eyes ... (have s)een they Salvation'.

Price had uttered those words on 4 November
1789, the anniversary of the Glorious Revolution, but that anniversary had quickly
been eclipsed by the anniversary of the fall of the Bastille, 14 July 1789, an event
which transcended all others in symbolising the triumph of Liberty. Two days
before the anniversary celebrations in 1791, Dent looked forward to the event in a
print entitled Revolutionary Anniversary or Patriotic Incantations. [Pl. 10]
In this poorly drawn though vigorous design, Priestley, Fox, Towers and Sheridan,
like the witches in Macbeth, prance around a cauldron. From the brew they call

up French spirits whose devilish features are obscured by a toppling crown. Pictures above the four sorcerers symbolise violence, mob rule, republicanism, or 'republicism' as it is called, and fanaticism through the figures of Jack Cade and Wat Tyler, a divine, resembling Towers, smashing a crown and sceptre with an axe, and a devil, resembling Priestley burning a church and trampling on a mitre. Priestley, holding a copy of 'Rights of Men' in his right hand, chants,

> O choice Spirit of dauntless Paine,
> Make, make our Cauldron blaze again.

Fox and Sheridan look forward to the chaos and opportunity afforded by the revolutionary spirit, but it is Priestley and Towers who have their targets marked out.

Gillray with his greater satirical genius commented on the celebrations in two prints, one of 19 July 1791 with the title, The Hopes of the Party Prior to July 14th 'From such wicked Crown & Anchor-Dreams, Good Lord Deliver Us', [4] the other of 23 July 1791 with the title, A Birmingham Toast, as given on the 14th of July by the -------- Revolution Society. In the former, Fox has taken on the role of a hesitant and somewhat cowardly executioner of George III. Priestley leans down and offers the king comfort in a parody of his religious and political philosophy:

> Don't be alarmed at your situation, my dear Brother; we must
> all dye once; and therefore what does it signify whether we dye
> today or tomorrow - in fact, a Man ought to be glad of the
> opportunity of dying, if by that means he can serve his Country,
> in bringing about a glorious Revolution:- & as to your Soul, or
> any thing after death don't trouble yourself about that; depend on
> it, the Idea of a future state, is all an imposition: &c. as every-
> thing here is vanity & vexation of spirit, you should therefore
> rejoice at the moment which will render you easy and quiet.

In his left hand, he holds the source of this stoic philosophy in a paper marked 'Priestley on a Future State'. As in many Gillray caricatures, there are no real heroes or even true victims. An irritable George III, impervious to his likely fate, asks, 'What! what! what's the matter now'?, while Pitt and Queen Charlotte hang in their last spasms from nearby lamp brackets.

The association of Priestley with king-killing is even more explicit in Gillray's second print on the anniversary theme. [Pl. 11]. Priestley offers a toast to 'The (King's) Head here!' and holds up an

empty communion dish for the purpose. To complete the blasphemous tenor of
the proceedings, he raises a brimming chalice for the toast. All except Sir Cecil
Wray, to the immediate left of Priestley, join in heartily, and Lindsey, on the far
left, in his cups, pronounces 'Amen! Amen!'. Behind him a group of
sanctimonious Dissenters pray for deliverance from 'Kings & Whores of Babylon'.
Their jaundiced view of the church is emphasized by a picture of St. Paul's as seen
from Hackney, where the Rational Dissenters had established an academy. It is
portrayed as a monumental pigsty. Note that there is not a mention of, nor an
allusion to, the real happenings in Birmingham. We may concede that Gillray had
engraved the first anniversary plate before he had heard of the goings-on there, but
not the second, published seven days after the riots began. Indeed, in it Gillray
accepted the inaccurate report, published in The Times on 19 July though subsequently
retracted, that Priestley had attended the Birmingham Bastille Day dinner and had
proposed the toast, 'The King's head on a charger'. There was here, undoubtedly
a failure of sympathy and percipience on the part of Gillray. Yet, given his
prejudices, it was only to be expected. At least he did not exult in Priestley's
misfortune.

For a while, Gillray left Priestley alone,
returning to the regicidal theme only when he heard the news of the assassination of
Gustavus III of Sweden. Six days after the news of that event reached London, he
published a print, Patriots Amusing Themselves;-or-Swedes Practising at a Post,
in which he took a swipe at both the King and the opposition. [Pl. 12] Fox,
corpulent as ever, is practising with a massive blunderbuss at hitting a post
resembling George III, capped by a wig and a hunting hat bearing the royal coat of
arms. The target is the king's behind. Sheridan, preparing his pistol, looks
forward to the moment when he has popped the post and can sing 'Ça-ira' merrily
with 'Dear Brother P.'. Priestley, for his part, holds two books, 'On the Glory
of Revolution' and 'On the Folly of Religion & Order', which he offers to Sheridan as
rather improbable wadding 'to ram down the Charge with, to give it force, & to make
a loud report'.

When Gillray next caricatured Priestley with
his foxy friends, he could not afford such a light touch, if the latter may be so-called,
for Louis XVI had been guillotined, France was in the grip of revolutionary turmoil

and Europe, including Britain, was engulfed by war. Entitled, Dumourier dining in State at St. James's on the 15th of May, 1793, it shows three familiar traitors, wearing bonnets rouges, serving up Britain's establishment garnished by frogs to the French general who occupies the King's chair at the royal dining table. [Pl. 13] Fox proffers Pitt's head, Sheridan the Crown, and Priestley a mitre. Dumourier, with the lean and hungry look of a modern Cassius, looks on the dishes with relish. In fact, the hero of Valmy had suffered a serious defeat at Neerwinden shortly before this print was published and was soon to defect to the Austrians. Gillray dedicated this print to the Association for the Preservation of Liberty and Property against Republicans and Levellers set up by John Reeves in November 1792. It was perhaps his fulsome inscription that led Dorothy George to suggest that the caricature represents an ironical attitude towards the loyalist society. But the irony, if such, is directed at the society and not at the values which it sought to preserve, for the dangers represented by the tatterdemalion Dumourier and the trio of fellow-travellers are not portrayed as chimeras of an overheated loyalist imagination. This is surely confirmed by another print published nine days later in which Gillray allowed himself the luxury of a hero. With an artistic economy which grows with the urgency of the situation, he pictures Britannia between Scylla & Charybdis: Pitt is navigating confidently the good ship 'The Constitution', with a fearful Britannia as a passenger, between the rock of Scylla topped by a liberty pole and bonnet rouge, and the whirlpool Charbydis ingeniously designed as an inverted crown. [Pl. 14] He is pursued unsuccessfully by three sharks, the dogs of Scylla, in the guise of Priestley, Fox and Sheridan. In the distance lies the haven of public happiness. Ironically, Priestley would have much approved of the navigation of the ship of state towards that goal.

Gillray portrayed Priestley twice more before he emigrated, in prints which showed that the caricaturist had recovered his sang-froid and sense of detachment. The first, Blue and Buff Charity; -or-The Patriarch of the Greek Clergy applying for Relief, satirizes the subscription to be raised to bale Fox out of his financial plight. [Pl. 15] The beggarly Fox is shown receiving gratefully worthless paper notes which tumble from the devil's talons into his bonnet rouge. His followers likewise wait in expectancy while keeping their daggers at the ready. Priestley is in the middle of the second rank.

Plate 1a   'Tale of a Tub'

Plate 1   B.M. 7887

Plate 2 B.M. 7632

Plate 3   B.M. 8114

Plate 4   B.M. 7628

Plate 5   B.M. 7629

Plate 6 B.M. 7630

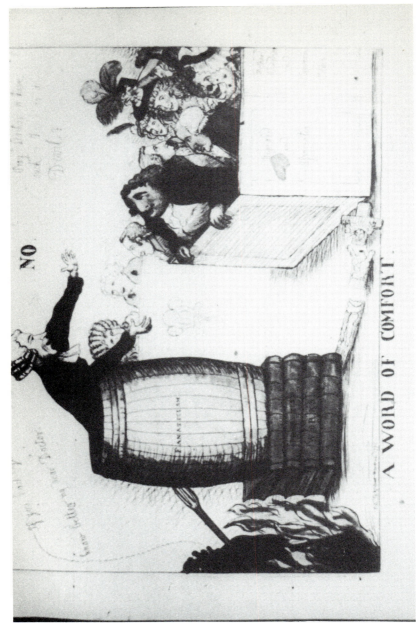

Plate 7   B.M. 7636

Plate 9   B.M. 7858

Plate 10   B.M. 7890

Plate 12   B.M. 8082

Plate 13 B.M. 8318

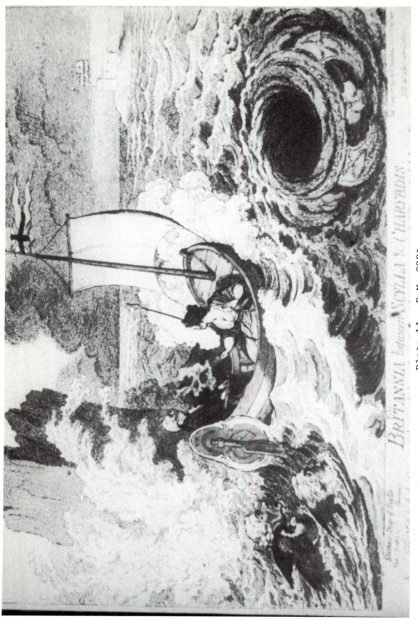

Plate 14   B.M. 8320

Plate 16 B.M. 8137

Plate 17 · B.M. 8131

Plate 18   B.M. 7896

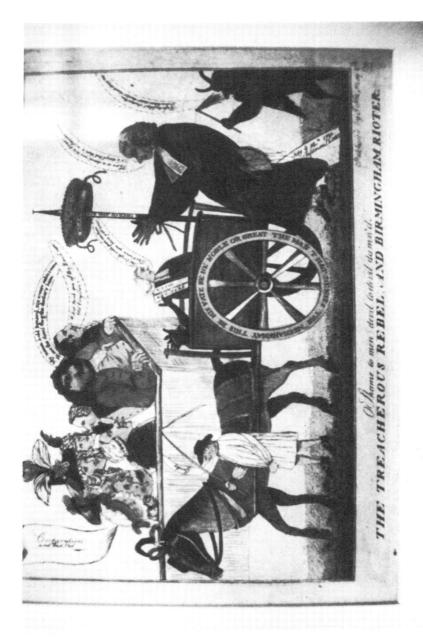

Plate 20 'The Treacherous Rebel and Birmingham Rioter'

Sheridan urges Charley to hurry up, for they have plenty of needy supporters, including 'old Phlogistick the Hackney Schoolmaster (who) expects some new Birmingham halfpence', an allusion to Priestley's claim for damages in the Birmingham riots which had been settled but which had yet to be raised. The final Gillray is a slightly lewd print on the theme of sedition, infidelity and the foreign foe.[5] It recalls the old unholy alliance between Catholic France and Infidel Turkey. Now, tricolour and crescent unite. The Turkish plenipotentiary presents his phallus-shaped credentials inscribed 'Powers for a new Connecion between the Port, England and France' (the print is subtitled, 'The final resource of French Atheists'). The king is alarmed by the object before his eyes, the queen hides her face but peeps through her fan, while the three elder princesses giggle and peep. Fox, Sheridan and Priestley follow in the train of the plenipotentiary. Priestley bows in obeisance but like the others keeps his eyes on the amazing proceedings. Only Pitt suffers from this caricature - he is portrayed as a fearful, naked manikin chained from his waist to a crown and orb - which introduces a farcical, and fanciful element into the debate on the French Revolution.

It is noteworthy that Gillray chose Priestley as the ideal subversive companion for Fox and Sheridan. In this trio, Priestley provided all the necessary resonances of regicidal, anti-clerical, anti-religious and anti-social philosophy and tradition. Thomas Paine, the bête noire of the government, did not feature much in their company, although in one Gillray which appeared in two versions in November and December 1792, Fox and Priestley are portrayed as his guardian angels. [Pl. 16] Paine was due to be tried on 18 December 1792 on a charge of seditious libel for the publication of Part II of the Rights of Man in February of the same year. Paine, who had in fact already escaped to France, is shown sleeping on a bed of straw. Despite his impassive countenance, his clenched hands and arched feet suggest that he is troubled by a brilliant vision of true justice represented by the scales of justice surmounted by shackles and a scroll inscribed, 'The scourge inexorable and the tort'ring hour awaits thee!'. This is Paine's 'Nightly Pest', the title of the print, which is so terrifying that the devil drops his bow and the music of Ça-ira, and flees fiddle in hand out of the window.

The association of Priestley with Painite republicanism was caricatured much more earnestly and graphically in a print published by Isaac Cruikshank at about the same time as the 'Nightly Pest'. [Pl. 17] Priestley is shown at a table, presided over by a devil, plotting with Paine. He shines a lantern to reveal a cluster of pistols, a musket and sundry knives. Paine, sitting on a barrel of gunpowder, waves his knives at Priestley, who offers him a dish full of phosphorus. In the background are emblematic pictures of regicide, republicanism and violence, while the room is littered with instruments for the same purpose. This is one of the few prints which makes play of Priestley's scientific eminence. To his right, there is a blunderbuss marked 'Royal Electric Fluid', and behind him, amongst tomes of treason from classical times on, there is a massive volume marked, 'Electrical Batteries so contrivd to Distroy any Assembly or Member at Pleasure'. Although the print is entitled, <u>Friends of the People</u>, which was the name of the moderate Whig society for parliamentary reform, Priestley is here firmly associated with the darker side of democratic reformism, and with the vile contemporary happenings in France (one of the pictures is of a guillotine). This is the only print which makes the connection with Painite, pro-French republicanism starkly and singularly. An earlier print by Sayers had, however, drawn attention to the involvement of one of Priestley's family in events in France. William Priestley, much to the concern of his father, had applied for French nationality in a speech before the French National Assembly in June 1792. In Sayer's satire on the occasion, William is being introduced as a good son of Dr. Priestley the great democrat. [6] He is held puppet-like before the members who have been given the heads of animals: two frogs, a mule and an owl. William declares, 'Papa sends me to you for Improvement I will bear true Allegiance'. In his left hand, he holds a firebrand and in his right, an electrical rod which is connected by a chain to a jar marked Phlogiston from Hackney College'. Although this was not one of Sayer's most convincing or weighty prints, it was nonetheless damaging to Priestley who was perhaps fortunate that the caricaturists did not develop this theme, given his reported trip to France later in the year, his election as a deputy for the Departments of Orne and Rhône-et-Loire in September 1792, and the increasingly panicky fears of republicanism at home.

It was neither Priestley's supposed association with Fox, nor that with Painite pro-French republicanism which brought out the full extent of the hostility of the caricaturists to him, but paradoxically, and perhaps this is a commentary on the art of the caricaturist, his own sufferings in the Birmingham riots. We have seen that Gillray's <u>A Birmingham Toast</u>, printed at the time, showed not a trace of sympathy towards Priestley. In the prints which relate directly to the riots, he is shown as a victim deserving his fate. One of these prints makes the point by naturalistic portrayal rather than by caricature proper. [Pl. 18] Its starting point is the story that at the time of the riots a local squire organised the burning of Priestley in effigy to avert more serious destruction. The print itself, however, shows the riots in full spate. A lifelike effigy of Priestley looks down from the stake with concern at the flames licking his legs, but, so it appears, with even more concern at the books and objects being used to fuel the flames: the <u>Rights of Man</u>, the <u>Bible (by) Priestley</u>, <u>Sermons</u>, and a picture of Oliver Cromwell. The mob carry banners of Church and State, they hurl stones at a meeting house, and flames in the distance depict the havoc that they have already wrought. The orgiastic mobsters are hardly heroes, but neither is Priestley. The print is entitled <u>The Exalted Reformer</u> and it appeared as an illustration to 'The Un-Priestley Divine in Effigy'.

Isaac Cruikshank, in his reaction to the Birmingham riots, drew Priestley as a wolf in sheep's clothing hanging from a gibbet. [Pl. 19] Priestley with his last gasp is uttering the words, 'Damn the Church and State I hate them both the hand bill ah! that is the Divel!!! '. This was a reference to the mysterious inflammatory handbills which circulated in Birmingham before the commemoration dinner on 14 July. But Priestley was not allowed thus to exculpate himself from a fate he clearly deserved; the print is entitled <u>Self Murder or the Wolf tried and Convicted on his own Evidence</u>. On the gibbet is a serpentine Dr. Price firing at the church from a canon marked 'J.P.'. Below, snakes spit out at Priestley venom labelled, 'Enthisism', 'Fanaticism', 'Athism' and 'Sedition'. To his left a whole rogues gallery of serpentine Dissenters is hatching out with the assistance of a little watering from a demon. The most prominent of these is Lindsey, who holds up his arms and exclaims, 'Believe me the Church of England which they thought they were supporting has

Received a greater Shock by their (the rioters) conduct than all our Brethren have aimed at'. The snakes remain deaf to his pleas, and his neighbouring minister, possibly Kippis, admits, 'To be sure we have had a Nock or two at it as well as the Constitution'. Above the Dissenters, a bishop warns a parson to beware of wolves in sheep's clothing, and the parson replies that it was Priestley, Price and Lindsey who had brought 'a stigma on all Dissenters in General tho' it is only the followers of those three Blasphemers who have made the Disturbance'. That assessment, which was at least somewhat discriminating, was not shared by Cruikshank. He suggests that ceaseless vigilance against <u>all</u> Dissenters was necessary. Under the serpentine divines, he wrote, 'If we are destroyed an Hundred will appear', and on the left he introduces John Wesley into the fray (Wesley had died recently on 2 March 1791). Appearing out of a cloud of smoke, the horned and tentacled evangelist announces, 'I now come forward in a Glorious Cause'. Cruikshank implied that Priestley deserved his fate, that the Rational Dissenters were particularly pernicious, but that Dissenters in general had to be watched.

The print which attacked Priestley most personally after the riots forms that nasty sequel, referred to earlier, to Dent's print, <u>A Word of Comfort</u>. The sequel is entitled <u>The Treacherous Rebel and Birmingham Rioter</u>, and is presumably also by Dent. [Pl. 20] Priestley appears before the same congregation, with the exception of the Prince of Wales and Mrs. Fitzherbert, although this time behind the poxy Mother Windsor and her ladies of easy virtue there is an inscription, namely, 'Corporation and Test Act'. Windham (or Hood), next to Fox, taunts Priestley, 'What think you of the Devil now old Conjurer? While Fox himself, looking foxier than ever, exclaims in bold print,

> My name's bold Renard, six scoure miles i come
> To see the devil flog the doctor's bum.

And that is indeed what is happening: Priestley tied to a maypole marked 'Church and State' on the pole and 'Treason' and 'Blasphemy' on the crown, and pulled along behind a horse and cart, is being whipped by the devil. His blood falls towards a paper on which there is written, 'July ye 14th. 1791 Infamous Libel'. He confesses,

> There is a Devil yes 'tis true
> No doubt but there's a Saviour too.

A cleric in the cart tells him,

> The cross with patience bear
> The crown you soon will wear.

But this is no martyrdom, rather Priestley is receiving his just deserts, as is indicated by the inscription on the cart wheel, 'The man that denieth the messiah may this be his fate be he noble or great'.

The Treacherous Rebel and Birmingham Rioter represents the savagest attack in caricature on Priestley. He is here not naive or misguided as in A Word of Comfort, rather he is evil, malign and thoroughly deserving of his fate as is indicated by the subtitle of the print, 'Shame to men: devil to devil damn'd'. His disreputable, if influential congregation are either amused or are indifferent to his condign punishment. Yet the print is more than a vicious caricature, it is a travesty. Priestley in his unorthodox way was a profound Christian, though he would never have wished such a fate upon anyone who denied the Messiah. The effect upon judgement of the prejudices which existed against him could not have been more graphically illustrated.

After Priestley's emigration to America, he naturally vacated the public stage in England and caricaturists ceased to bother him. He is depicted in only two extant prints of the time, both by Gillray, who may even have felt a touch of nostalgia at the departure of such an ideal subject. The first of these prints appeared in 1798 as a result of Gillray's association with George Canning, the editor of the Anti-Jacobin Review [Pl. 21]. Priestley plays a cameo role in a complex plot illustrating Canning's poem, New Morality. The targets are familiar: republicanism and atheism and a host of their associates and associations. Amongst a bewildering array of characters and symbols, there can be detected a delightful little caricature of Priestley standing by a 'Cornucopia of Ignorance'. He is holding a copy of his 'Political Sermons' and wears a bonnet rouge. Gillray alas does not allow him such a dignified appearance in his final caricature. In a satire on the European situation in 1808 with the picturesque title of The Apples and the Horse Turds; ----- or ----- Buonaparte among the Golden Pippins, Priestley has become, so it appears, but a name on a dunghill of republican horse turds. [7] He was in good company, for the turds include not only his old pals in caricature, Fox and Sheridan, but also a more real friend, Price, as well as Godwin, Rousseau, d'Alembert and Voltaire.

The caricatures of Priestley and his fellow Rational Dissenters demonstrate the extent to which in the eyes of the caricaturists they had become a danger to the community. At best, Priestley was a misguided divine, naive in his religious beliefs and a dupe of party politicians. At worst he posed an insidious threat to established values, was a dangerous ally of party politicians, an English Jacobin of the first water, and a preacher whose wicked message had to be countered if necessary by violence in order for the nation to be preserved. It was, however, only a unique conjunction of events which brought him to the forefront of affairs in the early 1790s. Until then, caricaturist generally ignored his rhetoric and his beliefs, which nonetheless they quickly turned to damaging effect against him. The depth of hatred which he then evoked is symptomatic of the threat which it was believed the Rational Dissenters posed to the existing establishment in Church and State and the extent to which Priestley himself symbolised their extensive and dangerous virtues at that most pressing time. It is remarkable that not one of the caricatures of Priestley shows him in a favourable light, and even more remarkable that no caricaturist chose to satirize the opponents of Dissent. Priestley's great opponent in controversy Samuel Horsley, whose attack on Priestley could so easily have been portrayed in terms of his ambition to achieve a bishopric and who was a ready-made target for counter-attack, did not appear in caricature until the mid 1790s and then for quite different reasons. Since Priestley and the Rational Dissenters were able and tireless propagandists with excellent contacts in the printing and publishing industry, one must ask why they failed to enlist at least one caricaturist for their cause. Without precise evidence, one can only conjecture that it is unlikely that they ever tried. They aimed to pursue truth by candid free inquiry and this excluded satire. For them, honest rational argument was the test of truth. It is certainly conceivable that a caricaturist might have stepped forward from their own ranks. There were a number of artists within or associated with the Rational Dissenting community. Of these, perhaps only Henry Fuseli had the ability to turn himself into a caricaturist of note. But inclination rather than artistic skill was the essential requisite of the caricaturist, and Rational Dissent offered its adherents and supporters no encouragement to turn to caricature to reply to its critics in kind.

There remains one final question to answer, which is, 'Why were the caricaturists themselves so unsympathetic to the Dissenters' cause in general and to Priestley in particular?' Dorothy George's excellent survey of political and social satire reveals that as religion began to lose its immediate and compelling political significance following the death of Queen Anne, the caricaturists took up a general posture of indifference towards religious dissent, but when provoked to comment did so only to reveal their deep hostility. Curiously, although the caricaturists were very free with their criticism of the Church, they could find nothing to commend in Dissent, and indeed opposed fiercely any moves towards the extension of toleration: in 1753 and in 1779/80 they were against any improvement in status for the Jews and Roman Catholics respectively. In contrast with their willingness to defend political dissent, they were not disposed to be open-minded about religious dissent. The latter conjured up all sorts of bogeys which threatened political stability, religious certitude, moral order and social well-being. Hogarth was probably typical in his unfussiness about the precise nature of the dissent in question. He altered one of his prints, Enthusiasm Delineated, which was an attack upon Popery, into a satire on Methodist enthusiasm with the title of Credulity, Superstition and Fanaticism. A Medley (15 March 1762).[8] It is true that caricaturists seemed more indifferent to the Dissenters than to other religious minorities, but too much should not be read into this for they tended to be slow to react. In the cases of the Act for the naturalization of foreign Jews of 1753 and of the Catholic Relief Act of 1778, they only began to comment when change was imminent or shortly after it had taken place. In this sense, caricaturists acted as a barometer of public opinion, turning their attention to religious dissent only when it began to play a significant role on the political stage. This is what happened to Priestley and the Rational Dissenters in the early 1790s. Unfortunately for them, the caricaturists also acted as indicators of public prejudice. Despite their willingness to serve a cause if paid well, one suspects that it would have been difficult for the Rational Dissenters to have obtained the services of a single caricaturist, especially once the disenchantment with the French Revolution set in. Unlike the cunning Fox or the oppressive Pitt, Priestley possessed no redeeming features for the caricaturist. But if he provided them with a marvellously rich image for obloquy, it now requires a good deal of historical knowledge to recapture the contemporary impact of their assaults upon his character and thought. If they helped to make his closing years in England uncomfortable, they did not seriously damage his reputation. It would no doubt have given Priestley pleasure to reflect that after all caricature is one of the more ephemeral arts.

ACKNOWLEDGEMENTS

Permission was kindly granted for the reproduction of caricatures by the following: Dickinson College Library, Carlisle, Pennsylvania (plate 20): Professor Derek Davenport (plates 9, 15 and 16): and by Dr. D.O. Thomas (plates 1a and 5). All the remaining plates were reproduced from the microfilm edition of English Cartoons and Satirical Prints in the British Museum with the kind permission of Mr. Charles Chadwyck-Healey and the Trustees of the British Museum. I am grateful to Martha C. Slotten and Marie Ferre for their assistance in reading plate 20. I wish to thank Professor Davenport for his most useful handlist of Priestley in caricature and for his kind provision of slides from his collection of Priestley caricatures. I have also benefitted greatly from the advice and assistance of Dr. D.O. Thomas, Professor P.D.G. Thomas and Mr. Richard Brinkley.

FOOTNOTES

Because of the constraints of space, it has been necessary to limit the footnotes to references to caricatures which have not been printed here. All except one of the caricatures of Priestley can be found in the British Museum collection and the museum catalogue number has been indicated on the plates and in the footnotes. Readers requiring further information should refer to the relevant volumes of M. Dorothy George, Catalogue of Political and Personal Satires Preserved in the Department of Prints and Drawings in the British Museum, or to her standard volumes, English Political Caricature to 1792 : English Political Caricature, 1793-1832 : A Study of Opinion and Propaganda (Clarendon, Oxford, 1959). A copy of the complete footnotes to this article may be obtained from the author.

[1] M. Dorothy George, English Political Caricature to 1792, plate 36.

[2] B.M. 5644; Dorothy George, English Political Caricature to 1792, plate 57.

[3] B.M. 7635; F.W. Gibbs, Joseph Priestley. Adventurer in Science and Champion of Truth (Nelson, London, 1965), plate 15.

[4] B.M. 7892; Draper Hill (ed), The Satirical Etchings of James Gillray (Dover Publications, Inc., New York, 1976) plate 21.

[5] B.M. 8356; Draper Hill (ed), The Satirical Etchings of James Gillray plate 33.

[6] B.M. 8108; Gibbs, <u>Joseph Priestley</u>, plate 16.
[7] B.M. 9522.
[8] M. Dorothy George, <u>Hogarth to Cruikshank : Social Change in Graphic Satire</u> (Allen Lane, London, 1967), plate 15.

Priestley in America: 1794 — 1804

By D. A. Davenport
DEPARTMENT OF CHEMISTRY, PURDUE UNIVERSITY, WEST LAYFAYETTE, IN. 47907, U.S.A.

The possibility that Joseph Priestley might one day emigrate to America occurred as early as 1772 when in a letter to Richard Price he wrote from Leeds:

> Indeed, my place is such, that, according to present appearances, the only motive I can ever have to remove is, that, agreeable as my situation is with respect to myself, it affords me no prospect of making any provision for a growing family. I have therefore thought that, if ever I do remove, it must be to America, where it will be more easy for one to dispose [of] my children to their advantage. (1)

That he would have been welcomed in the incipient republic is clear from a letter written less than a year later by John Winthrop in Massachusetts to Benjamin Franklin in London:

> I am extremely concerned to hear that Dr. Priestley is so poorly provided for, while so many are rolling about here in gilt chariots, with very ample stipends. I admire his comprehensive genius, his perspicuity and vigor of composition, his indefatigable application, and his free, independent spirit, and wish it were in my power to do him any kind of service. It would give me great pleasure to see him well settled in America; though indeed I am inclined to think he can prosecute his learned labors to greater advantage in England. A man of his abilities would do honor to any of the colleges. At present there is no vacancy among them; but if there were, I believe, Sir, you judge perfectly right, that his religious principles would hardly be thought orthodox enough. Indeed, I doubt, whether they would do at the Rhode Island College, any more than in the others. That college is entirely in the hands of the Baptists, and intended to continue so, and I never understood that Dr. Priestley was of their persuasion. However, I cannot but hope that his great and just reputation will procure something valuable for him, and adequate to his merit. (2)

There were perhaps few gilt chariots rolling about New England at that time but the warning about effects of Priestley's religious and even political unorthodoxy (3) was prophetic enough. More than twenty years later, in Northumberland, Pennsylvania, Priestley was to lament "for such is the bigotry of the people in this part of the country that tho, in every other respect, my reception here has been very flattering, their pulpits are all closed to me". (4) Not long afterwards President John Adams' Secretary of State, Timothy Pickering, was threatening Priestley with imprisonment under the Alien and Sedition Laws. (5) Not all things apparently had suffered a sea-change. However ever since John Donne's "O my America! my new-found-land" America had been as much a dream landscape as a geographical location and Priestley was to remain guardedly loyal to his adopted land until his death.

After Franklin's departure from England in 1775 and the subsequent seven year war for independence, Priestley's American connections understandably grew weaker. The two men were never to meet again though they carried on an affectionate if sporadic correspondence until Franklin's death in 1791. Priestley's scientific high-noon in the service of the Earl of Shelburne followed by his increasing involvement in political and theological controversy during the Birmingham years drove any thoughts of emigration from his busy mind. When these revived in the early 1790's it was to France rather than to America that his thoughts initially turned. The only surviving letter from Priestley to Lavoisier, written in June 1792, closes with the words:

> In case of more riots, of which we are not without apprehension, I shall be glad to take refuge in your country, the liberties of which I hope will be established notwithstanding the present combination against you. I also hope the issue will be as favourable to science as to liberty.... (6)

Shortly afterwards he was elected an honorary citizen of France. (7) Less than a year later with the revolutionary tide turning crimson in France he was to change his mind. As he wrote to William Withering:

> I wish the country was in a better state to invite your return to it, but it is far otherwise, at least with respect to myself and those who have generally passed for the friends of liberty. Such is the spirit of bigotry encouraged by the Court party, that great numbers are going to America, and among others all my sons, and my intention is, that when they are settled to follow them and end my days there. This will be a great mortification to me, after having replaced my apparatus, and recommenced my experiments, as I now have done. Indeed, to appearance, I have everything very comfortable about me, but I cannot get so much time to myself as I wish, and I have little intercourse with the members of the Royal Society.

> We have no intercourse now with France, and whether my
> son William has been able to leave it and go to America I
> cannot learn. Indeed, the prospect is very melancholy. The
> conduct of the French has been such as their best friends can-
> not approve; but certainly the present combination against
> them, which does not appear to have any other object than the
> restoration of the old arbitrary government, is as little to be
> justified.
>
> Wishing, tho' hardly hoping, for better times.... (8)

In that best of times and worst of times even the normally Panglossian Priestley could approach despair, though later he would have agreed with the verdict of his friend Thomas Cooper:

> They are a wonderful people; but in my opinion rather to be
> admired at a distance, than fit for a peaceable man to reside
> among....I look for happiness amid the attachments of friends
> and kindred; where the obligations of private society shall
> be inviolable; where I may talk folly and be forgiven; where
> I may differ from my neighbour in politics and religion with
> impunity; and where I may have time to correct erroneous
> opinions without the orthodox intervention of the halter or
> the guillotine. Such times may and will come in France, but
> I fear not before the present race shall die away. (9)

Early in 1794 the decision to sail for America was made:

> ...My own situation, if not hazardous, was become unpleasant,
> so that I thought my removal would be of more service to the
> cause of truth than my longer stay in England. At length,
> therefore, with the approbation of all my friends, without
> exception, but with great reluctance, I came to that resolu-
> tion [to leave England and settle in the United States].... (10)

His parting words were admirably in character:

> I do not pretend to leave this country, where I have lived
> so long, and so happily, without regret, but I consider it
> as necessary, and I hope that the same good providence that
> has attended me hitherto will attend me still. (11)

Joseph and Mary Priestley sailed from Gravesend on April 8 and arrived off Battery Point in New York on June 4, 1794. As usual he was fearsomely industrious during the long voyage:

> I read the whole of the Greek Testament, and the Hebrew Bible
> as far as the first Book of Samuel: also Ovid's Metamorphoses,
> Buchanan's poems, Erasmus' Dialogues, also Peter Pindar's poems,
> &c...and to amuse myself I tried the heat of the water at
> different depths, and made other observations, which suggest
> various experiments, which I shall prosecute whenever I get my
> apparatus at liberty. (12)

Priestley was given a hero's welcome to his adopted land. (13) He was attended by Governor Clinton, by Dr. Prevost, the Bishop of New York, by the

New York Association of Teachers, by the Democratic Society of the City of New York, by the yet respectable Tammany Society, by the Medical Society of New York and by the Republican Natives of Great Britain and Ireland. Their ponderous addresses of welcome and Priestley's almost equally ponderous replies have, alas, largely survived:

> While the arm of Tyranny is extended in most of the Nations of the world to crush the spirit of liberty, and bind in chains the bodies and minds of men, we acknowledge, with ardent gratitude to the Great Parent of the Universe, our singular felicity in living in a land, where Reason has successfully triumphed over the artificial distinctions of European policy and bigotry, and where the law equally protects the virtuous citizen of every description and persuasion. (14)

Newspapers joined in the praise:

> ....The name of Joseph Priestley will be long remembered among all enlightened people; and there is no doubt that England will one day regret her ungrateful treatment to this venerable and illustrious man. His persecutions in England have presented to him the American Republic as a safe and honourable retreat in his declining years; and his arrival in this City calls upon us to testify our respect and esteem for a man whose whole life has been devoted to the sacred duty of diffusing knowledge and happiness among nations.... (15)

It is all rather suffocating and one can sympathize with the anonymous reviewer in The Gentleman's Magazine for January 1795, approvingly quoted by William Cobbett, who remarked of "the dullness and ignorance of the addresses presented to the Worshipful Doctor, emulated by the same qualities in his answers". William Cobbett himself, under the pen-name of Peter Porcupine, was to prove much less generous. The Sunday following Priestley's arrival in New York happened to be Trinity Sunday. None of the city's many churches offered him the hospitality of its pulpit.

After two weeks in New York the Priestley family travelled to Philadelphia, at that time the intellectual and political capital of the young republic. Mary Priestley in particular found the city much less agreeable than New York. Though less fulsome than in New York, their welcome in the city, over which the spirit of Benjamin Franklin still reigned, was warm enough. Priestley had tea with President Washington and became acquainted with David Rittenhouse, Benjamin Rush and the other intellectuals who gathered around the American

Philosophical Society, of which Priestley had been a member since 1785. But the Priestleys were eager to move on and shortly they made the five-day journey to Northumberland in central Pennsylvania, not too far from the advancing American frontier and in Schofield's words a town "without any pretensions to liberalism, intellectualism or science." (16)

The late XVIII and early XIX centuries are replete with largely aborted utopian or communitarian experiments. (17) The pantisocratic community which Priestley hoped would assemble on the banks of the Susquehanna river was one of the most quixotic and least successful. In 1793 Priestley's oldest son, also Joseph, together with the Manchester-based radical Thomas Cooper, had visited America and had taken an option on a large tract of land in the Susquehanna valley some distance north of Northumberland. On his return Thomas Cooper wrote <u>Some Information Respecting America</u>, a book which adroitly mixes enthusiasm, observation, erudition and cupidity. (18) In this Cooper was a man very much of his times, for Americans in Paris, London, and Philadelphia - Gilbert Imlay, Joel Barlow, and Benjamin Rush for example - were all caught up in land speculation. The elder Priestley however saw in Joseph's and Thomas Cooper's actions the possibility of realizing his dream of a self-contained, emigré English colony, professedly Unitarian, on the banks of the Susquehanna. Many well-known figures - Coleridge, Southey and even the young Wordsworth - were caught up in his pantisocratic enthusiasm. Recognizing that eligible young women would be scarce on the American frontier Coleridge, Southey and a fellow-poet, Robert Lowell, made the decision to marry before leaving. They settled on three sisters and each proposed to, was accepted by, and married a Miss Fricker. But the schemes of poets as well as of mice and men 'gang aft a-gley'; Lowell died, Coleridge eloped with another woman, and Robert Southey was left with all three Misses Fricker on his hands. The plans of the Priestleys, Thomas Cooper, and their friends fared little better.

Perhaps because political tensions were easing in England few people joined the new colony. Addresses to the Federalist-dominated State Legislature for financial assistance for the establishment of a college predictably failed and when, after the elections of 1800, help was forthcoming it was too little and too late. The elder Priestley did not seem to mind:

> But of what importance is it where I was born, or whence I came... Here I am. Here is my family. Here is my property and everything else that can attach a man to any place. Let any man only view my house, my garden, my library, my laboratory, and the other conveniences with which I am surrounded, and let him withal consider my age, and the little disposition that I have shown to ramble any whither, and say whether any person among yourselves, or in the United States could remove with more difficulty, or with more loss, than I.

After first accepting and then reluctantly declining a call to become Professor of Chemistry at the University of Philadelphia (19) he resigned himself to the isolated life of Northumberland. In this resignation he was encouraged by the partly vain belief that his sons were now well-settled and far from the temptations of the big city, and by the contentment of his wife, Mary. Together they planned the fine two-storey house which still stands on the banks of the Susquehanna river but Mary Priestley was not to live to see its completion. She died in 1796 nine months after the death of Harry, the youngest of Priestley's three sons and seemingly his favorite. In spite of these twin blows Priestley decided to stay on in Northumberland though he made several extended visits to Philadelphia, the last in 1803. On that occasion he was invited to join members of the American Philosophical Society for a testimonial dinner. In his acceptance Priestley closed with the words:

> Having been obliged to leave a country which has long been distinguished by discoveries in science, I think myself happy by reception in another which is following its example, and which already affords a prospect of its arriving at equal eminence....

It would be many years, however, before that prospect was fulfilled.

Priestley was past sixty when he emigrated to America and it is clear that he sincerely hoped to stay clear of political entanglements and to spend his declining years in continuing his scientific investigations and writing his theological tracts and tomes. Left to himself he might possibly have succeeded but with William Cobbett as gratuitous enemy in Philadelphia and Thomas Cooper as friend, near-neighbour, and later house-guest, in Northumberland such hopes were bound to prove vain. Cobbett and Cooper make a strange pair, two crusading souls who somehow passed in the Pennsylvanian night. Cobbett, who wrote under the pseudonym Peter Porcupine, was in his early arch-Tory phase, damning with faint praise everything American and savaging with sulfurous damnations everything French or French-sympathizing. (20) In the usually unfair and always vituperative language of Peter Porcupine it is hard to foresee the measured prose of <u>Cobbett's Parliamentary Debates</u> or the bucolic resignation of <u>Rural Rides</u>. Thomas Cooper had earned his Jacobin spurs in the jousting-yards of Manchester politics and he arrived in America with his levelling spirit if anything more highly pitched than ever. As a young man he had spoken against the slave-trade to the Manchester Literary and Philosophical Society. In old age he was known as the Father of State's Rights in South Carolina. (21) Their fellow pamphleteers, Thomas Paine and Joseph Priestley, remained much more constant in their life-long beliefs.

Cobbett's first major work was <u>Observations on the Emigration of Dr. Joseph Priestley and on Several Addresses Delivered to Him, on His Arrival in New York</u> a magnificently unfair diatribe. (22) If only James Gillray had been around to illustrate it.

> It is no more than justice to say of these addresses, in the lump, that they are distinguished for a certain barreness of thought and vulgarity of stile, which, were we not in possession of the Doctor's answers, might be thought inimitable.

"In the lump", Cobbett is right. In a weak moment Priestley had bracketed "Marat, St. Paul and Jesus Christ". Cobbett leapt on this:

> If he did not foresee them [the horrors of the reign of terror] he must have an understanding little superior to that of an idiot; if he did, he must have the heart of a Marat. Let him choose."

He disapproves of Priestley's at times convoluted prose:

> This is neither the <u>style periodique</u> nor the <u>style coupé</u>, it is I presume the <u>style entortillé</u>: for one would certainly think the author had racked his imagination to render what he has to say unintelligible. This sentence of monstrous length is cut asunder in the middle by a semi-colon; which, except that it serves the weary reader by way of a half-way house, might be placed in any other part of the sentence to, at least, equal advantage.

Cobbett even anticipated Oscar Wilde's famous verdict on the death of Little Nell:

> There is something so pathetic, so irresistibly moving in all this that a man must have a hard heart indeed to read it, and not burst into laughter.

He goes on to question, on very dubious authority but with equal self-conviction, Priestley's scientific reputation:

> On this occasion the reader will please to bear in mind, that I am not pretending that we ought to dislike Dr. Priestley; for he is certainly as much entitled to our gratitude and esteem as [Benedict] Arnold was to that of the British.

The pamphlet closes with a flourish:

> The Doctor expresses a desire of one day returning to "the land that gave him birth" and, no offense to the New York addresses, I think that we ought to wish that this desire be very soon accomplished. He is a bird of passage that has visited only to avoid the rigour of an inclement season: when the re-animating sunshine of revolution should burst forth, we may hope to see him prune (sic) his wings, and take flight from the dreary banks of the Susquehannah to those of the Thames or Avon.

As Dumas Malone remarks from this "time forward the baying of the Federalist pack was never long silenced". Cobbett was to hound Priestley and Benjamin Rush, Priestley's friend and sometime physician, for the next five years. He was eventually sued for libel by Rush and was assessed the then enormous sum of five thousand dollars. He honored the debt but shortly afterward left to return to more peaceful pursuits in England. He could not resist one final attack, for in <u>The Rushlight</u> of April 1800 he addresses a long letter to Dr. Joseph Priestley. By the standards of Peter Porcupine, however, it closes almost elegiacally:

> Nay, Sir, go one step further with me: <u>confess your errors</u>; acknowledge that you were deluded, and were instrumental in deluding others. Make all the atonement in your power: return home, and tell what you have seen, Never was there a man, who had a fairer opportunity of evincing true greatness of mind, of exchanging contempt for respect, misery for happiness. You have a country that ever stands with open arms to welcome her wandering sons: turn not from her maternal embrace to the selfish hug of democracy: at any rate, dishonour not the name of Englishman by becoming the eleemosynary eulogist of a puffed up petty despot, whose person you must loath, whose actions you must reprobate, and whose character you must despise.

> With this admonition I bid you farewell, assuring you,
> that, while I rejoice at your public disappointments, there
> are very few who more sincerely regret your private calamities,
> and no one who more heartily despises your former panegyrists
> and present persecutors. (23)

Priestley made no public response to Cobbett's attacks until 1798 though he did complain privately to Lindsey of the phrase "I hope I shall see the malignant old Tartuff of Northumberland begging his bread through the streets of Philadelphia, and ending his days in the poorhouse, without a friend to close his eyes". On John Adams' election to the Presidency in 1796 the political climate in the young republic began to heat up. Priestley had first met Adams in 1784 when the latter was serving as the first American Ambassador to the Court of St. James's. The acquaintance was renewed in Philadelphia and Priestley dedicated the published version of his first series of Philadelphia sermons to President Adams. Somewhat later Priestley wrote a letter of support for his friend Thomas Cooper who had applied to Adams for an appointment in his Administration. Although the relationship between Adams and Priestley had by now clearly cooled there seems to have been no malice in the President's failure to reply. By 1798 the Federalist/Republican political confrontation - apart from the Civil War perhaps the most contentious period in the history of American politics - had polarized the country. Joseph Priestley was somewhat reluctantly drawn into the fray but Thomas Cooper clearly relished the opportunity. He took over the editorship of the <u>Sunbury and Northumberland Gazette</u> and quickly turned it into one of the most effective weapons in the Jeffersonian propaganda arsenal:

> I hope to afford some proof that I remain in this Country what
> I was in Europe, a decided opposer of political restrictions
> on the Liberty of the Press, and a sincere friend to those first
> principles of republican Government, the Sovereignty of the
> people and the responsibility of their servants. Having adopted
> these opinions on mature consideration, and the fullest convic-
> tion, I shall retain and profess them; but I am sorry to say
> they are likely ere long to become as unfashionable in this
> Country as in the Monarchies of Europe.
>
> The more we understand of the science of government the less
> necessity we find for governmental secrets. State-craft and
> priest-craft are fond of hidden mysteries; they delight in their
> esoteric and exoteric doctrines and measures; but hidden motives
> are always suspicious in a republican government. In such a

> government...secrecy is the child of misconduct and the parent of mischief.
>
> Is it a crime to doubt the capacity of a President? Have we advanced so far on the road to despotism in this republican country that we dare not say our President is mistaken? (24)

In fact they had. The U.S. Congress had recently passed "The Alien and Sedition Laws" which, though less draconian than those current in England, were none-the-less a powerful threat to civil liberties. Though no one was ever deported under the short-lived laws several Republican writers were arraigned for sedition. Timothy Pickering, who was Secretary of State in the Adams Administration, wrote complaining of one of "Mr. Cooper's mischievous addresses":

> As to Dr. Priestley, his conduct in this affair is wholly unpardonable. I once thought him <u>a persecuted Christian</u>: but I am now satisfied that <u>ambition</u> influences him, like the mass of seditions, turbulent <u>democrats</u>; and that no government which human wisdom could devise would ever make them contented, unless they were placed at its head. - I am sorry that Cooper like Priestley has not remained an alien. The <u>indecency</u> in these <u>strangers</u> - thus meddling with our government, merits a severe animadversion: I hope besides that Mr. Rawle will prosecute both. (25)

President Adams dissuaded Pickering from proceeding against Priestley with the scarcely complimentary words: "He is as weak as water, as unstable as Reuben, or the wind. His influence is not an atom in the world." He showed no such compassion towards Thomas Cooper and Cooper became one of the handful of people arraigned for sedition. After a palpably political trial he was sentenced to six months in jail and fined four hundred dollars. On his release he wrote with a composure worthy of Priestley and a truculence all his own of "a tedious imprisonment to which I had the honour of being sentenced for exposing some few among the errors of a weak, a wicked and a vindictive administration." (26)

On 17 February, 1801 Thomas Jefferson was, on the 36th ballot of the electoral college, declared the winner of the presidential election held the preceding December. The bitterness of the election campaign still rankled with John Adams and, after attempting to pack the judiciary, he rode out of town shortly before Jefferson's inauguration on 4 March, 1801. The two great men were not to be fully reconciled until 1813 when, thanks to the contrivance of Benjamin Rush, they began their autumnal correspondence which remains one of the chief glories of American letters. (27) But even then, almost ten years after his death, Priestley was to be caught up in controversy.

With Jefferson's assumption of the Presidency, Priestley found himself, perhaps for the first time, resident of a country of whose leaders and government he fully approved. One of the first letters Jefferson wrote from the Presidential Mansion, which in 1814 the British caused to be renamed the White House, was to an ailing Joseph Priestley:

> As the storm is now subsiding, and the horizon becoming serene, it is pleasant to consider the phenomenon with attention. We can no longer say there is nothing new under the sun. For this whole chapter in the history of man is new. The great experiment of our Republic is new. Its sparse habitation is new. The mighty wave of public opinion which has rolled over it is new. But the most pleasing novelty is its so quickly subsiding over such an extent of surface to its true level again. The order & good sense displayed in this recovery from delusion, and in the momentous crisis which lately arose, really bespeak a strength of character in our nation which augurs well for the duration of our Republic; & I am much better satisfied now of its stability than I was before it was tried. (28)

Priestley sent a copy of this letter to Theophilus Lindsey who ill-advisedly allowed its publication, much to annoyance of John Adams who understandably did not agree that the nation had narrowly avoided catastrophe. Twelve years later, Adams was to write of this letter in his correspondence with Jefferson:

> In your Letter to Dr. Priestley of March 21.1801, you "... disclaim the legitimacy of that Libel on legislation...." This Law, I presume was, the Alien Law....
>
> As your name is subscribed to that law, as Vice President, and mine as President, I know not why you are not as responsible for it as I am. Neither of Us were concerned in the formation of it. We were then at War with France: French Spies then swarmed in our Cities and in the Country. Some of them were, intollerably, turbulent, impudent and seditious. To check these was the design of this law. Was there ever a Government, which had not Authority to defend itself against Spies in its own Bosom? Spies of an Ennemy at War? This Law was never executed by me, in any Instance.
>
> But what is the conduct of our Government now [1813]? Aliens are ordered to report their names and obtain Certificates once a month: and an industrious Scotchman, at this moment industriously labouring in my Garden is obliged to walk once a month to Boston, eight miles at least, to renew his Certificate from the Marshall. And a fat organist is ordered into the Country. etc. etc.

etc. All this is right. Every Government has by the Law
of Nations a right to make prisoners of War, of every
Subject of an Enemy. But a War with England differs not
from a War with France. The Law of Nations is the same in
both. (29)

Adams was being somewhat ingenuous in trying to associate Thomas Jefferson with the passage of the Alien & Sedition Laws. The Laws were written by the Federalists in power and when passed Jefferson, as Vice-President and Chairman of the Senate, had no choice but to co-sign. Adams, however, does go on to say:

Scarcely anything that has happened to me in my curious life, has made a deeper impression upon me than that such a learned, ingenious, scientific and talented madcap as Cooper, should have influence enough to make Priestley my enemy.

In common with most of his other adversaries John Adams was never to forgive Thomas Cooper. On Cooper's headstone in the graveyard of Trinity Episcopal Church in Columbia, South Carolina we find the inscription: "Erected by a portion of his fellow-citizens, to the memory of Thomas Cooper, M.D. and L.L.D. ...." Cooper would have relished the qualifying "by a portion". (30) There are no qualifying phrases on the headstone of Joseph Priestley who, after initial burial in a Quaker cemetary, now rests on the banks of the Susquehanna River.

Robert Schofield characterizes Priestley's last ten years in America as 'Anti-Climax'. (31) As far as chemistry was concerned he is certainly correct for Priestley's many repetitive publications largely served to show that he was increasingly out of touch with the rapidly advancing science. But as for his other multifarious activities, they continued to the end and perhaps 'epilogue' is a fairer, and certainly a kinder, word than 'anti-climax'.

On Monday morning, the 6th of February, after having lain
perfectly still till four o'clock in the morning, he called
to me, but in a fainter tone than usual to give him some wine
and tincture of bark. I asked him how he felt. He answered,
he had no pain, but appeared fainting away gradually. About
an hour after he asked me for some chicken broth, of which
he took a tea cup full. His pulse was quick, weak, and
fluttering: his breathing, though easy, short. About eight
o'clock, he asked me to give him some egg and wine. After
this he lay quite still till ten o'clock, when he desired me

and Mr. Cooper to bring him the pamphlets we had looked out the
evening before. He then dictated as clearly and distinctly as
he had ever done in his life the additions and alterations he
wished to have made in each. Mr. Cooper took down the substance of what he said, which when he had done, I read to him.
He said Mr. Cooper had put it in his own language; he wished
it to be put in his. I then took a pen and ink to his bed-
side. He then repeated over again, nearly word for word,
what he had before said; and when I had done, I read it over
to him. 'That is right; I have now done.' About half an
hour after he desired, in a faint voice, that we would move
him from the bed on which he lay to a cot, that he might
lie with his lower limbs horizontal, and his head upright.
He died in about ten minutes after we had moved him, but
breathed his last so easy, that neither myself or my wife, who
were both sitting close to him, perceived it at the time. He
had put his hand to his face, which prevented our observing
it. (32)

The pamphlet mentioned here was Socrates and Jesus Compared and it was being revised at the express wish of Thomas Jefferson. Joseph Priestley viewed the two primarily as great teachers and on this his two hundred and fiftieth anniversary we must surely place him in their company.

References

1. Robert E. Schofield, A Scientific Autobiography of Joseph Priestley, 1733-1804. MIT Press, Cambridge, MA, 1966. p. 105.

2. Reference (1), p. 114.

3. The neatest comment on the word 'doxy' is rather surprisingly that of Bishop William Warburton: "Orthodoxy is my doxy; heterodoxy is another man's doxy." Even more surprisingly, the source cited in The Oxford Dictionary of Quotations is the Memoirs of Joseph Priestley.

4. Reference (1), p. 288.

5. James Morton Smith, Freedom's Fetters. Cornell U.P., Ithaca, N.Y., 1966. Chapters IV and V.

6. Reference (1), p. 263.

7. Derek A. Davenport, CHOC News, 1983, $\underline{1}$, #2, 1.

8. Reference (1), p. 265.

9. Dumas Malone, The Public Life of Thomas Cooper: 1783-1839, U. of South Carolina Press, Columbia, SC. 1961. p. 68.

10. Joseph Priestley, Autobiography of Joseph Priestley, Adam and Dart, Bath, 1970, p. 132.

11. Reference (10), p. 132.

12. Reference (10), p. 222.

13. Edgar F. Smith, Priestley in America: 1794-1804. Blakiston, Philadelphia, PA, 1920.

14. Reference (13), p. 22.

15. Reference (13), p. 20.

16. Reference (1), p. 278.

17. Arthur Bestor, Backwoods Utopias: The Sectarian Origins and the Owenite Phase of Communitarian Socialism in America, 1663-1829. U. of Pennsylvania Press, Philadelphia, PA. 1970.

18. Thomas Cooper, Some Information Respecting America, Dublin and London. 1794.

19. Reference (1), p. 281.

20. Mary E. Clark, Peter Porcupine in America: the Career of William Cobbett, 1792-1800. Dissertation, U. of Pennsylvania, Philadelphia, PA. 1939.

21. Reference 9, Chapter IX.

22. William Cobbett, Observations on the Emigration of Dr. Joseph Priestley, Philadelphia, PA. 1794. Reprinted in The Works of Peter Porcupine.

23. Peter Porcupine (William Cobbett), The American Rush-Light by the Help of which Wayward and Disaffected Britons may see a Complete Specimen of the Baseness, Dishonesty, Ingratitude and Perfidy of Republicans and of the Profligacy, Injustice and Tyranny of Republican Governments. Philadelphia, PA. No V., 30 April 1800.

24. Reference 9, p 92 et seq.

25. Reference (5), p. 310.

26. Reference (5), p. 331.

27. Lester J. Cappon, ed., The Adams-Jefferson Letters: The Complete Correspondence Between Thomas Jefferson and Abigail and John Adams. U. of N. Carolina Press, Chapel Hill, NC. 1959.

28. Dumas Malone, Jefferson the President: First Term, 1801-1805. Little Brown, Boston, MA. 1970. p. 27.

29. Reference (27).

30. Derek A. Davenport, The Chemistry Lectures of the Learned, Ingenious, Scientific and Talented Mad-Cap: Professor Thomas Cooper, Dickinson College 1811-1813. John and Mary's Journal, 1977, $\underline{3}$, p. 4-29.

31. Reference (1), Chapter VI.

32. Joseph Priestley, Jr. in Reference (10), p. 138.

'Fresh Warmth to our Friendship': Priestley and His Circle

By D. M. Knight
UNIVERSITY OF DURHAM, OLD SHIRE HALL, DURHAM DH1 3HP, U.K.

'I am sorry for the manuscripts, the Instruments &c of Dr. Priestley, but if popular operations of the kind in question must take place, it is just that it shou'd fall on those who are fond of appeals to the People': thus on September 4th 1791 John Baker Holroyd, later Earl of Sheffield, wrote to his friend Sir Joseph Banks.[1] Holroyd had been one of those responsible for suppressing the Gordon Riots of 1780, and was thus no believer in lynch law; his remarks probably indicate the general feeling among the improving landowners powerful in the Royal Society. Warmth of friendship was not what Priestley found in the London scientific community, who saw him as an agitator promoting materialism and democracy - a word which meant something more like communism does to us.

But at least Fellows of the Royal Society must have seen Priestley the chemist and Priestley the heretic politician as one person. This vision escaped Sir Leslie Stephen and his team, for in the <u>Dictionary of National Biography</u> there are two lives of Priestley following one another.[2] It is as though there were two persons of the same name living at the same time, one of whom was theologian and the other a chemist; and I know of no other such case in the <u>Dictionary</u>. The two cultures seem to have claimed an early victim in Priestley, whose posthumous reputation is uneasy: to most he is a chemist who wasted a lot of time, paper, and ink on theological and political controversy, while to others he is an important Dissenter who did some interesting or fortunate chemical experiments. There does not seem to be much intellectual connection between the two Priestleys, as one might hope there would be; the science and the theology had little obvious effect on each other, for connections between philosophical materialism and actual experiments on gases do not seem close, though they can be teased out.[3] Indeed, Maurice Crosland has suggested recently that Priestley took up chemistry mainly because it was cheap.[4]

Priestley's entry, or entries, in DNB fills nearly twenty pages; and it is interesting that the portrayal of the radical theologian (by Alexander Gordon) is more sympathetic than that of the chemist, by P.J. Hartog. Hartog refers to Priestley's 'many wrong conclusions, and ... records of unsatisfactory experiments' and to his 'blind faith in the phlogistic theory' which meant that the significance of his isolation of oxygen was 'lost upon him'. Hartog does however recognize that Priestley's references to chance in his discoveries should not be taken at face value. Posterity has not been altogether kind to Priestley, or has not quite known what to make of him; as John Passmore wrote of him in 1965, 'our age is suspicious of versatility' and indeed the age of specialization began in the nineteenth century about when Priestley died.[5]

'Fresh warmth to our friendship' was a phrase used by Priestley,[6] who was indeed supported through most of his life by a circle of friends. He had the gift of making friends as well as a gift for experimental chemistry, and the secret of this was probably, to quote Passmore again, that he 'was one of those very rare people who not only approve in principle of free discussion but also rejoice in the actual practice of it'. He needed to be, as he moved to Calvinism to Unitarianism and thereby outraged both the religious (who believed in the Trinity) and the irreligious (who rejected providence); for Unitarianism was characterized as blasphemous (and thus if taught even illegal) in generally tolerant England, until Priestley and his contemporaries gradually gave it respectability. They fought much the same battle as the agnostics a century later. There are anecdotes of the orthodox refusing to talk to the profane Priestley, but being won round by his charm. His voluminous and disorganized works seem like the work of a talker rather than a writer.

This was not how they struck William Hazlitt, one of Priestley's pupils and an admirer in the next generation upon which I mean to concentrate, who has given us the most vivid pen-portrait of Priestley that I know.[7] Irritable in little things, but calm on great occasions 'his frame was light, fragile, neither strong nor elegant; and in going to any place he walked on before his wife (who was a tall, powerful woman) with a primitive simplicity, as if a certain restlessness and hurry impelled him on with a projectile force before others ... His feet seemed to have been entangled in a gown, his features to have been set in a wig.' A scholar who sat at breakfast with a folio volume and a notebook, with 'the prim formal look of the Dissenter', he 'stammered, spoke thick, and huddled words ungracefully together. To him the whole business of life consisted in <u>reading and writing</u>; and the ordinary concerns of life were considered as frivolous or mechanical interruption to the more important interests of science and of a future state.' Clearly, Priestley wrote whereas other men talked.

Priestley's writing, for Hazlitt, showed him to be 'one of the very few who could make abstruse questions popular', for there is not 'an obscure sentence in all he wrote'. Here Hazlitt put him ahead of Paley, and considered him to be 'certainly the best controversialist of his day, and one of the best in the language', which is a compliment indeed from a skilled and fearless controversialist and journalist. Hazlitt saw in these writings ingenuity, vigour, logical clearness, and also verbal dexterity and artful evasion of difficulties. It is hardly surprising that Priestley was a hero to the Unitarians of the next generation like Lucy Aikin, who remembered 'the benignant smile with which he greeted us little ones', Charles Lamb, and S.T. Coleridge.[8] His influence in Unitarianism persisted down to the middle of the nineteenth century, to the rise of W.E. Channing and James Martineau.

Lamb wrote to Coleridge[9] on January 9th 1797 of Priestley 'whom I sin in almost adoring' and his idea that real friendship was the true balsam of life, keeping firmness of mind where the world would otherwise warp and relax it. He had earlier, on May 31st 1796, written: 'Coleridge, in reading your R[eligious] Musings I felt a transient superiority over you, I <u>have</u> seen priestly. I love to see his name repeated in your writings .... You would be charmed with his sermons, if you ever read 'em,-- You have doubtless read his books, illustrative of the doctrine of Necessity.' Priestley's <u>Doctrine of Philosophical Necessity</u>, 1777, was based on Hartley's notions, and indeed Coleridge became a convert and named his first child after Hartley in consequence; he also comtemplated joining Priestley on the Susquehanna. As a

Unitarian preacher, so Hazlitt tells us, 'Mr. Coleridge once threw a respectable dissenting congregation into an unwonted forgetfulness of their gravity, be reciting a description, from the pen of the transatlantic fugitive, of the manner in which the first man might set about making himself, according to the doctrine of the Atheists. Mr. Coleridge put no marks of quotation either before or after the passage, which was extremely grotesque and ludicrous; but imbibed the whole of the applause it met with in his flickering smiles and oily countenance'.[10]

Hazlitt abandoned Unitarianism, and indeed any religion; Lamb remained at least nominally a member, but Coleridge moved back to the Anglican doctrines of his father. These eminent younger friends or disciples of Priestley thus diverged from his views and practices, though perhaps preserving some of his tolerance and reasonableness. But Coleridge was not the only one to hear from the transatlantic fugitive; he also wrote to Humphry Davy. Davy in 1810 said of him:[11] 'To theory Dr. Priestley paid but little attention; and his hypotheses were rapidly formed, and relinquished with an ardour almost puerile. His chemical writings are principally narrations of facts; and though the style and arrangement are defective, from hasty composition, yet it is impossible not to be amused and interested by his details. They are copious, distinct, and satisfactory; and the manner in which they are pursued leaves a very favourable impression of the simplicity, the ingenuousness, and candour of his mind. Dr. Priestley was a discoverer before he was a chemist. In a letter, which I received from him a few months before his death, he makes this statement, in his usual unaffected manner'. He believed that Priestley would be remembered as an illustrious chemical discoverer, for permanent glory was to be gained 'by pursuing the immutable in nature [rather] than the transient and capricious in human opinion.'

Davy was the protégé of Gregory Watt and Thomas Beddoes, and like Priestley and Coleridge benefitted (though in his case indirectly) from Wedgwood money, which supported Beddoes' Pneumatic Institution in Bristol.[12] There Davy in his research on the oxides of nitrogen, and especially nitrous oxide, carried on Priestley's tradition of experiments on gases. Davy's book, published in 1800, is explicitly following on Priestley's work;

and Davy shows the same manipulative skills, preference for working with the gaseous state, and feeling for the qualitative rather than the quantitative which had characterized Priestley. Thus Davy missed the discovery of multiple proportions because he set out his weights of oxygen and nitrogen as percentage compositions instead of (like Dalton) the different weights of one which combined with a standard weight of the other. Davy was also to show an opportunism with apparatus and hypotheses which is perhaps again like Priestley.

Davy went on to direct his most successful work against Lavoisier's doctrine of acidity, demonstrating that the supposed generator of acidity, oxygen, was in fact a component of the strongest alkalies, but not of hydrogen chloride. This brings us back to Priestley, and to Davy's remarks about his paying little attention to theories. This seems a paradoxical remark to historians like Hartog, for whom the most striking feature of Priestley is that he clung through thick and thin to the phlogiston theory, dying as an unrepentant phlogistonist in exile in Pennsylvania, and presumed to be excommunicated by the chemical community. This does not suit the picture of a man who rapidly formed and easily relinquished hypotheses, and is indeed surprising in one whose religious opinions changed from Calvinism through Arianism to Unitarianism, and who was above all undogmatic. He was also in poltical matters pro-French; so that the famous riot directed against him and others was occasioned by his associates celebrating Bastille Day. In Germany,[13] conversion to French chemistry and to French politics sometimes went together, and conversion to oxygen also came more easily to those on the fringe of the chemical community there, as Priestley was by the 1790s.

What should be remembered is that Priestley was not the only phlogistonist to survive into the nineteenth century. Work on the Voltaic cell led to a number of phlogistic explanations of the apparent decomposition of water as nothing of the kind; and indeed Davy at the height of his fame following the decomposition of potash in 1807 toyed publicly with the phlogiston theory, in his Bakerian Lecture before the Royal Society.[14] He returned to phlogiston in another paper read to the Royal Society in June 1808 where he described the isolation of Alkaline Earth metals, and discussed the so-called 'ammonium amalgam' formed when ammonium salts are electrolysed with a mercury cathode - by now the word 'phlogiston' had moved from the footnote to the text. In his Bakerian Lectures of 1809 and 1810 comes the same emphasis that the question of whether the phlogistic ot the antiphlogistic theory be true, is open. The tone is of faint praise for Lavoisier, with arguments against him.

For Davy, hydrogen was probably the cause of both inflammability and of metallization; hydrogen after all seemed to convert nitrogen into a metal not unlike potassium, and was very likely to be a component of other metals. Davy hoped that this phlogistic theory would lead to a really simple chemistry, in which oxygen, hydrogen, and a few unknown bases were the only real elements. With his work on chlorine, it seemed to the young Faraday that it would not be surprising 'if the old theory of Phlogiston should be again adopted as the true one', though it would probably not entirely set Lavoisier's aside.[15] In fact this was not what happened: for Davy, rather, set aside the 'materialistic' doctrine that properties were the result of a constituent. He turned instead to a Newtonian doctrine of powers and forces, with perhaps only one true element, all the others being stable configurations of hydrogen corpuscles or something even simpler.

The last resurrection of phlogiston was also at the Royal Institution, by William Odling the translator of Laurent and later the Professor of Chemistry at Oxford. In 1871 he lectured on 'The Revived Theory of Phlogiston'.[16] Pointing out that Cavendish as well as Priestley had gone on believing in phlogiston, Odling suggested that there must be something in a theory which had appealed to so many great chemists of the eighteenth century. He believed that in talking about phlogiston, they had had a vision of chemical potential energy -- hazy no doubt but not false. For him, therefore, 'the phlogistic and the antiphlogistic views are in reality complementary and not, as suggested by their names and usually maintained, antagonistic to one another.' He saw the dynamical chemistry of the later nineteenth century having its roots in the theorizing of the previous century.

This sophisticated interpretation or revival of the phlogiston theory bears only a general relationship to Priestley's. As the Unitarianism of the nineteenth century moved away from Priestley's

Biblical emphases, so Davy and Odling in chemistry moved away from his materialism. But ripples from Priestley's circle continued to spread along after his death; the man who in his life found it easy to make friends, continued to influence people after his death.

References

[1] 'The Sheep and Wool Correspondence of Sir Joseph Banks', ed. H.B. Carter, B.M. (Nat. Hist.), London, 1979, p.211.

[2] 'Dictionary of National Biography', Compact Edn., Clarendon Press, Oxford, 1975, pp. 1709ff. The standard introduction to Priestley's science is 'A Scientific Autobiography of Joseph Priestley', ed. R. Schofield, M.I.T. Press, Cambridge, Mass.

[3] J.G. McEvoy, 'Joseph Priestley, Aerial Philosopher', *Ambix*, 1978, 25, 1-55, 93-116, 153-175; 1979, 26, 16-38.

[4] M.P. Crossland, 'Priestley Memorial Lecture', *Br. J. Hist. Sci.*, 1983, 16, 223-238.

[5] 'Priestley's Writings on Philosophy, Science and Politics', ed. J.A. Passmore, Collier, New York, 1965, pp. 37, 39.

[6] J. Priestley, 'Autobiography', ed. J. Lindsay, Adams and Dart, Bath, 1970, p.118. Unitarians in those days had something in common with Catholics; see E. Duffy, 'Ecclesiastical Democracy Detected', *Recusant History*, 1970, 10, 197, 320-322.

[7] W. Hazlitt, 'Complete Works', ed. P.P. Howse, Dent, London, 1934, Vol. 20, pp. 236-239. *Cf.* L.S. Macaulay, 'Hours in a Library', Smith Elder, London, 1899, Vol. 2, pp. 363ff.

[8] H. McLachlan, 'The Unitarian Movement in the Religious Life of England', Allen and Unwin, London, 1934, pp. 276, 286, 294.

[9] 'The Letters of Charles and Mary Lamb', ed. E.W. Marrs, Cornell University Press, Ithaca, 1975, Vol. 1, pp. 12, 88.

[10] W. Hazlitt, ref. 7, p. 238.

[11] J. Davy, 'Memoirs of the Life of Sir Humphry Davy', Longman, London, 1836, Vol. 1, pp. 223ff.

[12] T.H. Levere, 'Dr Thomas Beddoes (1750-1808); Science and Medicine in Politics and Society', to appear in *Br. J. Hist. Sci.*, 1984, 17. R.E. Schofield, 'The Lunar Society of Birmingham', Clarendon Press, Oxford, 1963. H. Davy, 'Researches Chemical and Philosophical chiefly concerning Nitrous Oxide', Johnson, London, 1800, p. xi; see also p. 565.

[13] K. Hufbauer, 'The Formation of the German Chemical Community (1720-1795)', University of California Press, Berkeley, 1982.

[14] D. Knight, 'The Transcendental Part of Chemistry', Dawson, Folkestone, 1978, Ch. 3,4. H. Davy, Phil. Trans., 1808, 98, 33, 363ff; 1809, 99, 103; 1810, 100, 69. J.H. Brooke, 'Davy's Chemical Outlook: the Acid Test', in 'Science and the Sons of Genius: Studies on Humphry Davy', ed. S. Forgan, Science Reviews, London, 1980, pp. 121-175.

[15] 'The Selected Correspondence of Michael Faraday', ed. L.P. Williams, Cambridge University Press, 1971, p. 16.

[16] W. Odling, 'The Royal Institution Library of Science', Elsevier, London, 1970, Vol. 2, pp. 282-291.

From Chaos to Gas: Pneumatic Chemistry in the 18th Century

By R. G. W. Anderson
THE SCIENCE MUSEUM, EXHIBITION ROAD, LONDON SW7 2DD, U.K.

Although I cannot claim to have a rigorous statistical basis for this contention, I would suggest that Joseph Priestley has been responsible for more commemorations, celebrations and symposia than any other British scientist. For example, this year I am aware of three meetings, three exhibitions, the inauguration of a centre for the history of chemistry and a commemorative postage stamp. And celebrations like these have been held at regular intervals for some time. To mark the centenary of the discovery of oxygen in 1874 commemorative meetings were held at Birmingham, Leeds, Paris and Northumberland, Pennsylvania. A volume published to describe these events notes "The inhabitants of Birmingham performed a great act of retributive justice to the illustrious memory of Dr. JOSEPH PRIESTLEY, by erecting a marble statue" and the Birmingham Morning News reported this by printing an extensive ode which begins:

"Seer of the late won renown
Lo! we have crown'd thee;
Stand with the heart of the town
Throbbing around thee;
Stand 'mid the fashion and pride,
Traffic and barter,
God-lit apostle and guide,
Champion, martyr."

How can such enthusiasm for the memory of Priestley be explained? Why have we continually celebrated Priestley, but forgotten all about Henry Cavendish's 250th anniversary two years ago? Partly it must be that Priestley's breadth of intellectual interests was so wide, and he meant, and means, so many different things to different groups of people. I described him earlier as a scientist, but I might well have referred to Priestley as a preacher or a teacher; a progressive theologian or a notorious heretic; a political irritant or an inspiration to revolutionary movements; and many other things besides. We are celebrating Priestley's 250th anniversary here at Imperial College mainly because of his role as a scientific discoverer. The exhibition earlier this year at the Bodleian Library was set up by the unitarian Manchester College, and concentrated on Priestley's theology. And although I was not present to hear it, I am certain from its title that the opening lecture at the Joseph Priestley celebration held at Bucknell University, Lewisburg, Pennsylvania in April dwelt on Priestley and the American War of Independence.

Who, then, was Joseph Priestley? Perhaps a few indisputable facts would now be appropriate after the complexities and uncertainties I have raised. He was born in March 1733 at Birstal in Yorkshire, the son of a cloth dresser. After attending a grammar school he went, at the age of nineteen, to the newly established Daventry Academy where he read widely, including mathematics, physics and philosophy, and where his dissenting religious views developed. In 1755 he was appointed minister to a congregation at Needham Market in Suffolk which proved disastrous partly because of what his congregation considered to be his extreme theology.

In 1758 he moved to Nantwich in Cheshire, where his doctrines proved more acceptable, and he also started a school. Teaching developed into a passion and in 1761 he was asked to act as tutor in languages and belles lettres at the dissenting academy at Warrington. A complex chart he designed and published, depicting historical periods, was rewarded by the degree of doctor of laws from Edinburgh University. His scientific interests were growing. He visited London at Christmas, 1765, met Benjamin Franklin who was then renowned for his electrical researches, and attended a meeting of the Royal Society on 9 January 1766. Priestley decided to gather material for his own book on electricity which was published in 1767. During the course of this he was elected a fellow of the Royal Society. In September 1767 he moved to Leeds to be minister of the Mill-Hill chapel and his scientific interests turned to optics and gases, while maintaining a regular output of theological works. His opportunity for research was improved when, in 1773, Lord Shelburne offered him a generously salaried post as literary companion on his estate at Calne in Wiltshire. After another fruitful period in his life, Priestley moved to the New Meeting house at Birmingham in 1780. Here he interacted with a brilliant circle which included Matthew Boulton, James Watt, Josiah Wedgwood and Erasmus Darwin, who with others formed the Lunar Society, which met for dinner once a month to discuss scholarly matters.

However, Priestley's radical views on politics became widely known and during the anti-French Church and King riots on Bastille Day, 1791, his house and laboratory were wrecked by a riot and Priestley had to flee the city. After a rather disturbed period during which he taught at a dissenting college at Hackney he decided, at the age of sixty-one, to emigrate to America and he settled in a remote English community at Northumberland, Pennsylvania. Though he turned down the chair of chemistry at the University of Pennsylvania, he continued to experiment in chemistry. He died in February 1804.

Priestley published on everything which interested him, from his first book, Rudiments of English Grammar of 1761, to his last paper, "The Discovery of Nitre in Common Salt frequently mixed with Snow" of 1804. It has been estimated that he published 50 works on theology, 13 on education and history, 18 on political, social and metaphysical subjects and 12 books and 50 papers on scientific topics. When Priestley was once asked how many books he had written, he replied "Many more, Sir, than I should like to read." It would be impossible for me ever to summarise his output in this paper, and in any case this is not my theme. But when considering Priestley and the chemistry of gases, it must be remembered that this is but a part of one aspect of the work of a complex character. Nevertheless, this area of interest constituted Priestley's most sustained effort in scientific research. His first experiments were conducted in Leeds in a brewery adjacent to his house. This was in the late 1760s. His investigations continued, on and off, to the end of his life. The period of most intense activity was the 1770s. Priestley investigated the gases which are now called carbon dioxide, carbon monoxide, nitrogen, nitric oxide, ammonia, hydrogen chloride, silicon tetrafluoride and, of course, oxygen. Some of these he characterised chemically for the first time. His results were published mainly in the Philosophical Transactions and in his book Experiments and Observations on Different Kinds of Air which came out in three volumes from 1774 to 1777.

However the main concern of this paper is not to survey Priestley's scientific work or even to consider his own investigations into the chemistry of gases. Rather, it is to see where Priestley's work on gases stands in the context of developing techniques and discoveries in the 18th century. To be able to understand the tradition within which he was working we should attempt to set aside our preconceptions of the gaseous state in terms of atoms, molecules and the kinetic theory. Probably the best place to start is in the previous century by considering the work of Johannes Baptista van Helmont.

Van Helmont was born in Brussels in 1579. He studied widely, before taking a medical degree when he was twenty years old. He was unorthodox in his views and was frequently at loggerheads with the Church. He managed to combine mystical ideas with true scientific research, and spent much of his life as a wandering courtier and scholar. A major part of his research was devoted to examining the products of the combustion of solids and liquids. He found that these were different from atmospheric air and water vapour and that they were specific to the substance of origin. He referred to the "specific smoke" by the "new term gas". "Gas" was a word which he coined himself. It almost certainly was derived from the Greek word "chaos", meaning the confusion which was the first state of the universe.

Up to van Helmont's time, and for many scholars for decades later, different substances in the gaseous state were held to be atmospheric air in different state of purity. Van Helmont's achievement was to identify a number of gases as individual substances, though some which he differentiated were identical gases from different sources. Thus he spoke of gas carbonum formed by burning charcoal, of gas pingue created by dung and of gas sulphuris made by burning sulphur. However he distinguished between gas carbonum and the gas which collects in mines, the gas which is evolved in bubbles from Spa water and the gas evolved in eructations. All these must have substantially been carbon dioxide.

Van Helmont did not collect gases. Indeed he thought this to be impossible and that they could not be contained because they possessed an untameable, wild spirit, or <u>gas sylvestre</u> which would burst a sealed vessel in its efforts to escape. Gases were therefore uncontrollable, and he wrote:

> "I call this spirit, hitherto unknown, by the new name of gas, which can neither be retained in vessels nor reduced to a visible form, unless the seed is first extinguished."

If it can be considered a meaningful thing to say, van Helmont can be considered the discoverer of gases.

During the latter part of the 17th century efforts were successfully made to manipulate gases. The air pump was perhaps the most important apparatus developed at this time. Otto von Guericke built the earliest example and his famous experiment at Magdeburg in 1657 demonstrated that teams of horses could not separate hemispheres from which the air had been evacuated.

Apart from other things, von Guericke's experiments showed that empty spaces could exist. In his own development of the air pump, Robert Boyle stated that he was stimulated by von Guericke's "way of emptying glass vessels". In 1658 or 59 Boyle's assistant Robert Hooke built a single-barrelled air pump. Robert Boyle was a great demystifyer of science and he worked to present a mechanical picture of chemical reactions, exploring the behaviour of substances by analogy with machines. He considered air to be composed of corpuscles. To explain its compressibility, he likened the particles to small coiled springs. Boyle was the first to collect hydrogen, and here it is best to repeat his own description of the experiment:

> "We took a clear glass bottle (capable of containing by guess about three ounces of water) with a neck somewhat long and wide, of a cylindrical form; this we filled with oil of vitriol and fair water, of almost a like quantity, and casting in half a dozen small iron nails . . . and speedily inverting the phial, we put the neck of it into

a small wide-mouthed glass . . . with more of the same liquor in it . . . soon after we perceived the bubbles produced by the action of the menstrum upon the metal, ascending copiously . . ."

Thus the gas was collected in the same vessel as it was prepared.

Boyle was also the first to describe the combustion of hydrogen, which he described as: "Upon the approach of a lighted candle, it would readily enough take fire and burn with a blueish and sometimes greenish flame at the mouth of the vial."

He was greatly interested in the nature of fire and in calcination, and oxidation of metals, which he interpreted in terms of his corpuscular theory. He sealed various metals in glass flasks and heated them strongly to convert them to their calces, or oxides. He then opened the flasks and weighed the product. The increase in weight was attributed to fire particles which had penetrated the glass and combined with the metal. An alternative, and even more exotic interpretation of this phenomenon will be mentioned later.

Also working on similar problems at Oxford, but a few years later, was John Mayow. In particular, he experimented on respiration, fermentation and combustion, and interpreted his results in a novel way. Mayow considered air to be impregnated with what he called "nitro-aerial spirit", a volatile substance which is contained in nitrates. The elastic power of air was due to nitro-aerial spirit. When a flame burned in a confined space nitro-aerial particles were removed and at the same time the air lost its elastic force, which therefore seemed due to the same aerial particles as those by which a flame was supported. Mayow's main experiments on combustion and respiration were described in his five tracts published in 1674. His apparatus was ingenious, consisting of an inverted cupping glass (usually used by surgeons for bleeding patients) over a pan of water. By this means he could generate hydrogen by lowering a piece of metal into a vessel of acid and he could ignite material by means of a burning mirror strategically placed outside the apparatus. To investigate respiration, Mayow put a mouse in a cupping glass sealed with a bladder, and noticed that after breathing for some time the bladder bulged into the glass; in another experiment, the mouse was in a cage in the glass over the water in the pan. This time the water rose upwards. In both cases the diminution in volume of air was explained by respiration causing the removal of nitro-aerial particles. Mayow adopted Boyle's view that air particles are like little springs, but claimed that they were not all alike. His experiments suggested to him that there were two kinds of particles which made up air. Some were branches and hooked together while others, the nitro-aerial particles, were very "subtle, solid, smooth, agile and fiery."

The first phase of experimentation with gases can be concluded with a brief account of the work of Stephen Hales, whose main work was published in 1727. Hales, who combined inventive scientific research with his professional work as a parish priest, is always referred to as the 'Perpetual Curate of Teddington', which post he held from 1709 until his death in 1761. He performed many quantitative experiments with gases, burning measured weights of natural substances and recording the volumes generated. For the first time, an apparatus was derived which separated the generator of the gas from the receiver: materials were heated in a bent iron gun-barrel in a furnace which led to an inverted glass bell, filled with water and suspended over a barrel. Another apparatus was devised specifically for measuring the volume of gas evolved in distillation or combustion.

Hales recorded results of heating 24 gas-producing substances including coal, oyster shells, gall stones, tobacco and hog's blood. He did not, however, distinguish between the gases evolved: he simply recorded the quantity of air and discarded it. His conclusions were vague: he stated that "air abounds in animal, vegetable and mineral substances" and he considered that air "is very instrumental in the production and growth of animals and vegetables, both by invigorating their several juices, while in an elastic state, and also by greatly contributing in a fix'd state to the union and firm connection of the several consituted parts of those bodies, viz. their water, salt, sulphur and earth." Thus Hales supposed that air could exist in a non-elastic form in many substances, from which it could be set free again. Air itself he considered to be an element, and the differences between one gas and another could be accounted for by impurities. He stated that "our atmosphere is a Chaos, consisting not only of elastick, but also of unelastick particles." In an interesting experiment on mineral waters, he found that Pyrmont water contained about twice as much 'air' as rain water, and he explained that this "air contributes to the briskness of that and many other mineral waters".

If we take stock of the position by the mid 18th century, it can be seen that experiments with gases had claimed the attention of various chemists. Techniques for conducting experiments had been refined, from the pessimistic efforts of van Helmont to the development of the pneumatic trough by Stephen Hales. Although quantitative results had been obtained, there was less consideration given to qualitative experiments. What thought there had been to the subject was largely conjectural, with interpretations given in terms of prevailing theories of elementary matter. In general, gases prepared or collected from different sources were not considered to be different species in the same way in which different solid substances were held to be.

Although widely varying attitudes to the nature of airs or gases were adopted in the 18th century, there was one theory which was almost universally agreed until the end of the period: this was the phlogiston theory developed from earlier ideas by the German chemist Ernst Stahl, who lived from 1660 to 1734. This provided, amongst other things, a theory of combustion which became the central doctrine of chemistry. Stahl argued that when combustion occurred, the inflammable principle, phlogiston, was lost. When, for example, a metal was heated it lost phlogiston and was converted to the calx, or what is now understood to be the oxide. The metal was therefore a more complex substance than the calx. Phlogiston lost from a metal, when heated, was dispersed through the air which surrounded it. When air had become completely phlogisticated it would not serve to support combustion of any material: it would not allow a metal heated in it to yield a calx, and it would not support life, for the role of air in respiration was to remove phlogiston from the body. Air was thus essential as the medium to carry away the phlogiston, but was a mechanical aid rather than playing any other role in the process. Plants could absorb phlogiston from the air, and animals could obtain it from plants. Hence plant and animal substances were rich in phlogiston and converted them to metals again. Phlogiston was never found in a perfectly pure state: the purest form was soot formed from burning oil, because it burned away without leaving a residue. Charcoal was also rich in phlogiston, which explained its role in the production of metal from its calx.

Stahl himself recorded that he began to think about the principle of inflammability in 1679. He named it phlogiston in a work published in 1718, explaining that he derived the term from a Greek word meaning 'combustible'. In general the phlogiston theory provided a satisfactory framework to explain many observable processes, but there were some worrying inconsistencies. When an organic substance burned, the apparent products weighed less than the original substance. However when a metal was calcined, which was recognised as being basically the same process, the calx seemed to weigh more than the starting material. This was unimportant to Stahl, who probably considered phlogiston to be weightless. To later chemists elaborate explanations had to be constructed. One was that phlogiston had negative weight, or buoyancy, so that when phlogiston was lost, the calx would weigh more than the original metal. Early in the 18th century air was assumed to be a substance which could not play a chemical role: later, when it was suspected that gases played a part in combustion, it was assumed that although phlogiston was lost on combustion, another substance with greater weight was taken up, which explained the observed quantitative phenomenon.

When, then, can be found the first evidence that gases could participate in chemical reaction and could have chemical identities of their own? If it were necessary to produce a single piece of evidence of changed attitude, that document would probably be a thesis submitted by a student for a medical degree in 1754. The thesis, printed in Latin according to the requirements of Edinburgh University, was De Humore Acido a Cibis Orto et Magnesia Alba, on the acid humours arising from food, and magnesia alba. The author was Joseph Black, a Scot of Ulster parentage who was born in Bordeaux in 1728. Black's work was quantitative, though he did not use volumetric measurements, but weighed his substances routinely throughout, using a simple balance. The substance used in his first experiments was magnesia alba, or basic magnesium carbonate. In the course of his experimental scheme, he found that on heating magnesia alba a gas was expelled, which he suspected had originated in the pearl ashes, potassium carbonate, used in its preparation. By means of a cyclic series of experiments, Black showed that the original weight of magnesia alba could be recovered by dissolving the product of heating, magnesia usta (magnesium oxide) in sulphuric acid and reconstituting the magnesia alba as a precipitate by the addition of a solution of fixed alkali (sodium carbonate), suggesting that the origin of the gas, which Black called fixed air, was indeed the alkali. Black continued his experiments for a further year in Edinburgh, practising medicine and waiting for an academic job-vacancy. On 5th June 1755 he read a revised account of the experiments described in his thesis to the Philosophical Society, but the work was extended to include further investigations into the relationship between chalk and lime. It was in this paper that Black snowed that chemical test whereby atmospheric air was a mixture of gases and that carbon dioxide could be differentiated chemically using lime water. He wrote:

"Quick-lime does not attract air when in its most ordinary form, but is capable of being joined to one particular species only . . . to this I have given the name of fixed air . . . the particles of quick-lime which are nearest the surface gradually attract the particles of fixed air which float in the atmosphere. But at the same time that a particle of lime is thus saturated with air, it is also restored to its state of mildness and insolubility."

There is a record that Black demonstrated the role of carbon dioxide in respiration on a massive scale. His colleague, John Robison, wrote about an experiment which Black performed in Glasgow in the winter of 1764-65. A considerable quantity of caustic alkali was rendered "mild and crystalline by causing it to filter slowly by rags, in an apparatus which was placed above one of the spiracles in the ceiling of a church, in which a congregation of more than 1500 persons had continued near ten hours." This experiment says more about the Kirk-going habits of the Scots than it does about Black's experimental techniques. But the more usual laboratory procedures of Black were as inventive, using a balance to weigh the reactants and products at every stage in his cycle of experiments.

Henry Cavendish was a near contemporary of Black's. In certain ways they were contrasting figures. Cavendish studied at Cambridge, though he did not take a degree. As a gentleman of means (being related to the Duke of Devonshire) he did not have to work for his living but devoted most of his time to scientific research. Black's career was spent teaching chemistry in the medical faculties of Glasgow and Edinburgh. Cavendish had a retiring disposition (there is only a single likeness of Cavendish, and that was sketched surreptitiously), Black was clubbable, and several portraits exist. But there were similarities. Both published little of their work, and both were deeply concerned in the study of what they called 'factitious airs', that is, gases which were contained inelastically in other bodies, but which were capable of being freed and made elastic. While, as already been described, Black characterised fixed air, or carbon dioxide, Cavendish characterised inflammable air, or hydrogen. This was his first paper read to the Royal Society, on 29th May 1766.

Cavendish found that from the same weight of a particular metal, zinc, iron or tin, the same volume of inflammable air was obtained when it was added to hydrochloric or sulphuric acid. He suggested that the gas came from the metal rather than the acid, and that the inflammable air was pure phlogiston. Inflammable air formed an explosive mixture with air, the most violent detonation being when the volumes were mixed in the ratio 3:7. He found that inflammable air differed from fixed air in that it was insoluble in water and alkali. Moreover Cavendish measured the specific gravity of inflammable and fixed air by weighing samples on a balance. Weighing gases from different sources had been tried before: Francis Hauksbee in 1710, Stephen Hales in 1727 and Isaac Greenwood in 1729, but none of the experiments showed significant variation in weight from a sample of atmospheric air. Cavendish's superior technique and apparatus showed that fixed air was 1.57 times heavier than common air and that inflammable air was about eleven times lighter. He collected his gases by inverting a bottle filled with water in a trough of water. For gases soluble in water, he used mercury. Cavendish's balance had to be capable of more accurate weighings than those hitherto used. If the balance preserved at the Royal Institution is indeed the one he used - and it would be a very difficult, if not impossible, matter to prove one way or the other - then it was clearly purpose built for him and it incorporates innovative features. Its most likely constructor is Samuel Harrison of Birmingham. The work of Black and Cavendish together provided strong evidence against the traditional notion of a single, universal air.

Exactly a fortnight after Cavendish's paper on inflammable air was read to the Royal Society, Joseph Priestley was elected a Fellow - on 12th June 1766. It is fair to say that he was in the early stages of self-education in science, though by the Spring of 1767 he had completed the work of compilation and research for his History of Electricity. He moved from Warrington to Leeds later that year. Recalling this period in later life, Priestley wrote:

". . . nothing . . . engaged my attention while I was at Leeds so much as the prosecution of my experiments relating to electricity, and especially the doctrine of air . . .

When I began these experiments I knew very little of chemistry . . . But I have often thought that upon the whole, this circumstance was no disadvantage to me; as in this situation I was led to devise an apparatus, and processes of my own, adapted to my peculiar views. Whereas, if I had been accustomed to the usual chemical processes, I should not have so easily thought of any other; and without new modes of operation, I should hardly have discovered anything materially new."

Priestley's first work was conducted on fixed air, available from his conveniently located next door brewery. He devised an apparatus for dissolving the carbon dioxide in water under slightly enhanced pressure. Five years later a description of the technique was published as a pamphlet. Impregnation of liquids with the gas could produce artificial mineral water (which obviated the need to travel to spas), and could also improve beer, wine and cider which had gone flat and might even,

it was thought, produce a beverage to combat scurvy. Priestley's
apparatus was not very easy to manipulate: it incorporated a
pig's bladder, and was soon improved by others. We might
remember that a well-known soft drinks company is celebrating
its bicentenary this year, having connections with one Jacob
Schweppe who came from Geneva, and that the current multi-
million pound industry can trace its origins to Priestley's work.

Priestley conducted many experiments on the production of
gases and their properties over a period of several years. A
list of those which he prepared for the first time or which he
helped to characterise had already been given. It would be much
too complex a task to summarise his work here and, moreover, it
would be totally confusing. Hence only particular studies on
nitrous air and the production of dephlogisticated air are here
considered.

Stephen Hales had prepared a colourless air by adding pyrites
to nitric acid, and had noticed that on mixing this with common
air, a red turbid gas had resulted. After taking advice from
Cavendish, Priestley repeated the experiment on 4th June 1772 by
adding brass filings to nitric acid and, on exposure to common
air, the same red fumes were observed. Over water, a large
diminution of the total volume of gas was observed, though over
mercury there was only a minor change. Thus the water had dissolved the red product. Priestley concluded that he had found
a direct means of measuring the "air fit for respiration", which
he argued must be proportional to the volume change. His
earlier technique for determining this property was to introduce a live mouse into the vessel of gas being tested, which he
disliked doing, and he wrote "the goodness of air may be distinguished much more accurately than it can be done by putting
mice to breathe it" which "was a most agreeable discovery to me".
Priestley devised a quantitative instrument which could conveniently be used to analyse the 'fitness' or 'goodness' of air
by this means, which came to be called a eudiometer, meaning
'fine weather measure'. Some keen investigators even undertook
'eudiometrical tours' in the Alps, measuring the quality of air
as they progressed from valley to mountain peak.

Priestley's most significant experiment of his series can be
seen, in retrospect, to be that performed on 1st August 1774 at
Calne in Wiltshire, whilst in the employ of Lord Shelburne.
Priestley had been heating a variety of substances over mercury
by focussing the sun's rays on them by means of a twelve inch
diameter lens in order to discover whether any factitious gases

might be released as a result. On this particular day he heated <u>mercurius calcinatus per se</u>, what is now known as mercuric oxide, and noted that "air was expelled from it very readily". He continued:

> "what surprised me more than I can well express, was, that a candle burned in this air with a remarkably vigorous flame . . . I was utterly at a loss how to account for it."

At first he thought he had found a species of nitrous air, even though he found that the gas was insoluble in water. Further experiments showed that his mice lived longer in a limited volume of it than in common air, and that by applying his nitrous air 'goodness test', the new gas was between five and six times as 'good' as common air. Priestley's enthusiasm flowed over into his account, and he wrote "I wish my reader be not quite tired with the frequent repetition of the word <u>surprize</u> . . . but I must go on in that style a little longer."

Priestley interpreted his newly discovered air in terms of the prevailing phlogiston theory. Because of its ability to receive more phlogiston than common air, shown, for example, by its superior capacity to sustain life by accepting more phlogiston from respired air, Priestley called it 'dephlogisticated air'. Later he prepared the same gas by heating red lead and also from nitre.

In October 1774 Priestley travelled to Paris with his patron Lord Shelburne, and while there he met Antoine-Laurent Lavoisier and other French scientists. Lavoisier was a member of the Academy of Science. His lucrative job of helping administer the tax gathering system allowed him to indulge his scientific interests, but it was also to be the cause of his death during the reign of terror which followed the Revolution. Lavoisier was already involved in experiments on gases. He had assessed the state of the art for the Academy, repeating and developing the work of others. In January 1774 Lavoisier published these results in his first book, the <u>Opuscules physiques et chimiques</u>.

Exactly what transpired between Priestley and Lavoisier at dinner will never be known. This occasion must rate highly of those at which historians of science would like to have been the proverbial fly on the wall. Certainly Lavoisier was already aware of the peculiar properties of <u>mercurius calcinatus per se</u>, the red calx of mercury. He set about his own experiments. To a meeting of the Academy in 1775 he referred to the gas as "not only common air . . . but even more pure than the air in which we live" and three years later he called it "the purest part of the air" or "eminently respirable air". He also showed that it combined with carbon to produce fixed air and with phosphorus to form phosphoric acid. Because he showed the "eminently respirable air" was a constituent of so many acids, Lavoisier considered it to play the role of the "acidifiable principle". Hence he termed the air "principe oxygine", the latter word being derived from Greek, meaning 'acid producer'. In 1787 the gas was referred to as oxygène.

Meanwhile in England, Priestley and Cavendish were experimenting on the explosive reaction between dephlogisticated air and inflammable air. For this devices called 'new eudiometers' were used, cylinders of brass or glass which incorporated wires across which an electric charge could be discharged. In 1781 it was recorded that after firing, a dewy deposit was formed which was shown to be water. This too, was reported to Lavoisier in Paris, during the visit of Cavendish's assistant, Charles Blagden, in 1783. Further experiments followed which convinced Lavoisier that water was simply composed of the two gases, a compound of the 'principe oxygine' with what he called the 'principe inflammable aqueux', which when turned into Greek became 'hydrogène', the principle of water. Shortly afterwards Lavoisier decomposed water over heated iron, showing that the inflammable air obtained was identical with that formed by the action of some metals on sulphuric acid.

It was clear to Lavoisier and other French chemists at this time that if further progress were to be made and confusions between chemists minimised, a new and universally used nomenclature was needed. In 1787 a scheme was proposed which essentially survives to the present. Each simple substance was to have a definite name which, where possible, would express its character. The names of compounds should indicate their composition in terms of their simple constituents. Lavoisier's widely read and highly influential final book dealt with explaining his new system of chemistry in simple terms. It was published in Paris in 1789. By 1790 it had been translated into English, by 1791 into Italian and Dutch, 1792 into German and 1797 into Spanish (published in Mexico). Dover Publications keep it in print to this day.

Lavoisier's Traité élémentaire de chimie described in detail the experimental basis for the rejection of phlogiston and for the new theory of combustion in which oxygen held the central role. The antiphlogistic theory had been gradually accepted by many chemists over the previous decade or so. Joseph Black himself had somewhat hesitantly adopted Lavoisier's views and had been teaching them to his students from the mid-1780s.

Lavoisier's work heralded what has become to be called 'the chemical revolution'. In his Traité he wrote that "Chemistry, in subjecting to experiments the various bodies in nature, aims at decomposing them so as to be able to examine separately the different substances which enter into their composition." By studying carbon compounds by burning them in oxygen and determining the carbon dioxide formed, the basis of organic analysis was laid by Lavoisier. In 1868 the French chemist C. A. Wurtz, in reviewing the development of chemistry, made the famous pronouncement "Chemistry is a French science. It was constituted by Lavoisier, of immortal memory."

Joseph Priestley was the only major chemist who refused to abandon the phlogiston theory in favour of Lavoisier's new chemistry. In 1796 when in America he published Consideration on the Doctrine of Phlogiston and four years later his last scientific work The Doctrine of Phlogiston Established. In a letter of 1800 referring to his belief he wrote "I feel perfectly confident of the ground I stand upon . . . though nearly alone, I am under no apprehension of defeat." This was the year in which Volta's pile was introduced into England and experiments were almost immediately conducted into the electrolysis of water. Priestley believed that the production of oxygen and hydrogen was not due to decomposition, but to gases dissolved in the water, writing that the "modern hypothesis of the decomposition of water . . . I consider as wholly chimerical."

It is perhaps surprising that Priestley who in so many areas was willing to suggest the overthrow of the established order, dug his heels in when it came to displacing an unsatisfactory theory with one which his own work had helped to establish. However by doing so he remained as he had always been throughout his life - one of a small, embattled minority.

Finally, I want to add a brief coda to the story of these pneumatic investigations, one which brings us up right to the last year or two of the 18th century. It concerns the efforts of Thomas Beddoes, sometime reader in chemistry at Oxford University, to treat disease by the inhalation of gases. In 1797, Beddoes proposed to the Lunar Society that a Pneumatic Medical Institute should be established. The idea was not exactly a new one - many experiments had already been carried out on the medical uses of gases - but Beddoes' concept was to be carried out on a greater scale than before, establishing a joint laboratory and hospital where the possible curative effects of gases could be clinically tested.

Beddoes was supported financially by Josiah Wedgwood and technically by James Watt, who helped design and make the apparatus for him. Eventually the Pneumatic Institution was opened at Hotwells, a suburb of Bristol, on 21st March 1799. Beddoes' assistant was a young man of twenty called Humphry Davy. There were high hopes that a cure would be found for consumption, but it was soon realised that the experiments with gases failed: neither oxygen, nor carbon dioxide, carbon monoxide, nitrogen, ether, nor any or the other gases tried, in any combination, seemed to have an efficacious effect. Perhaps the experiments with greatest impact were those conducted with nitrous oxide. The gas did not effect a cure, and its anaesthetic effects were not fully appreciated, but the hallucinatory and euphoric effects were widely tested and enjoyed.

Priestley knew of Beddoes' aspirations: in October 1795 he wrote to William Withering:

"I am glad to hear of the fair trial that will be made of Dr Beddoes's theory at Birmingham. I want Mr Watts apparatus; I could use it for various purposes . . . I must get everything of much value from England."

Remarkable changes in chemistry had taken place during the course of Priestley's lifetime. He was born only shortly after the first practical method of collecting gases had been devised. It was not until he was twenty one years old that any gas was recognised as having specific chemical properties. By the time he left England the phlogiston theory had been widely discredited (though not by him) and the role of oxygen in combustion and respiration was recognised. When established in America, the first steps had been taken which were to lead to the use of gases as anaesthetics.

But that is not all. On 6th September 1803, five months to the day before Priestley's death, a north country teacher jotted down in his notebook a list of elements (and compounds), alongside which were written a series of numbers. John Dalton had compiled the first table of atomic weights. Joseph Priestley had truly lived through and had made his own contributions to a chemical revolution.

Bibliography

A. J. Berry, "Henry Cavendish. His Life and Scientific Work", Hutchinson, London, 1960.

F. W. Gibbs, "Joseph Priestley. Adventurer in Science and Champion of Truth", Nelson, London, 1965.

Henry Guerlac, "Antoine-Laurent Lavoisier, Chemist and Revolutionary", Charles Scribner's Sons, New York, 1975.

Walter Pagel, "Joan Baptista van Helmont. Reformer of Science and Medicine", Cambridge University Press, Cambridge, 1982.

J. R. Partington, "A History of Chemistry", volumes 2 and 3, Macmillan, London, 1961 and 1962.

Robert E. Schofield, "A Scientific Autobiography of Joseph Priestley, 1733-1804", M.I.T. Press, Cambridge Mass., 1966.

The Professional Work of an Amateur Chemist: Joseph Priestley

By R. E. Schofield
IOWA STATE UNIVERSITY OF SCIENCE AND TECHNOLOGY, AMES, IOWA, 50010, U.S.A.

Joseph Priestley was thirty-four when his first scientific publication, A History of Electricity, with original experiments, appeared. By that time he had already served two congregations as a dissenting minister, taught six years in a dissenting academy, published six books, received an LL.D. degree, and become a Fellow of the Royal Society. He was thirty-nine before his first paper on chemistry was published. During the five intervening years he had published a second edition of History of Electricity, with additions; a student text on electricity, another on perspective, a pamphlet on artificially carbonated water, a History of Optics, and five papers on electricity in the Philosophical Transactions of the Royal Society. In the same period, while serving as minister to a large dissenting congregation in Leeds, he published twenty-one other books and pamphlets, on theology, politics, and education, and at least twelve papers in The Theological Repository, a journal of controversial theology which he founded and edited. By his death in 1804, he had increased his scientific publications to a total of twenty volumes, in first editions, and approximately fifty-eight scientific papers (thirty-nine of these after 1794). In his lifetime, he published one hundred and thirty-four pamphlets and books (many in several volumes) and at least forty-two papers on subjects other than science.

During the forty-three years of this incredible career as a publishing scholar, less than fifteen percent of his book publications, in first editions, clearly related to science. Until his exile to the United States in 1794, only one-third of his published papers were on scientific subjects.[1] Even accepting Priestley's own caveat that a paragraph of science writing might take as long in preparation as "whole sections, or chapters" of his other works took in writing, it is hard to see, in this record, the career of a professional scientist.[2]

Of course, the term "professional scientist," in the sense of a person who earned his living by doing science, can scarcely be said to have meaning in eighteenth-century Britain.[3] Most of Priestley's scientific contemporaries earned their livings as teachers, physicians, clergymen, etc., doing their original scientific investigations as Priestley did his, in the time they could spare from remunerative activities. But Priestley was not a "professional" in the alternate meaning of this term either. He never "professed" a commitment to scientific study and investigation of an intensity which could justify its being defined as a calling, a vocation he forsook only to earn the living required to continue it. His preference was always for the ministry and science was, for him, chiefly a support to his personal theological study and to his public theological debates.[4] More than half of his adult life Priestley served as a minister and over sixty percent of his published writings were on religious or theological issues.

Three times during his career, from 1761 to 1767, from 1773 to 1780, and from 1794 until his death in 1804, Priestley was not to be active as a minister. During the first of these periods, he taught at Warrington Academy. Conscientious in his obligations as a teacher, during those years he did nearly all his writing on educational subjects. Not until he had "composed all the lectures . . . [he] had occasion to deliver" was he to turn his attention to science and even then the major emphasis, in the History of Electricity, was educational.[5] During the second period, he acted nominally as librarian and actually as literary companion to William Petty, Lord Shelburne. He had no formal obligations and could expend his efforts more or less as he pleased. It was then that the better part of his scientific research was done -- and during this period his scientific writing constituted no more than twenty percent of his publications; he wrote and published more on theology during these years and as much in metaphysics as he did on chemistry. After

1794, in exile in the United States and subsisting on the generosity of his brother-in-law, on donations from admirers, and on an annuity from Lord Shelburne, he had, again, no obligations to anyone but himself; he could not even collect a congregation from among the orthodox Calvinists of the region in which he had settled. He established a laboratory at his home and wrote four scientific pamphlets and nearly twice as many scientific papers during these last ten years of his life as in the previous sixty. For the first time, the number of scientific items he published exceeded those on other subjects.[6] In number of pages, though, the sum of his science writing was less than a single volume of one of the twenty books on theology he published during the same period while the quality of the science papers -- generally repetitive and uninspired attacks on the new chemistry of Lavoisier -- could not, this time, make up for their brevity. Clearly, even when given the opportunity, Priestley would not concentrate his efforts on scientific research.

Because of the successes of his scientific activities, Priestley's "philosophical" (i.e. scientific) friends wished him to confine his attention to science and he had, gently, to lecture them that he was a better judge than they, theology having been his "original and proper province," as to what were the "most important pursuits, those from which . . . mankind at large will finally derive the greatest advantage."[7] Nor, so far as the majority of his contemporaries were concerned, was he entirely wrong about the relative significance of his various works. No doubt not everyone agreed with Coleridge when he wrote: "I regard every experiment that Priestley made in chemistry, as giving wings to his more sublime theological works."[8] Nonetheless, in the wider reception given his non-science work, one can see the comparative value the general public assigned his science. Excluding modern reprints of his scientific writings, only the History of Electricity reached a fifth edition in English; volumes of the Experiments and Observations on Air reached but third editions. There were twenty-two fifth editions of non-science works, nine tenth editions, and one theological tract reached a twentieth edition. When, between 1817 and 1831, an edition of Priestley's collected works was published, it was the Theological and Miscellaneous Works, excluding the science. On the Continent, the balance of interest was, understandably, the other way, but even there, in German, French, Dutch, and Italian translations, theological, educational, and political works number nearly half of those of science.

Now Priestley would be the last person to argue that contemporary public interest was a measure of merit. He had, after all, in successive moves from Calvinism to Baxterianism to Arianism to Socianism (or unitarianism), deliberately chosen religious opinions which appealed to smaller and smaller minorities of his countrymen. Yet surely he would be surprised that the judgment of posterity -- that judgment he himself frequently invoked -- should so readily ignore not only his own sense of the innate value of his non-scientific work but also that of his age in so nearly confirming the opinion of his science friends and critics. Today, it is primarily Priestley's scientific work, and specifically his pneumatic chemistry, which is remembered and celebrated. Few people remember, and fewer care, that he considered himself an amateur in chemistry and a professional, from personal conviction and for livelihood, in education and theology.[9]

It can be argued that, in style and in substance, Priestley's science is informed by, as it informs, his educational and political theories and his theological opinions.[10] During this 250th anniversary celebration of the year of his birth, however, it seems appropriate, for once, to consider Priestley's non-scientific works in their own right, without more than peripheral reference to the science. As at least one of the other papers of this conference will be devoted primarily to Priestley's theology, this paper will chiefly concern itself with his work in education, politics, and metaphysics, though some attention will, necessarily, be paid to the theology, especially in connection with his political activity. Not only was that work the greater part of his active life, it gave him the most satisfaction -- and produced the most dramatic impression on his contemporaries. He was, after all, not mobbed from his home in Birmingham and thrust into unwanted exile from his country because of his chemical experiments.

Excepting two bits of juvenilia (a poem and a book review), Priestley's first publication was a theological work, The Scripture Doctrine of Remission, written about 1757 while in a "low and despised situation" at Needham Market, Suffolk. That book was published anonymously in 1761 and until its reappearance as papers in the Theological Repository contributed little toward his scholarly career. By the time it was published, however, Priestley had moved twice: in 1758 to a still low but less despising congregation at Nantwich, Cheshire and, in 1761, to become teacher of languages and belles lettres at Warrington Academy, Warrington, Lancashire. He carried with him, to Warrington, the manuscript of The Rudiments of English Grammar, originally prepared for the use of pupils at a school he had begun in Nantwich. The Rudiments was first published in 1761, revised for an edition of 1768, and went through at least seven further English editions in Priestley's lifetime, as well as appearing in French translation.

Priestley was employed to teach English grammar and composition, Latin, Greek, French, and Italian, and the theory and practice of rhetoric. Apparently there was no problem in finding texts for languages other than English and Priestley wrote none. He did, however, prepare A Course of Lectures on the Theory of Language and Universal Grammar, printed at Warrington in 1762. These two books the Rudiments and Lectures . . . on Language mark Priestley's formal introduction to the world of scholarship. Examination of the texts show that, this early in his career, Priestley was already demonstrating the essential qualities of all of his future work. These are not simply texts of grammar, they are, as well, reasoned justifications of his approach to language study. As such, they have elicited, from modern students of linguistics, the judgment that he "was one of the great grammarians of his time."[11] Yet neither work is, except in its final impact, notably original; every leading idea can be traced to an earlier source. What the books illustrate throughout is a vigorous common sense and an ability to select, from a variety of sources, sets of significant ideas which he linked together into a coherent argument, followed to its logical conclusion, and stated with conviction.

That English should be taught independently of irrelevant Latin grammar was, for example, argued from at least as early as the 17th century. The notion that languages changed with time was expressed by John Locke and by Isaac Watts, in books Priestley read as a youth. To these, he added a dissenter's distaste for authoritative establishments to set standards of belief or taste. In consequence, Priestley was one of the very few English grammarians of the century to declare, and consistently to follow that declaration, that the only standard for the English language was that of usage. He did not mean, as many modern language relativists have taken him to mean, that fixed rules for English were not possible. Indeed, he argued that the function of language as a means of communication must mean that there are general rules of language, to be learned by comparison of varieties of them. These rules, however, could not yet be established and could only become known through competition of different forms in a marketplace of usages over time. This, then, required careful observation of past and contemporary usage. Because he believed that languages developed naturally, out of speech (the "imperfections of all alphabets . . . seems to argue them not to have been the product of divine skill"), examples of usage should not be taken simply from cultivated and celebrated writers, but also from those who illustrate "the natural propensity . . . general custom and genius of the language as it is commonly spoken." [12]

Priestley's examples of usage have provided students of language with many of their illustrative forms and are one of the reasons they cite his work so frequently. But for all their historical significance, and their many editions, Priestley's works on grammar were superseded by still more popular contemporary texts -- by Robert Lowth's Short Introduction to English Grammar, first published within a month of the first edition of the Rudiments, by Lindley Murray's English Grammar of 1795, and, in the United States, by Noah Webster's Grammatical Institute of the English Language (1782-1785). The influence of Priestley's grammatical writings was, however, not lost. Later editions of Lowth and all editions of Murray cite examples and quote approvingly from Priestley while Webster makes his debt to Priestley's ideas quite clear. And as late as 1827, the reputation of the Rudiments was still sufficient to justify its exploitation in a Dr. Priestley's English Grammar Improved.

Priestley next prepared his Course of Lectures on Oratory and Criticism. Although he made his manuscript lectures available to students to be copied, the Oratory and Criticism was not published until 1777 and had only that one English edition during his lifetime, though it was pirated for a Dublin edition in 1781 and was translated into German. It has, however, recently been reprinted as a "Landmark in Rhetoric and Public Address," a "notable contribution to rhetoric," and been described by a principal historian of British logic and rhetoric as being "of special interest to historians of rhetoric."[13]

These lectures, as published, say little about the problems of elocution -- which is probably just as well, as Priestley spoke always with a Yorkshire accent and a barely controlled stammer. They are primarily concerned with the design of speeches and compositions for effective argumentation and persuasion, with rhetoric, in fact, and for this purpose he could call upon a long tradition of texts. He made liberal use of these texts, and his debt to writers such as John Ward, Alexander Gerard, and Henry Home, Lord Kames is obvious. His development of their ideas was, nonetheless, not traditional.

The proximate origin of this originality was his use of David Hartley's associationist (physiological) psychology as a principle of organization for the diverse materials from which the lectures were composed. Indeed, the eventual publication of the Oratory and Criticism was probably part of Priestley's campaign, between 1774 and 1778, to establish Hartley's Observations on Man his Frame, his Duty, and his Expectations (1749) as a viable alternative to Thomas Reid's Scottish Common Sense Philosophy and David Hume's pyrrhonic scepticism. In the Lectures on Oratory and Criticism, Priestley used the association of ideas to create a system of topical invention, argumentative structure, and analysis at considerable variance with the classical theory of the rhetorical texts from which he borrowed. Associationism was still more significant in Priestley's opposition to the "faculties psychology" of Scottish Common Sense aesthetics. The conclusion he drew from Hartleyian analysis was strikingly parallel to that he had drawn respecting language standards. Although, in time and with the spread of a common culture among "all the nations of the earth . . . a uniform and perfect standard of taste will at length be established. . . ," there was not yet sufficient similarity in people's associations to afford a foundation for such a standard.[14]

Had Priestley's Oratory and Criticism been published when first written, it might have been as influential as his works on language and grammar. The fifteen-year delay in publication gave the lead to George Campbell's Philosophy of Rhetoric (1776) in placing rhetoric on a psychological basis, while Archibald Alison's Essays on the Nature and Principles of Taste (1790) took the lead in applying associationism to aesthetics. Yet Wordsworth, who is supposed to have drawn his concept of poetic invention from Alison's Essays, is alternately described as reading Priestley's edition of Hartley in the bookshop where he "foregathered with his malcontent friends," while a recent analysis of Wordsworth's Preface to the Lyrical Ballads cites Priestley's Oratory and Criticism and the Lectures on Language as frequently as any other of the possible influences on that work and more often than Alison's Essays.[15]

Having taught at Warrington Academy for all of four years, Priestley next determined to change the curricular thrust of the entire institution -- and succeeded in doing so. In 1765 he published An Essay on a Course of Liberal Education for Civil and Active Life, in which he recommends a new set of courses in history and law for students not intending to enter one of the learned professions. The first part of the Essay criticizes the state of learning in English schools and universities "as wholly unfit to qualify men to appear with advantage in the present age" and goes on to propose changes both in method and curriculum to remedy the problem.[16] These proposals were supplemented and amplified in Priestley's Miscellaneous Observations relating to Education, published in 1778 but "written at different times, as particular occasions suggested," and based upon experience with liberal education which had "been the business of a great part of . . . [his] life to study and conduct . . . ."[17] His general treatment of education, in the Essay and Miscellaneous Observations, is the basis for claims that Priestley was the most significant English writer on educational philosophy in the period between Locke and Herbert Spencer.[18]

The declaration that Priestley "did more perhaps than any other man to modernize the curriculum of the dissenting academies" is a consequence of the second part of the Essay, its emphasis upon history, and its printing of three course-syllabi on: The Study of History and General Policy, The History of England, and The Constitution and Laws of England.[19] Although Priestley did not then publish the full lectures themselves (indeed only those on History and General Policy were ever published and they not until 1788), the courses were adopted at Warrington and several other of the dissenting academies. And to aid in teaching them, Priestley published history teaching aids. His Chart of Biography and Description of a Chart of Biography (1765) and the New Chart of History and Description of a New Chart of History (1769) appear to have been early and important examples of the now-familiar time-line diagrams accompanied by explanatory handbooks.

These works of educational reform, like those earlier on grammar, rhetoric, and aesthetics, were soundly based on earlier writers: on Locke and Isaac Watts and David Hartley; on Philip Doddridge, who had developed the teaching methods Priestley had studied under at Daventry Academy, and, for the emphasis on history, on Bolingbroke and perhaps even on King George I, who had endowed Regius professorships in modern languages and history at Oxford and Cambridge in 1724. Even the Charts of biography and of history were modifications of a French historical chart, published in English translation in 1750. But that chart had been clumsy and ill-conceived, the Regius professorships in modern history had become sinecures by the end of their first decades, and the recommendations of Locke, Watts, and Doddridge had remained fragmentary and unused. Priestley made of his materials something that was systematic, operational, and widely noticed. The Essay on a Course of Liberal Education was one of the most frequently published of his writings, in separate printings, appended to early editions of the Miscellaneous Observations relating to Education, and prefixed to the Lectures on History and General Policy, for at least sixteen printings, including translations into Dutch and French. The Miscellaneous Observations was printed at least six times by 1831. Each of the Charts was revised and reprinted at least twice and their Descriptions went through at least nineteen editions, including those in the United States, Holland, Austria, and Italy, before their final appearance around 1840. And the Lectures on History, for all their late appearance, were to be republished eight times by 1826, including Dutch and French translations, were used as texts at Princeton and Harvard Universities in the United States and, after John Symonds and Matthew Arnold rescued the Regius professorships in modern history at Cambridge and Oxford, were recommended for use in the English universities.

By 1766 Priestley seemed fully embarked on a distinguished career as a teacher, yet early the next year he accepted a position as minister to the congregation at Mill Hill Chapel, Leeds. Neither his publications, nor the LL.D. or F.R.S. obtained because of them, could have obtained for him a better position in teaching; the income at Warrington was always uncertain and, still more important, the invitation from Leeds gave him the opportunity of returning triumphant to the region of his youth, in a position where he might dominate over the provincial chapel which had refused him membership at the age of nineteen. The position was too tempting to refuse.

He responded with enthusiasm to the challenge of his return to the ministry. During the next six years, he was active in preaching, visitation, the religious instruction of children, and, of course, in religious and theological writing and publication. He prepared a Catechism for Children (1767, with fourteen further editions to 1817), published pamphlets on family prayer, church ritual and discipline, and many sermons, including one, An Appeal to the Serious and Candid Professors of Christianity (1770), which went through at least twenty printings by 1836. By the time he went to Leeds he had nearly completed his theological itinerary to humanistic Unitarianism, but much of his writing was intended to be independent of any specific doctrinal position; his theological argumentation, which also begins seriously during this period, was a separate occupation as, to some degree, was his writing on politics.

Now Priestley was always to deny that he particularly concerned himself with politics and, insofar as the majority of his political writings related also to religion, that is perhaps a defendable stance. But no dissenting minister, concerned for the social as well as spiritual well-being of his co-religionists, could well avoid some involvement in political issues.

Although the Toleration Act of 1689 had freed dissenters from some of the worst consequences of the Test and Corporations Acts, they still suffered under some legal disabilities, there was frequent petty persecution, and Unitarianism, itself, remained a penal offence as late as 1813. Most of Priestley's political writing discuss the legal status of dissenters and of dissenting ministers. His Remarks on Some Paragraphs in the Fourth Volume of Dr. Blackstone's Commentaries on the Laws of England, relating to Dissenters (first published in 1769, with some ten additional printings in London, Dublin, and Philadelphia) was of that class, while his View of the Principles and Conduct of the Protestant Dissenters, with respect to the Civil and Ecclesiastical Constitution of England of 1769 marked the beginning of his campaign for the repeal of the Test and Corporations Acts. That campaigning was to continue, with a brief interlude during his politically-sensitive service with Lord Shelburne, into his ministerial service at New Meeting House, Birmingham and included his Letter to the Right Honourable William Pitt (1787) and his sermon on The Conduct to be observed by Dissenters in Order to procure the Repeal of the Test and Corporations Acts of 1789. By that time, however, he had become more extreme and thought repeal was only the beginning of necessary politico-religious reform. As his Letters to the Right Honourable Edmund Burke indicate, by 1791 he wanted disestablishment of the Anglican Church.

Arguments for religious freedom shade easily into arguments for other freedoms, and particularly did so during the years of the Wilkes exclusion crisis and the Townsend Acts. No doubt Priestley was encouraged by his friend Benjamin

Franklin in the writing of such pamphlets as: The Present State of Liberty in Great Britain and her Colonies (1769) or the later An Address to Protestant Dissenters . . on the approaching Election . . . , with respect to the State of Public Liberty in General, and of American Affairs in Particular (1774). Even the more opulent members of his congregation were, however, unlikely to object to the pamphleteering, for they were cloth-merchants suffering from the colonists' non-importation agreements.

Although some of the pamphlets were published anonymously (the Address of 1774 to protect Lord Shelburne), Priestley was soon identified as a leader in the group of English reformers who worked, in vain, during the eighteenth century, for redistribution or even extension of the franchise, easing of the penal codes, abolition of slave trade, and parliamentary accountability, as well as religious reforms. Every serious scholar of eighteenth-century English political reform movements refers frequently to Priestley and quotes from his publications.[20] Yet he was not a stereotypical Enlightenment "natural-rights" theorist. In his first, and most general, purely political work, An Essay on the First Principles of Government; and on the Nature of Political, Civil, and Religious Liberty (1768), he does, in making what henceforth was to be a widely-employed distinction between civil and political liberty, emphasize that each individual has certain rights which are natural, equal, and inalienable. When he treats of political liberty, however, he does so from the nature and object of government. As he declares the object of government to be: "the goodness and happiness of the members, that is, of the majority of the members of any state," Priestley transcends natural-rights doctrine for a utilitarian, relativist radicalism. The First Principles was only to be published in four editions by 1835 (and a Dutch translation), but its central doctrine was to have a major influence, for it was avowedly the inspiration for Jeremy Bentham's Utilitarian formula: "The Greatest Good to the Greatest Number."[21]

In 1773 Priestley was lured from his position at Leeds to become literary companion to William Petty, Lord Shelburne. Shelburne had recently lost his wife and was at loose-ends, being out of political office and favor. A notable patron of the arts and of liberal political philosophers, he was persuaded by Richard Price to take Priestley under his protection. Priestley, making but 100 guineas a year and concerned to insure some long-term security for a growing family, accepted Shelburne's offer of £250 a year. It seems, on the whole, to have been a mutually satisfying relationship. Shelburne had distinguished himself politically in support of measures in which Priestley believed, while Priestley's growing reputation in politics and science (he presented his first paper on chemistry to the Royal Society during 1772) brought Shelburne into contact with a number of persons, in Britain and on the Continent, he might not otherwise have met. The two men were, however, very different in temperament and it is doubtful if they ever became really friendly. When Shelburne remarried in 1779, the blunt middle-class virtues of the Priestleys were increasingly out-of-tune with an aristocratic household, and Shelburne's intermittent negotiations for a return to political office (he joined the Rockingham Administration as secretary of state in 1782 and became First Lord of the Treasury at Rockingham's death) were surely impeded by too close an association with Priestley's circle of radicals. The connection was, therefore, broken off in 1780 and Priestley retired to Birmingham, with an annual pension of £100 which was continued until his death.

The seven years of Priestley's connection with Shelburne were years of scientific advance and intellectual consolidation. He had a chemical laboratory at Bowood, Shelburne's estate in Wiltshire, wrote there five of the six volumes of his major work on pneumatic chemistry, and in Wiltshire's summer sun of August 1774, performed the experiments leading to the discovery of oxygen. He had also the leisure to bring to publication manuscripts previously left unfinished (e.g. Oratory and Criticism and Miscellaneous Observations on Education), to continue his theological investigations with, for example, a Harmony of the Evangelists in Greek (1777) and English (1780) and his Institutes of Natural and Revealed Religion, in three volumes (1772,

1773, 1774), and finally, to begin to put together a philosophical justification for his theological and scientific opinions.

Priestley was never to write a single, comprehensive and coherent, exposition of a philosophic doctrine. In a series of publications, beginning in 1772 and essentially completed by 1778, he did, however, outline a metaphysical position so different from that of his contemporary countrymen that one wonders at its neglect by historians of philosophy. The series commences with the first volume of his Institutes of Natural and Revealed Religion (1772), continues through An Examination of Dr. Reid's Inquiry into the Human Mind on the Principles of Common Sense (1774) and his edition of Hartley's Theory of the Human Mind, on the Principle of Association of Ideas (1775), and culminates in Disquisitions relating to Matter and Spirit (1777), The Doctrine of Philosophical Necessity illustrated (1777), and A Free Discussion of the Doctrines of Materialism and Philosophical Necessity . . . between Dr. Price and Dr. Priestley (1778). Of the several lines of thought followed in these volumes, perhaps the most interesting, at least in its connecting Priestley's science and theology, is that which develops a philosophical monism. By denying any essential, knowable, distinction between matter and spirit, he, at once, attacks the independent existence of spirit -- and, therefore of a pre-embodied Christ, the possession of distinct faculties of mind, and the explanation of scientific phenomena by separate and distinct imponderable fluids of heat, electricity, and magnetism.

The unitary materialism apparent in this position required a discussion of the nature of matter. Priestley declared that it was impossible to determine ultimate properties of matter lying behind our experience of its properties, but that this experience was actually one of direct interaction with forces of attraction and repulsion. So far as man could know, then, matter was nothing but these forces, designedly organized by the laws of nature and of nature's God. These laws could be discerned by man through reason, observation, and experiment, for God was benevolent and a benevolent God would structure the world to be understood. Contrary to Hume, therefore, man could know, a priori, that the appearance of causality was a reflection of reality.[22]

There are curious resemblances between this metaphysics of matter and that extractable from the work of Immanuel Kant, but there is no possibility of any direct relationship. Kant could not read English and was aware of Priestley's work only at second-hand and incorrectly. His Critique of Pure Reason (1781) and, more significantly, Metaphysical Foundations of Natural Science (1786) each appeared after Priestley's metaphysical writings were essentially completed and were not easily available in England until after Priestley's departure for the United States. The most that can then be said is that Priestley constructed a materialist philosophy of science, out of the writings of the Cambridge neo-Platonists, Isaac Newton, John Locke, David Hartley, and the Dalmatian nature philosopher, Roger Joseph Boscovich, which more nearly approaches the idealist philosophy of science of Immanuel Kant than any other Briton was to achieve during the eighteenth century.

Less than six months after his arrival in Birmingham, Priestley was invited to become senior minister to the congregation of New Meeting House and happily accepted. Resuming the duties of his favorite occupation, he again preached regularly and conducted classes for the younger members of the congregation. He also resumed a vigorous schedule of political and theological publication; during his eleven years in Birmingham, he was to publish forty-six books and pamphlets and twenty-one articles on theology and politics compared to three books and eleven papers on science.

Most of his scientific work consisted of objections to the new chemistry being proposed by Antoine Lavoisier, though his membership in the Lunar Society of Birmingham increasingly involved him in technological consultations which did not lead to publication. He had long corresponded with physicians and with Josiah Wedgwood on practical applications of his science and now, to that continuing correspondence, he added letters to Arthur Young on agriculture and to his brother-in-law, John Wilkinson, on smelting iron. He also talked with, and wrote to, fellow members of the Lunar Society on the chemical composition of inks, on alkalies, balloon gases, food preservation, and even on the economics of substituting condensable gases for steam in Watt's atmospheric-pressure steam engine.[23] Except in comparison with his previous productivity, this would seem a respectable spare-time scientific achievement for a busy minister, but, for Priestley it indicates that his interests had clearly shifted toward fulfilling his theological responsibilities.

Something must, therefore, be said here about those theological studies which run, like a leit-motif, through all his scholarly activities. For Priestley, there were two supports for a belief in Unitarian Christianity: that to be derived from natural religion, on which he had written in his metaphysical works and continued to defend until his death; and that to be derived from revelation. The Scriptures which transmitted that revelation were, however, not the result of plenary inspiration. Certainly the writers were inspired, but they wrote of their inspiration as men of their time, in the languages of their time, and in response to particular occasions. The books of Scripture were assembled from previously scattered devotional and disciplinary writings and they had since given rise to interpretations which increasingly corrupted the text and the original intention of the writers. Priestley and his theological friends occasionally concerned themselves with textual criticism, primarily in articles written for Priestley's Theological Repository. But most of his writing, from the Scriptural Doctrine of Remission of 1761 to the end of his career, involved the application of historical analysis to cleanse Christianity of its corrupting accretions and to penetrate through the language of the Jews, and especially the Hellenicized Jews, to a primitive Christianity directly reflecting what had been revealed.

During his Birmingham years, in the History of the Corruptions of Christianity (1782) in two volumes, the History of Early Opinions concerning Jesus Christ (1786) in four volumes, the General History of the Christian Church to the Fall of the Western Empire (1790) in two volumes, and the many volumes of controversy which these excited, Priestley made a substantial contribution to the development of what is called the higher, or historical, criticism of the New Testament. It is a measure of the continuance of doctrinal parochialism that the material contained in these more than six thousand pages should be so generally ignored by historians of the higher criticism. Most of them fail to mention Priestley, skipping lightly from Richard Simon and John Locke to the German critics of the late eighteenth and early nineteenth centuries. Of the few who do mention him, most adhere to a nineteenth-century romantic historical positivism, discounting Priestley's work because of his Unitarian bias while praising later more orthodox critics for arriving at the same conclusions.[24]

Unhappily, Priestley's English contemporaries did not take his work that lightly. George III named one of Priestley's adversaries, Dr. Samuel Horsley, a bishop for his efforts to silence him. And he had to be silenced, for his arguments were attracting favorable attention. A number of promising clergymen, such as Theophilus Lindsey, Gilbert Wakefield, John Disney, Thomas Fysshe Palmer, John Jebb and William Frend, even left or had to be forced out of positions in the Church or universities and then established themselves as Unitarian ministers or teachers. Still, had Priestley not combined theological heresy with politics, he might have escaped trouble. But in one of his attacks on religious establishments, he injudiciously compared reasoned argument to the laying, as it were, a trail of gunpowder to the foundations of the Church; now it could be claimed that the Church was in danger. In the face of Burke's impassioned denunciation, Priestley defended the French Revolution; now it could be claimed that the King was in danger. In 1791, as the aftermath of a quiet celebration, which Priestley did not attend, of the Glorious Revolution of 1688, a Church-and-King riot destroyed the New Meeting House, Priestley's home and laboratory, and sent him fleeing for his life.

There were contemporary claims by Birmingham dissenters that the riot had been staged by local tories and clergymen, with the active encouragement of the national conservative establishment. These claims were not then investigated and certainly cannot now be substantiated. It can be said, however, that deliberately or not the establishment incited the riot. From pulpit and Parliament, they had played the part of Henry II toward Thomas à Becket, asking in effect: "Who will rid us of this turbulent Priest?"[25] And they had their way. Priestley settled temporarily in London, where he replaced the deceased Richard Price as minister and taught history and natural philosophy at Hackney New College. But he continued to be threatened, his wife was unhappy and his sons unable to find jobs. In 1794 the Priestleys sailed to the United States and settled in Northumberland, Pennsylvania. Though his friend, Benjamin Franklin, was dead and he found that prejudice and persecution were not unique to the old world, he was befriended by Thomas Jefferson and regarded as an elder statesman of science by members of the American scientific community. He attempted to continue his life of scholarship and research, adding comparative religion to his armory in support of "rational Christianity," but his long period of creativity was essentially over. His writings of this period are mostly compilations or variations on old themes. His wife and youngest, favorite, son predeceased him and, when he died in 1804, he was buried with them in a Quaker burial ground in Northumberland, as no orthodox church would accept his body. His house, still standing (and now beautifully restored) on the banks of the Susquehanna River, has become a museum and a Mecca for the pilgrimage of chemists.

This paper has attempted to show the quality and quantity of Priestley's work in areas other than science. It would be hard to summarize its argument better than in the words of Frederick Harrison, the English apostle of Comte's positivist religion, when he nominated Priestley for the Positivist Calendar of secular Saints:

> If we choose one man as a type of the intellectual energy of the eighteenth century we could hardly find a better than Joseph Priestley, though his was not the greatest mind of the century. His versatility, eagerness, activity and humanity; the immense range of his curiosity in all things, physical, moral or social; his place in science, in theology, in philosophy and in politics, his peculiar relation to the Revolution, and the

pathetic story of his unmerited sufferings, may make him the hero of the eighteenth century. [26]

No more in grammar, aesthetics, history, politics, education, metaphysics, or theology, than in science, was Priestley *the* foremost scholar of his day; no one came close to his level of achievement in *all* of these. Even without his science, Joseph Priestley was the major figure of the English Age of Enlightenment.

Notes

1. The evidence for these enumerations is derived from my own bibliographic listings, but can, generally, be confirmed by examining Ronald E. Crook, A Bibliography of Joseph Priestley 1733-1804 (London, the Library Association, 1966). Note that a few of his last publications were, in fact, posthumous, but this does not bear on the significance of the comparison of scientific to non-scientific publication.

2. Joseph Priestley, Experiments and Observations, relating to Various Branches of Natural Philosophy, vol. iii (Birmingham, for J. Johnson, 1786) p. ix.

3. The terms "scientist" and "science" are anachronistic as well each being of nineteenth-century English coinage. The correct terms: "natural philosopher" and "natural philosophy" are, however, somewhat misleading today and are longer and more difficult to use. This paper will, therefore, continue to use the anachronisms: scientist and science.

4. See, for example, his declaration that he chiefly valued the success of his scientific studies "from the weight it may give my attempts to defend Christianity . . .," in J. T. Rutt, ed., Theological and Miscellaneous Works of Joseph Priestley, "Memoirs and Correspondence- 1733-1787," vol. 1, Part I (New York, Kraus Reprint Co., 1972, from the 1831 edn.), p. 200. Hereafter cited as Priestley, Memoirs, i, I.

5. Priestley, Memoirs, i, I, p. 55.

6. Enumeration of the scientific papers Priestley published during his decade in the United States is confused by the fact that many of them are essentially copies of one another. He would send a paper to the American Philosophical Society, send another version for earlier publication to the New York Medical Repository, and then send the same paper, or a slight variation from it, to England for publication in the Monthly Magazine or Nicholson's Journal.
7. See op. cit., note 2, pp. vii-viii.
8. S. T. Coleridge to John Prior Estlin, 16 January 1798, in Leslie Griggs, ed., Collected Letters of Samuel Taylor Coleridge (Oxford, Clarendon Press, 1971), vol. i, p. 372.
9. "Not being a professed chemist and attending only to such articles in that branch of knowledge as my own pursuits are particularly connected with . . . illustrations of chemical processes are not so likely to occur to me, as they are to others, who by their profession give a general attention to every thing within the whole compass of chemistry," Priestley, Experiments and Observations relating to . . . Natural Philosophy (London, for J. Johnson, 1779), p. 39.
10. Indeed, I have so argued, in my "Joseph Priestley: Theology, Physics, and Metaphysics," to appear in a forthcoming issue of Enlightenment and Dissent.
11. The quotation is that of Ivan Poldauf, On the History of Some Problems of English Grammar before 1800 (Prague, Karlovy University, Prague Studies in English, 1948), p. 231, but the sentiment is repeated in almost every other study of the history of English grammar. See, for another example, Scott Elledge, "The Naked Science of Language," in Howard Anderson and John S. Shea, eds., Studies in Criticism and Aesthetics, 1660-1800 (Minneapolis, Univ. of Minnesota Press, 1967), pp. 266-295.
12. Priestley, A Course of Lectures on the Theory of Language and Universal Grammar (Warrington, by W. Eyres, 1762), p. 29; The Rudiments of English Grammar (London, T. Becket and P. A. DeHondt, J. Johnson, 1768), p. xii.

13. Joseph Priestley, A Course of Lectures on Oratory and Criticism, Vincent M. Bevilacqua and Richard Murphy, eds. (Carbondale, Ill., Southern Illinois University Press, 1965), p. xxxi; Wilbur Samuel Howell, Eighteenth-Century British Logic and Rhetoric (Princeton, N.J., Princeton University Press, 1971), pp. 632-633.

14. Priestley, Oratory and Criticism, pp. 134-135.

15. Willard L. Sperry, Wordsworth's Anti-Climax (Cambridge, Mass., Harvard University Press, 1935), pp. 126-127; W. J. B. Owen, Wordsworth's Preface to Lyrical Ballads (Copenhagen, Rosenhilde and Bagger, vol. ix of Anglistica, 1947), passim.

16. Priestley, Essay on a Course of Liberal Education ([London], C. Henderson, T. Becket and deHondt, J. Johnson and Davenport, 1765), pp. 23-24.

17. Priestley, Miscellaneous Observations relating to Education (Bath, for J. Johnson, 1778), pp. xi, v-vi.

18. See, for example, H. M. Knox "Joseph Priestley's Contribution to Educational Thought," Studies in Education. The Journal of the Institute of Education 1 (1949), 82-89; the sentiment is repeated in different terms in many histories of English education.

19. Frederick Elby and Charles Flinn Arrowood, The Development of Modern Education (N.Y., Prentice-Hall, 1934), p. 606; the claim is the more significant as curriculum reform in dissenting academies led that in the universities, often by decades.

20. See, for example, two recent works which cover the period: Colin Bonwick, English Radicals and the American Revolution (Chapel Hill, Univ. of North Carolina Press, 1977), which has fifty-six separate index-references to Priestley; and Albert Goodwin, The Friends of Liberty: The English Democratic Movement in the Age of the French Revolution (Cambridge, Mass., Harvard University Press, 1979), which has eleven.

21. Priestley, An Essay on the First Principles of Government (Dublin, James Williams, 1768), p. 17; the origin of Bentham's "formula" has been argued by scholars who claim closer parallels in the writings of Helvetius or

Beccaria, than in this phrase of Priestley. Bentham, however, explicitly acknowledges the origin in Priestley's First Principles, see Mary P. Mack, Jeremy Bentham: An Odyssey of Ideas 1748-1792 (London, Heinemann, 1962), pp. 102-103.

22. The textual citations which support this summary are to be found in my "Joseph Priestley: Theology, Physics, and Metaphysics," loc. cit., supra n. 10.

23. For an account of Priestley's work in applied science, see Robert E. Schofield, The Lunar Society of Birmingham: A Social History of Provincial Science and Industry in Eighteenth-Century England (Oxford, Clarendon Press, 1963), especially pp. 200-202.

24. I have found no reference to Priestley's contributions in over twenty books or articles relating to the higher criticism and its history. Only Unitarian historians, such as Alexander Gordon, Heads of Unitarian History (London, Philip Green, 1895), pp. 117-118, generally remark favorably on Priestley's scripture criticism. F. C. Conybeare, History of New Testament Criticism (London, Watts & Co., 1910), pp. 93-96, describes Priestley as "a bold innovator in the domain of Church History," whose "main heresy was the entirely correct opinion that the earliest Christians neither knew anything of Trinitarian doctrine nor deified Jesus . . .;" Conybeare declares that Priestley developed a "fertile method of inquiry," and then concludes that he was "a mixture of enlightenment and superstition."

25. The question is a paraphrase of a variously-quoted statement supposedly made by King Henry II; the version most recently offered: "Who will get rid of this turbulent priest for me?", is that of Christopher Fry in Curtmantle A Play (London, Oxford University Press, 1961, 1965), 2nd. edn.

26. Quoted by T. E. Thorpe, Joseph Priestley (London, J. M. Dent & Co., 1906), p. 1.

'A Sower Went Forth': Joseph Priestley and the Ministry of Reform

By J. H. Brooke
DEPARTMENT OF HISTORY, UNIVERSITY OF LANCASTER, BAILRIGG, LANCASTER LA1 4YG, U.K.

In his celebrated satire of the early Royal Society, Jonathan Swift spoke of the fevered projects of the 'Grand Academy of Lagado':

> The first man I saw was of a meagre aspect, with sooty hands and face, his hair and beard long, ragged and singed in several places. His clothes, shirt, and skin were all of the same colour. He had been eight years upon a project for extracting sun-beams out of cucumbers, which were to be put into vials hermetically sealed, and let out to warm the air in raw inclement summers. He told me, he did not doubt in eight years more, that he should be able to supply governors gardens with sun-shine at a reasonable rate; but he complained that his stock was low, and intreated me to give him something as an encouragement to ingenuity, especially since this had been a very dear season for cucumbers. I made him a small present, for my Lord had furnished me with money on purpose, because he knew their practice of begging from all who go to see them.

The satire was too early to be directed at Priestley. It is usually thought to have been penned with Stephen Hales in mind, the inquisitive vicar of Teddington who had been investigating the effect of sunlight on the growth of vegetables.[2] But one finds oneself thinking of Priestley, not least because, had the project been successful, he would have made good use of the sunbeams. "I have had the most pleasing results", he wrote to Alessandro Volta, "from my experiments on the exposure of various substances to the sun beams in water. Flesh, and most vegetables, turn green... and, notwithstanding their becoming putrid, emit great quantities of the purest air".[3] Dependent as he was on a burning glass for a focus of heat he was frequently stalled by the raw, inclement summers of Birmingham. "Having at length got sunshine", he reported to Benjamin Franklin in June 1782, "I am busy in prosecuting the experiments about which I wrote to you, and shall soon draw up an account of them for the Royal Society".[4] But there are other reasons why one thinks of Priestley. If Professor Schofield is correct in stressing the physical rather than chemical theory that lay behind Priestley's work on gases, then Priestley emerges as a latter-day Hales whose vocabulary as well as experimental technique reflected those of his predecessor.[5] More immediately perhaps, Priestley got his revenge against Swift whom he regarded as proud and malevolent.[6] He got his revenge because spinach, groundsel, and doubtless cucumbers too, could be made to release that "purest air" if not the sunbeams themselves. And Priestley did not doubt in a few years more that he would be

able to supply governors' gardens at a reasonable rate. His dephlogisticated
air was promised as a "fashionable article in luxury".[7] And I find myself
linking Priestley with Swift and Hales for one last reason. In his Cambridge
days Hales had been party to the chemical experiments of William Stukeley who
had sometimes surprised his College with a sudden explosion.[8] To read the
letters and papers of Priestley is also to be assaulted by explosion after
explosion. When he began to suspect that his dephlogisticated air was a
distinctive species, he conjectured that it might explode more violently with
'inflammable air' (hydrogen) than would common air. He was far from disappointed. His report was about fifty times as loud.[9]

In one respect, however, Priestley and Hales were poles apart. With his
Cambridge background and vicarage garden, Hales was likely to be perceived as
a pillar of the establishment. Born into a dissenting family, against whom
the gates of Oxbridge would have been barred, Priestley was to become a
prolific and notorious critic of political and theological orthodoxies. Explosions which had been a reality in his laboratory passed into metaphors which
erupted through his prose. Passive obedience to political authority he would
describe as an "exploded doctrine".[10] Well before the French Revolution he had
enriched his metaphor with gunpowder, predicting the fall of the English
hierarchy and the blasting of religious orthodoxy.[11] The chemical metaphor
mediated between Priestley the intellectual reformer and Priestley the
revolutionary, certainly the <u>perceived</u> revolutionary. Metaphors, when taken
literally, can be very costly. "Now, prithee Mr. Priestley", enquired one
of his detractors, "how would you like it yourself, if [church people] were
to send you word that they had laid trains of gunpowder under your house or
meeting-house?"[12] It didn't happen quite like that but, as is well known,
Priestley was to lose both to the mob.

For the large part of his career Priestley had almost certainly perceived
himself as a reformer rather than revolutionary. And principally a religious
reformer whose duty, as a historian, was to expose the manner in which
Christianity had been corrupted by the intrusion of alien philosophies, and
how it would remain so as long as the voice of reason and the liberty of
religious expression were suppressed. This is one of the reasons why Priestley
has been so attractive to historians, as rationalist and reformer in an age
of reason. But when one turns to the beliefs which sustained and animated his
campaign one is confronted by a set of paradoxes which few commentators have
been able to resist. What is one to make of a man who proclaimed himself a
Christian and denied the divinity of Christ; an apologist who considered this
the best of all possible worlds and yet one which man could improve; a theist
who denied that God could act directly on the human mind and yet who insisted
that his God was more in control of human affairs than the God of religious

orthodoxy; a Scriptural exegete who accepted the reality of certain biblical miracles as part of an argument to show that miracles did not occur; a philosophical determinist who believed that a denial of the autonomy of the will made men more, not less, responsible for their actions; an advocate of toleration for Roman Catholics whilst denouncing Catholic religion as "properly anti-Christian" and "a system of abomination little better than heathenism"[13]; a materialist who did not believe in matter, certainly not solid matter as usually understood; an empiricist who, having discovered oxygen, not only considered it a compound but supplied Lavoisier with an important clue for establishing it as an element[14]; and over-riding all, it would seem, a radical in politics and religion and yet so conservative in his chemical theory that he was left picking nits in the new French system?

Priestley scholars have resolved and transcended these paradoxes in various ways. To portray him as a reactionary in science because he would not part with phlogiston is to miss the point that he was not alone in resisting the nomenclature of the French, and the even more important point that Lavoisier's system was not inviolable[15]: the very name oxygen, or 'acid-producer', testifies to an oversimplification which the young Humphry Davy was to expose[16]. One scholar has even gone so far as to suggest that Priestley was no more committed to the truth of the phlogiston theory than he was to the falsehood of the oxygen: each depended on theoretical constructs not ultimately verifiable.[17] The anti-theoretical thrust of Priestley's writings can then turn him from a conservative into a radical. It was the new system which in a few years, so he told one of his correspondents, would be "universally exploded"[18]. This same preoccupation with facts rather than theoretical constructs has allowed John McEvoy to present Priestley as a scientific radical in a further respect. In his writings on scientific method and in his account of his own discoveries, Priestley would introduce a kind of egalitarianism into the conduct of a scientific investigation. One did not require the theorizing genius of a Newton to make advances in science, and, as Priestley's own experience confirmed, one's theoretical preconceptions were often as not dashed by unforeseen results.[19] Like an 18th century Peter Medawar, Priestley insisted that the typical scientific paper, sanitized of the real route to discovery, was a fraud. The image of Priestley as a reactionary in matters scientific has been transformed still further by two recent suggestions from Simon Schaffer.[20] I should like to mention these briefly because they will serve as a more substantial introduction to my theme: the main components of Priestley's theology and their bearing on his science.

Dr. Schaffer has argued that a radical interpretation of Priestley's science was made possible by the disruption of a tradition in which lecturers on natural philosophy had amused and amazed their audience with the display of powers, such as electricity, which were separately and immediately ascribed

to the power of God himself. The light which passed from nature to nature's
God was transmitted, as it were, instantaneously - the whole performance
geared to entertainment and the moral exhortation that went with a vicarious
exposure to the arm of God. It was a practice in which Priestley to some
degree participated. But with his participation the rules of the game were
changed. Indeed the popularization of science ceased to be a game, an amuse-
ment; it took on a new earnestness as a means of displaying not God's immediate
power but the rationality of His creation. It was now the interconnections
and the mutual adaptations of the powers of nature, not the powers themselves,
that bore witness to the divine plan. Nature still pointed towards nature's
God, but since an understanding of those connections required a modicum of
science, the inference was no longer spectacular and direct, but sober and
mediated by scientific knowledge. The effect, in Schaffer's own words, was to
break "the connection which natural philosophers had hitherto exploited between
God's direct power and that of the performer". It was a break which allowed
other lecturers, Adam Walker, Erasmus Darwin, and Thomas Beddoes to invest
their science with radical political purpose.

The second of Schaffer's suggestions is that because Priestley saw nature
as an ordered system rather than as a plethora of separate powers, his scientific outlook and the way it was interpreted can only be comprehended where
there is an appreciation of that system which, for Priestley, conferred significance on individual facts. Priestley himself would certainly have agreed.
Even the most trifling observations, on plant respiration for example, could
acquire, in his own words, "the greatest dignity and importance; serving to
explain some of the most striking phenomena in nature, respecting the general
plan and consitution of the system, and the relation that one part of it bears
to another.[21]" And again, scientific discoveries "give us a higher idea of the
value of our being, by raising our ideas of the system of which we are part".[22]
A system, a general plan, a constitution of nature designed to promote human
happiness.[23] There is no escaping the fact that the nature of nature was, for
Priestley, grounded in his theology.[24]

This means that if we were only interested in Priestley's scientific
output, we would still have to take his religious beliefs into account. It is
clear, for example, that his attitude towards the controversial issue of
spontaneous generation was informed by his strong theistic principles. His
contemporary, the poetic evolutionist, Erasmus Darwin was not only arguing for
the generation of micro-organisms from inanimate matter, but invoking some of
Priestley's observations to prove it. The green matter produced in stagnant
water which Priestley finally ascribed to the presence of seeds or germs in
the air, Darwin wanted to see emerge from more primitive organic, but not
organized, particles. True to form, Priestley accused the doctor of

harbouring an 'exploded doctrine'.[25] Empirical evidence was, of course, marshalled by both sides. But the key line of argument for Priestley went like this. The gap between organized and unorganized matter was so great that there could be no natural connection between them. Consequently, the affirmation of spontaneous generation was to affirm an effect without a cause. For an oak to emerge from an organic particle rather than an acorn would be as miraculous is if it emerged from a bean, a cucumber or nothing. But the advocates of spontaneous generation were in the business of denying miracles. That, from Priestley's angle, was part of their motivation. The upshot was that all the beautifully adapted animals and plants would be destitute of an intelligent cause. And if they were, so might be the whole system. Priestley could agree with Darwin that "there is more dignity in our idea of the Supreme Author of all things, when we conceive Him to be the cause of causes, than the cause simply of the events which we see", but the logic of Darwin's position was to dispense with the Author altogether.[26] Priestley's theism also steered him away from the transformation of species on which Darwin had been versifying.[27] Of even greater interest, perhaps, it regulated his enquiry into the diminution, vitiation and restoration of common air.[28] This is a point to which I shall return, but there seems little doubt that his belief in nature as a system, shot through with benevolence, sustained a long interest in the means which nature employed for replenishing the air. Thus, in August 1771, he could write to his friend and fellow unitarian, Theophilus Lindsey,: "I have discovered what I have long been in quest of, <u>viz.</u>, that process in nature by which air, rendered noxious by breathing, is restored to its former salubrious condition".[29]

An analysis of Priestley's <u>scientific</u> mentality would, therefore, be incomplete if it failed to take his theology into account. But there are more pressing reasons why any portrait must bring his religious mission to the fore. In his prolific theological writings he made a conspicuous and unique contribution to Unitarian doctrine. He was closely involved in setting up the first avowedly Unitarian chapel in England, with Lindsey as its first incumbent.[30] During his own career he was minister to five dissenting congregations and, from a reading of his private correspondence, there is no question but that he placed a far higher value on the identity and purification of Christianity than on the identity and purification of gases.[31] It was only on theological subjects that he was prepared to work for nothing.[32] A scientific reputation was certainly to be prized but only insofar as it lent weight and authority to the defence of his faith.[33] It was the winning of converts to his rational Christianity which, in the words of one Priestley scholar, "gave meaning to his very existence".[34] To convert an intellectual was the highest prize. As he wrote, following his visit to Paris in 1774:

> I can truly say, that the greatest satisfaction I receive
> from the success of my philosophical pursuits, arises
> from the weight it may give to attempts to defend Christianity,
> to free it from those corruptions which prevent its reception
> with philosophical and thinking persons, whose influence
> with the vulgar and unthinking is very great.[35]

The vision which sustained the whole enterprise was recorded in his General History of the Christian Church. There would come a time when every remaining corruption of Christianity would be removed, such that nothing would be found in it that any unbeliever, Jew or Mahometan could reasonably object to. Since "whatever is true and right will finally prevail", rational Christianity would, in due time, be the religion of the world.[36] No discovery in science could begin to compare with the discovery of life and immortality that was to be found in the gospel.[37]

In the remainder of this essay I shall try to sketch the rudiments of Priestley's Christianity and how they interlocked with his conception of science.

The rejection of Calvinism

It has been said that Priestley's whole life was a pilgrimage from the most conservative to the most liberal form of Protestant Christianity.[38] Such a description implies a gradual process of emancipation which, in several respects, it was. Having assimilated the doctrines of a strict Calvinism from early childhood and in the home of his aunt, he was clearly finding them uncomfortable by the early 1750s when he chose to attend the Dissenting Academy at Daventry rather than the Calvinist Academy at Mile End proposed by his relatives.[39] The doctrine of Christ's atonement for man's sin and especially the doctrine of double predestination - whereby the fate of the individual soul was determined only by the initiative of God's Sovereign Will - created an atmosphere of "gloom and darkness" against which his spirit rebelled.[40] Enjoying the freedom of enquiry which Daventry offered, he emerged in 1755 disenchanted with one facet of the doctrine of the Trinity. Describing himself at that stage as an Arian he still believed in the divinity of Christ, that He had been begotten of the Father. But he could not believe in the eternal coexistence of Father and Son in the form of one substance.[41] It was not until 1767 that he finally became a Socinian, committed to the view that Christ was essentially human, albeit with a divine mission.[42] During the period of his first ministry in the late 1750s he had finally been persuaded of the falsity of the doctrine of atonement, but only after "much pains and thought".[43] The full inspiration of the authors of Scripture was a casualty of the same period.[44] Although a convinced Socinian by the late 1760s several years were still to elapse before his notorious critique of the duality between matter and spirit - a critique which eventuated in the conclusion that materialism, Socinianism and the doctrine of philosophical necessity (or determinism) were mutually supportive strands in a coherent intellectual system.[45] Although the emancipation was gradual it was structured throughout by the reaction against Calvinism which had marked the beginning of his independence:

When he appealed to the serious and candid professors of Christianity to use their reason in matters of religion, he listed five doctrines endorsed by Calvinists which he considered insupportable.[46] These were that unregenerate man had no power to do God's will; that through the Fall of Adam all men were born into a state of original sin; that only a predetermined elect would enjoy salvation; that Christ was fully divine, and that through his sacrificial death had made atonement for man's iniquity. In what became a characteristic, though by no means original method, Priestley argued that the general tenor of the Scriptures was against each of these doctrines. Thus the countless exhortations to repent and to turn to God made no sense if man was so totally depraved that he was bereft of the power to obey.[47] Similarly he marshalled his favourite texts I Corinthians 8 v. 6 and I Timothy 2 v. 5 to establish that "There is one God, the Father; and one mediator the MAN Christ Jesus". Another characteristic move was to expose what, for Priestley, were the intrinsically immoral aspects of the Calvinist creed. To expect a man to repent of the sin of Adam, or to feel anything like remorse for it, was blatantly unreasonable when, in Priestley's words, "he cannot but know that he never gave his consent to it".[48] The doctrine of election which placed the fate of individuals at the mercy of God's arbitrary decree might pretend to be a doctrine of tender mercy, but in reality it was one of cruelty. Priestley's technique was to invoke analogies drawn from the human family. Surely no man of real goodness or compassion would wish to have children, knowing they were predetermined to misery.[49] The doctrine of atonement fell by the same token. Christ's injunction to forgive a penitent brother would not deserve the name of forgiveness if one were to insist on any atonement.[50] Priestley's God was the Father of the prodigal son, not a God of wrath.[51] However much he may have caricatured them, the doctrines of Calvinist orthodoxy failed, too, for practical reasons. He considered them extrinsically as well as intrinsically immoral. How, he would ask, can a man love his fellow creatures if the majority of them have to be regarded as objects of divine abhorrence? By contrast, the belief that God wants the happiness not the destruction of his creatures encourages good works and the love of others.[52] Theological beliefs, in other words, were to be judged by their moral implications and moral effects. The danger with Calvinist orthodoxy was that it could induce intense anxiety about one's spiritual condition, a fatal sense of despair and impotence, even a life of licentiousness.[53]

'A Sower Went Forth': Priestley and the Ministry of Reform

Given that the Gospel was concerned with the reformation of character, any doctrine which promised sudden acceptance with God was bound to be facile. Death-bed conversions were simply not on. "Some, indeed, are said to have been called at the <u>eleventh</u> hour," Priestley conceded, "but none at the twelfth".[54] True religion was born of serious resolution and sustained effort. Ideally it should be inculcated and practised from childhood. Otherwise, Priestley wrote, "there is very great danger that the thorns, briars, or bad soil, will prevent the good seed from ever coming to maturity".[55] The parable of the sower would serve him well on many occasions.

The suggestion that Priestley's religious pilgrimage was structured by a reaction against Calvinism receives further support from a work which he published even earlier than his <u>Appeal to the Serious and Candid Professors of Christianity</u>. In 1769 he had published some <u>Considerations on differences of Opinion among Christians</u>, in which he presented a sharp antithesis between the idea of God according to the <u>orthodox</u> system (which he rejected) and the <u>rational</u> system (which he preferred).[56] It took Priestley far more space to characterize the orthodox system than his own. One sentence sufficed to expound his doctrine which, by its independence of Trinitarian metaphysics and substitutionary concepts of Atonement, had the great advantage that it could actually be understood by the plain man.[57] The rational Christian, Priestley affirmed, considers the Divine Being as having produced all creatures with a view to making them happy, bearing a "most intense, and absolutely impartial affection to all his offspring ... inflexibly punishing all wilful obstinate transgressors, but freely pardoning all offences that are sincerely repented of, and receiving into his love and mercy all who use their best endeavours to discharge the duty incumbent upon them".[58] It was an egalitarian theology in those two respects: it preached God's impartial affection and in a way that the common plowman could grasp. His Socinian theology of repentance, as Martin Fitzpatrick has recently explained, was concerned with the reconciliation of man to God and not, through Christ's propitiation, the reconciliation of God to man.[59] The point that requires emphasis is that Priestley's antithesis between the rational and the orthodox was, and was perceived to be, an antithesis between incommensurable positions. If, for example, one renounced the doctrine of the Trinity, one could accuse those who worshipped Christ as idolators, in breach of the first great commandment: Thou shalt have no other Gods besides me.[60] When Priestley made that accusation it came as a very hard saying.

Given the antithesis in his mind between rationality and orthodoxy, given that he frequently used the word 'orthodoxy' when deriding the very existence of an <u>established</u> church,[61] it is tempting to ground his doctrinal rebellion in a political rebellion, particularly when one knows of his efforts to secure greater liberties for dissenters.[62] Was the 'orthodoxy' he so berated merely the religious symbol of all that was wrong with a society in which dissenters had to pay taxes in support of an established church they had renounced?[63] It is a tempting hypothesis but I think there is something of the devil in it. The clergy of the established church certainly saw him as a political activist.[64] A unitarian theology was bound to appear subversive to those who saw the marriage between Anglicanism and the state as essential to social and political stability. Nor did Priestley himself refrain from incriminating court and clergy on an issue as grave as the American war of independence.[65] His theology he did regard as an agent of revolution. But, and it is an important qualification, the revolution he had in mind was primarily intellectual. An internal revolution within Christianity was required before any external, political revolution could have any effect.[66] So, although he could warn that "a small change in the political state of things ... may suffice to overturn the best compacted establishments at once",[67] his emphasis usually fell on the emancipation of the mind. The present system might vanish like an enchanted castle but in the meantime what was required was the reformation of the mind.[68] As late as 1790 he reminded his readers that his opinion had always been that dissenters should patiently acquiesce in their exclusion, that "the true Christian maxim is patiently to bear every kind of persecution till it shall please God to put an end to it".[69] His activism, as he perceived it, was of the literary mould: "in proportion as a Christian will be patient in suffering, he will be bold to speak and to write".[70]

There is another reason why to ground Priestley's theology in an essentially political rebellion would be just too neat. The Calvinism against which he reacted was, in large measure, the orthodoxy of the dissenters themselves. One early critic even thought that for all his censure of Canterbury and Rome, Priestley had saved his severest blows for "poor orthodox dissenters".[71] Recent scholarship too has stressed the extent to which Priestley was a-typical of those who campaigned for the repeal of the Test and Corporation Acts.[72] His explosive theology, his sympathy for the cause of Catholic emancipation, and his dislike of hierarchies could make him a too conspicuous and embarrassing ally. Perhaps the right question is the one voiced by that same early critic:

> ... in the name of common sense, and rational religion, what have the orthodox dissenters done to the Rev. Dr. Priestley, that he should set them forth in such a disagreeable point of view?[73]

I don't know whether that question has ever been fully answered. One could point perhaps to experiences of rejection and exclusion in his early life, such as the celebrated occasion when he was refused membership of the chapel attended by his aunt because he had not evinced a compelling enough repentance for the sin of Adam.[74] One wonders too about his relationship with his brother Timothy who, as an upright Calvinist himself, was not averse to spreading injurious tales of Joseph's deviation.[75] What is clear is that a dialogue with orthodoxy was acted out before his very eyes whilst a student at Daventry. Dr. Ashworth, he recollected, took "the orthodox side of every question, and Mr. Clark, the sub-tutor, that of heresy, though always with the greatest modesty".[76] Having been exposed to the pride of those who were sure that they at least were among the elect, Priestley was peculiarly susceptible to a modest heretic.

Before leaving his dialogue with and rejection of Calvinism, there is the question whether it had any bearing on his science. The answer seems to be that there was a fundamental respect in which it did. A crucial weapon in Priestley's vocabulary was the adjective 'arbitrary'. Whenever Calvin's doctrine of predestination was under discussion Priestley would disparage the 'arbitrary decree'.[77] Not surprisingly in one who was to be sympathetic to the French Revolution the same epithet spilled over into his political comment. Thus he would refer to the "late arbitrary government of France".[78] One suggestion is that Priestley's presuppositions about the natural world were structured by the same renunciation of the arbitrary. If nature could be manipulated by a Sovereign Will, what guarantee was there of her uniformity? At the highest level of abstraction one might say that in Priestley's mind the very possibility of science was linked with the possibility of his rational religion. The uniformity of nature was more securely grounded in a Deity who did not intervene than in one who did.[79] In this he parted company with those giants of the 17th century, Boyle and Newton, who had been content to see in nature's laws the constancy of a Will which could move matter as it wished.[80] But Priestley, as we have already seen, was not above changing the rules of the game. The laws by which nature was ruled became part of a preordained nexus, inviolable and sacrosanct. This connection between his science and his religion is not as tenuous as it might seem. When discussing the doctrine of divine influence he would insist that all that God had done to reclaim mankind was to send the gospel among them.[81] Whatever moral good had come of it had been by natural means. Even the weeping of Christ over Jerusalem had not prompted a divine interposition. "So sacred with him", Priestley wrote, "are his established laws of nature" that there would be no supernatural action to rescue the impenitent.[82] Priestley's theology of repentance and his theology of the natural order were twin facets of a common mentality.

The authority of Scripture

A charge commonly levelled against Priestley by his critics was that he was setting up reason in place of revelation. It was a painful charge because far from undervaluing the Scriptures there were few men, he volunteered, who had "given more attention to them than myself".[83] He claimed it was through reading the Scriptures that he had become a unitarian.[84] In marshalling his proof texts he knew his Bible backwards. But he was to make no bones about its fallibility:

> The Scriptures ... were written without any particular inspiration, by men who wrote according to the best of their knowledge, and who from their circumstances could not be mistaken with respect to the <u>greater facts</u>, of which they were proper <u>witnesses</u>, but (like other men, subject to prejudice) might be liable to adopt a hasty and ill-grounded opinion concerning things which did not fall within the compass of their own knowledge 85.

On this reasoning, the doctrine of Christ's resurrection was safe since there were so many witnesses. More than safe, it was, as we shall see, the linch-pin in his theological system. But what of the Virgin Birth? In the nature of the case there were no witnesses to a miraculous conception. And if Christ were so different from us, how could his resurrection be a precedent for our own?[86] This is a fairly typical example of the way in which rational criteria did obtrude on a doctrine apparently sanctioned by revelation. But for Priestley it was a perfectly legitimate procedure. After all, to renounce one's reason was to surrender one's protection against the "gross delusions" of the papists.[87] And was it not written in the Scriptures themselves that one should "Prove all things, and then hold fast that which is good"?

What the Scriptures proved to Priestley was that, when read with an unprejudiced eye, they spoke one simple message: that God's free mercy was available to all who showed penitence. That had been Christ's message too - the religious truth for which he had died. It was this to which Priestley held fast as he began the laborious task of excising and reinterpreting biblical texts habitually cited in defence of orthodoxy. One example of his technique will have to suffice. In I Corinthians 2 v 14 we read that "the natural man receiveth not the things of the spirit of God". It was a proof text for those who considered unregenerate man powerless to do the will of God. But Priestley made short work of it. When St. Paul had used the same word for 'natural' in other contexts he had been referring to man's animal, sensual nature.[88] The proof text therefore contained a mistranslation and should have read "the <u>sensual</u> man (who had no higher aims than the gratification of his animal desires) receiveth not the things of the spirit of God". And that was incontestable.

I shall not dwell on this aspect of Priestley's programme because it belongs to a more specialized history of biblical exegesis. But one cannot fail to note that he found himself trapped in the same poignant situation as so many reformers who presumed to find the basis of a true, primitive Christianity in the Bible shorn of corruption. "I cannot help wishing", he once wrote, "that persons of all sects and parties would study their bibles more and books of controversy less". The only remedy for religious differences was to get back to first principles "without the help of commentators".[89] But, as Priestley's own exegesis showed, if a common plowman would understand St. Paul, a commentator could not be dispensed with. Commentaries designed to eliminate commentaries had the disturbing effect of adding to their numbers. Priestley may even have felt the frustration of this as tracts proliferated in response to his own. And that might help to explain why for his purge on contemporary Christianity he turned to history as well as Scripture.

But where there any connections between Priestley's attitude to Scripture and his attitude to science? I think there were, even though they were not always direct. In his own understanding of Scripture, Priestley detected a sense of progress. He was convinced that, as biblical scholarship advanced, the unitarian cause would be even further consolidated. The reformation of Christianity had not ceased at the Reformation and it would not cease in his own day. This vision of progressive reformation, even if it owed nothing to science was justified by it. "In nature", Priestley wrote, "we see no bounds to our inquiries. One discovery always gives hints of many more, and brings us into a wider field of speculation. Now why should not this be, in some measure, the case with respect to knowledge of a moral and religious kind."[90] The analogy actually went deeper than that. Observing that "learned unitarians increase, while learned trinitarians decrease", he drew an explicit parallel between the spread of Newtonian science towards universal acceptance and the prospective universality of his own creed.[91] More specifically his admission of the limitations under which the biblical authors had performed clearly pre-empted any possibility of an appeal to Scripture to block a scientific innovation. From the secular perspective of the 20th century, one might expect to find him rejecting biblical miracles on the ground that they were scientifically impossible. That line of thought, however, he barely entertained, if at all. Miracles which had been publically known and acknowledged were not called into question. They served the purpose of authenticating a divine message which rendered further miracles redundant.[92] In that respect there were limits to Priestley's radicalism. But his conviction that miracles did not belong "in this age of the world"[93] was certainly of a piece with his scientific determinism.

The purification of Christianity

On his visit to France where he met the scepticism of the <u>philosophes</u>, one episode stuck in his mind:
> When I was dining at ... Turgot's table, M. de Chatellux ... in answer to an inquiry said the two gentlement opposite me were the Bishop of Aix and the Archbishop of Toulouse, "but", said he, "they are no more believers than you or I". I assured him I was a believer; but he would not believe me.[94]

It is a telling anecdote because much of Priestley's motivation for the purification of Christianity stemmed from his desire to recast it in a form which would win the assent of intellectuals. Whether it was his friend, Benjamin Franklin, the French free-thinkers, or the historian Edward Gibbon, Priestley wished to persuade them that the Christianity they were rejecting was a bastardized form.[95] A pure unitarianism was, by contrast, unassailable.[96] In fact, if Christianity was to become the universal religion it had to be unitarian for how else could Jews and Mahometans be expected to withdraw their objections?[97] For such pragmatism he was not surprisingly dubbed "half a Mahometan".[98] But, as with so much of the abuse he had to endure, it missed the point. Once the trinitarian obstacle were removed a rational dialogue between the two religions would establish the one and destroy the other. It is perhaps a measure of Priestley's optimism that he thought "less than a century" might suffice.[99]

Under these pressures for purification Priestley made his distinctive contribution to the academic study of religion. His scriptural exegesis was reinforced by a historical critique, designed to show how Christianity had come to be so corrupted. As Professor Schofield has pointed out, the depth of this historical critique, anticipating as it did the Higher Criticism of the German historians, is still not fully appreciated.[100] On the one hand he set more rigorous standards for the evaluation of historical evidence. That the evangelists Mark and John gave no hint of a miraculous conception over-ruled the testimony of Matthew and Luke who did.[101] On the other hand he developed the challenging historical thesis that it was through contamination with Platonism that primitive Christianity suffered its most seminal corruption. It was the Platonist belief in the pre-existence of souls which did the damage.[102] Once the eternal pre-existence of Christ was countenanced, it was only a step or two to create the Trinity. The thesis itself was not original. It had been argued by no less a scientific precursor than Newton.[103] But Priestley developed it with such assiduity that each successive corruption seemed the most natural thing in the world. He managed to convey the dynamics of the process. As Jews and heathens had been so scandalized by discipleship of one who had been crucified as a common malefactor, so Christians, in general, he suggested, would be disposed to adopt any opinion which would wipe away the reproach.[104] And one corruption did naturally flow from another. Once the souls of the Greek philosophers were imported, it became natural to think in terms of the immortality of the soul rather than the biblical doctrine of the resurrection of the dead.[105] And with immortal souls on tap, it was not surprising that some began the practice of praying to and for the dead.[106] After that the deluge: praying to the saints, belief in transubstantiation ... and the doctrine of the atonement. For, as Priestley pointed out, if Christ had remained a "mere man" there would never have been any question of his sufferings placating the divine wrath.[107] The doctrine of the free mercy of God would not have been impeached.

Priestley's purpose was, then, to show that the causes of religious error could be located in specific cultural conditions. Religious truth, on the other hand was either transparent to reason, or, like the doctrine of the resurrection transparent through revelation. As Margaret Canovan has recently observed, there is something rather startling about this asymmetry;

> The reverse of Priestley's theory that religious errors could be explained in historical terms, was his logical, if startling, conviction that religious opinions for which there was no historical explanation must be true[108].

But what if a historical explanation were offered for a central doctrine like the Resurrection? Priestley simply had to reply to Gibbon who had done precisely that. The promise of eternal happiness, in Gibbon's opinion, had been so advantageous that it was not in the least surprising that it had been so widely accepted. Priestley's retort was that it was not enough that a doctrine should be inviting. It must also have appeared credible.[109] This was a typical response from one who never doubted that his own creed bore the marks of eternal truth. The irony of his position, as Canovan shows, was rooted in the unconscious assumption that his own interpretations, like the original gospel, were exempt from the cultural distortions peculiar to his own age.[110]

One of those peculiarities was an optimistic belief in progress, coupled in Priestley's mind with the central fact of Providence that out of evil good would always come. Thus, in an engaging, and entirely characteristic manner, he could argue that the very corruption of Christianity against which he strove had been permitted by Providence because it would enable a purer version to triumph through criticism.[111] It was entirely characteristic because Priestley had a keen sense of the dynamics of history in which actions for the worse invariably provoked reactions for the better. Many of his political comments were based on that pattern. Thus the clergy of the established church were contributing to their own downfall by supporting a war against America which enlarged the national debt.[112] And they would contribute to their own downfall if they, be it ever so slightly, became more exacting in their extraction of tithes.[113] The formula that 'everything that is is right' was not good enough for Priestley. His keen sense of the dialectics of history meant that even everything that was wrong was right.

And what of the bearing of all this on science? Heaven bent on the purification of Christianity, Priestley clearly saw in scientific knowledge a means of enlightenment that could only assist his cause:

> This rapid progress of knowledge ... will, I doubt not,
> be the means under God of extirpating all error and
> prejudice, and of putting an end to all undue and usurped
> authority in the business of religion as well as of science.[114]

In the preface to his <u>Observations on Air</u> he deliberately pitted the growth of rational knowledge against the Papacy, provoking a predictable response from the French censors.[115]

The disposal of spirit

A perennial theme in Priestley's writings was that Christianity had been cluttered with immortal souls - spirit substances which had, as it were, a life of their own, detachable from the matter with which, during this earthly existence, they were conjoined. It was the assertion of this separate spirit world which had been responsible for most of the corruptions to which Christianity had been subjected, whether it was the pre-existence of Christ or the paraphernalia of purgatory. His quest for a pure religion therefore took on another dimension when, in addition to exegetical and historical arguments, he enlisted the power of metaphysics. In his <u>Disquisitions relating to matter and spirit</u> (1777), he protested that on the conventional view, matter and spirit were such incommensurable things that they could not conceivably interact at all.[116] With that train of reasoning set in motion he went on to abolish the spirit world altogether. Or, more accurately, he replaced the matter/spirit dichotomy with a unified picture of nature in which every kind of property, whether physical or mental, could be reduced to a single category of attractive and repulsive powers. From this point of view it was simplistic to define matter in terms of solid, impenetrable atoms. For as Priestley explained, if there were no "powers of mutual attraction infinitely strong ... between the parts, the atom could not hold together... that is, it could not exist as a <u>solid atom</u>[117] ...". Drawing on the work of David Hartley,[118] he tried to show that mental attributes - thoughts and ideas - could be explained by similar powers of attraction and repulsion operating in the brain. Minds and spirits as <u>things</u> were thus disposed of.

The argumentation in support of this monism was extremely complex.[119] But the religious impulse that lay behind it is quite easy to grasp. In the first place the dualistic view was clogged with religious difficulties. If the soul was immaterial and the body material, there was the old chestnut whether the two came together at conception, at birth or whenever[120] ... But far more seriously the doctrine of the immortality of the soul made the doctrine of the resurrection superfluous. Priestley belonged to that mortalist tradition which proclaimed that when you're dead you're dead. But, by the grace of God one had the hope and promise of resurrection in God's own time. That, for Priestley, was the biblical view and it was rendered otiose by conceptions of immortality and purgatory. This was an absolutely critical point. It allowed him to turn the tables on those who accused him of surrendering revelation to reason. It was the Trinitarians and the immaterialists who were grafting the inventions of reason on the simple biblical view.[121] And it was a critical point because he happened to believe that the only essential article of Christianity was the doctrine of resurrection.[122] Without it, and the concomitant rewards and punishments, there could be no social control[123] and no ultimate rationale for the reformation of character.[124] He even went so far as to speculate on the chemistry of resurrection:

> Death, with its concomitant putrefaction and dispersion of parts, is only a decomposition; and whatever is decomposed, may be recomposed by the being who first composed it; and I doubt not but that, in the proper sense of the word, the same body that dies shall rise again '[125].

All this was just too much for one contemporary who was driven into verse:

> So when we view a castle-wall
> Rent by a pond'rous cannon-ball,
> We must conclude from your new rules
> (Our reas'ning fathers being fools),
> That 'tis not solid brick or stone
> Which solid iron has o'erthrown;
> But that a Nothing did attract,
> And had not strength to counteract,
> By its repulsive force the thing,
> (The Thing! your pardon Sir, the Nothing)[126]

But this was only half. Suppose some poor fellow sank in the Thames to be eaten by eels, the eels subsequently gracing the palate of an alderman. What on Priestley's view would happen at the resurrection?

> Poor Thomas in the Thames was drown'd
> And though long sought could not be found ...
> At the last trumpet's solemn sound,
> How mangled will poor Tom be found! [127]

Because Priestley's matter, comprising attractive and repulsive powers, was unlike conventional matter, or conventional spirit, he was happy to say that it might as leave be called spirit as matter.[128] He would have agreed with the remark of a later materialist, John Tyndall, who after rebuking the philosophers who had made it solid, impenetrable and inert, complained that matter had been "much maligned". But most of Priestley's contemporaries were mystified, even if they were not angry. Richard Price kept asking "what is it that attracts and repels, and that is attracted and repelled?"[129] Nor was he happy with Priestley's account of the resurrection:

> It is ... implied, that the men who are to be raised from death, will be the same with the men who have existed in this world, only as a river is called the same, because the water, though different, has followed other water in the same channel ... Did I believe this to be all the identity of man hereafter, I could not consider myself as having any concern in a future state.[130]

Priestley's materialism was as much maligned as the matter he sought to activate. But in his own mind it made sense philosophically, scientifically and theologically. So tightly are these strands woven together in the Disquisitions that one cannot tease out one without pulling on all the rest.[131] Even the history of chemistry contributed to the argument. Chemistry, like Christianity had been infected with 'spirits'. But Priestley's century, and Priestley's own work, had shown that a vocabulary of 'spirits' could be replaced by a vocabulary of 'airs'. By analogy, the time was right for disinfecting Christianity.[132]

If Priestley's materialism mediated between his religion and his science, there is the question whether it had a direct impact on his scientific work. This has been a matter of dispute among Priestley scholars - not least because Priestley himself declared that one could never gain access to the inner constitution of matter.[134] Professor Schofield has urged the strong claim that the attractive and repulsive powers, which Priestley associated with the name of Boscovich, stamped on his science a concern for physical rather than intrinsically chemical investigation. And that might help to explain some of his resistance to Lavoisier who built his science on the primacy of the chemical elements.[135] The least that can be said is that Priestley advertised an ontology of attractive and repulsive powers which was to have vibrations in the subsequent researches of Davy and Faraday.

Divine Activity in the World

It was not only Priestley's materialism that mediated between his religion and his science. There was also his determinism, or the doctrine of philosophical necessity as he preferred to call it. Since there was now no ultimate distinction between mind and matter, Priestley was committed to the view that the workings of the mind were determined by fixed causal laws in just the same way as any physical phenomenon. He did not demur at this. It was in keeping with his resolve to publicize the works of David Hartley who had suggested that the association of ideas in the mind could be correlated with patterns of vibration in the brain. But it still seems odd at first sight that Priestley did not demur. What happened to free will and moral responsibility if human decisions were locked in predetermined sequences? This intuitive reaction was enough for most of his critics. But with characteristic independence Priestley rejected the intuitive view. In common with other 18th century radicals he saw in determinism a necessary presupposition if campaigns for moral and educational reform were to be effective. He put it like this:

> ... one principal reason why I reject the doctrine of
> philosophical liberty, is that exactly in the degree
> in which we suppose the mind not to be determined by me
> motives, in that very degree do rewards and punishments
> lose their effect, and a man ceases to be a proper subject
> of moral discipline.[136]

He did not deny that men had power over their own actions.[137] But unless that power operated according to fixed causal laws their behaviour would be indeterminate and <u>un</u>accountable. Moral responsibility, in other words, assumed determinism.

Far from being theologically embarrassing, Priestley's doctrine seemed to him infinitely consoling. As a minister he knew only too well that there were men and women who rejected the gospel. But they did so because their minds were not yet in the right state of preparation to receive it. For want of proper instruction early in life their vibrations had, as it were, got into the wrong grooves. The duty and heavy responsibility of the minister was to ensure that the right experiences and instruction were cultivated which would produce in due course a <u>guaranteed</u> response. "Without this persuasion concerning the uniformity of the laws of nature respecting our minds", Priestley wrote, "minister and people will both be subject to great occasional despondence".[138]

The burden of responsibility was so great because the minister could not call upon God to work some miracle on the mind. That was the God of vulgar Christianity whom Priestley had discarded. When the sower went forth, whether the seed grew or not depended on the prior condition of the ground where it fell. It was a favourite parable because it underlined the point that "all the benefit we are authorized to expect from the gospel arises from the <u>natural</u> effect that the great truths and motives of it are calculated to produce upon the mind".[139] And because, in that sense, religious responses were predetermined, it was vital to guide the young in their search of the Scriptures.[140] His catechisms therefore contained questions to be researched rather than answers to be learned by rote. "What became of Ishbosheth?" gives one pause for thought. But the ulterior design was clear enough:

> What cautions does Solomon give concerning bad company?
> What description does Solomon give of the artifices of
> an harlot?
> Repeat some of the advice which Solomon gives to young
> persons in the book of Ecclesiastes.[141]

The great beauty of Priestley's determinism was that it allowed the whole of nature and the whole of history to be subsumed under God's ulterior design.[142] The doctrine of necessity, he wrote, ultimately ascribes all to God. It allowed the integration of general and special providence without any violation of the natural order.[143] And it had one consequence for a Christian minister that was the most consoling of all: "my sphere, and degree of influence on other beings and other things, is his influence".[144]

The bearing of his determinism on his perception of science was equally direct. Scientific discoveries, particularly of covering laws, helped to disclose the interconnections that were built into nature.[145] For Priestley, unlike Hume, they were necessary connections, part of the fabric of creation not merely expectations of constant conjunction in the human mind.[146] And since nature was the product of design one would expect to find examples of economy and benevolence. There was economy, for example, in that a single principle, phlogiston, conferred common properties on all the metals.[147] It was the quest for economy that led him to speculate in January 1788 that water might be the basis of all kinds of air.[148] And there were many such speculations in his letters and papers.[149] As for benevolence, the vitiation and restoration of air provided a scientific parable of the transformation of evil into good. Even in stagnant waters, the unwholesome effect of putrid matter was offset by the "purest dephlogisticated air, supplied by aquatic plants".[150] In an age which had seen Hume's emasculation of natural theology, Priestley's science gave it new vigour. The message was clear enough to Sir John Pringle who presented Priestley with the Copley medal in November 1773:

> From these discoveries we are assured, that no vegetable grows in vain, but that ... every individual plant is serviceable to mankind; if not always distinguished by some private virtue, yet making a part of the whole which cleanses and purifies our atmosphere.[151]

This does not, of course, mean that Priestley somehow deduced the <u>details</u> of his science from theological assumptions. On specific details his metaphysical preconceptions may even have led him astray. It took him a long time to appreciate that there could be an air purer than common air. And he was inclined to overestimate the restorative effect of shaking noxious airs with water, believing as he did that his experiment was simulating a beneficent interaction between atmosphere and sea.[152] The interaction between his science and religion was nevertheless pronounced and operated on many different levels. His rational Christianity provided a presupposition for science in the shape of a pre-determinate uniformity of nature. It provided a sanction in that advances in scientific knowledge spelled the retreat of religious superstition. It provided a motive because science could assist in the exploding of corruptions; - a motive too in that applied science was a vehicle, under Providence, for improving man's lot.[153] And it also regulated his scientific reflections in the sense that the search was on for examples of beneficence and economy. The interaction was clearly a two-way process. He wrote as if the same objectivity could be obtained in the pursuit of religious as of scientific truth.[154] Criteria of validation appropriate to the sciences became useful implements when tackling a religious dogma such as the immediate influence of God on the human mind - an influence Priestley dismissed, partly because there was no way it could be proved.[155]

In popular essays in the history of science one often comes across references to the 'separation' of science and religion - a process usually lodged in the Scientific Revolution or, failing that, in the Enlightenment. The case of Joseph Priestley shows how simplistic that formula can be. In a letter from America, dated April 3 1800, he confided that one of his primary objects had been to join (natural) philosophy to Christianity from which it was "too much separated".[156]

His case has also been cited, with good reason, in support of a correlation between enthusiasm for science and religious dissent. With good reason because the dissenting academies found a place for science in their curricula;[157] because non-conformists were proportionally more active in careers which departed from the traditional professions[158] - as, in large measure, they were obliged to be - and because in the early literary and philosophical societies of industrial centres such as Manchester and Newcastle there was a strong unitarian presence.[159] There is, however, a danger in the correlation if it glosses over the fact that different forms of science have been both promoted and obstructed in almost every religious tradition. If there is a general correlation perhaps it is between science and a degree of heterodoxy within the religious tradition to which the man of science belonged. Of that correlation Priestley would be just as perfect an exemplar. In dissenting from 'orthodox dissent' his individualism found expression in a theology that was rationalist even by the standards of 'rational dissent'. It has recently been said of him that he came "close to arguing that freedom of religion was, if one had to choose, more important than the security of the state".[160] On the issue of religious liberty he still strikes a chord as one who believed it was a case of "every man for himself".[161] But we must not be deceived by his apparent modernity.[162] It is the greatest irony of Priestley's rationalism that he believed (almost believed)[163] every man would, eventually, agree with him.

References

1. J. Swift, "Gulliver's Travels", Part 3, Chapter 5.
2. M. Nicolson, "Science and Imagination", Cornell U.P., Ithaca, 1962, Chapter 5, p.147.
3. J. Priestley to A. Volta, 5 August 1779, in R. E. Schofield, "A Scientific Autobiography of Joseph Priestley", M.I.T. Press, Cambridge Mass. and London, 1966, p. 174.
4. J. Priestley to B. Franklin, 24 June 1782, ibid., p.208.
5. R. E. Schofield, "Joseph Priestley and the Physicalist Tradition in British Chemistry", in L. Kieft and B. R. Willeford (eds.), "Joseph Priestley: Scientist, Theologian and Metaphysician", Bucknell U.P., Lewisburg, 1980, pp. 99-102.
6. J. Priestley to W. Morgan, 23 October 1802, in Schofield (note 3), p. 316.
7. H. Hartley, "Studies in the History of Chemistry", Clarendon Press, Oxford, 1971, Chapter 1, p.12.
8. S. Hales, "Vegetable Staticks", with foreword by M. A. Hoskin, Scientific Book Guild, London, 1961, p. xi.
9. F. W. Gibbs, "Joseph Priestley", Nelson, London and Edinburgh, 1965, p. 122.
10. J. Priestley, "An Examination of Dr. Reid's Inquiry into the Human Mind on the Principles of Common Sense; Dr. Beattie's Essay on the Nature and Immutability of Truth; & Dr. Oswald's Appeal to Common Sense on Behalf of Religion", London, 1774, in J. T. Rutt (ed.), "The Theological and Miscellaneous Works of Joseph Priestley", 25 vols., London, 1817-35, Vol. 3, pp. 101-2.
11. J. Priestley, "Reflections on the Present State of Free Inquiry in this Country", Birmingham, 1785, pp. 40-1. See also J. Priestley, "Letters to the Rev. Edward Burn of St. Mary's Chapel, Birmingham", 1790, p.ix.
12. Cited by Gibbs (note 9), p. 183.
13. M. Fitzpatrick, The Price-Priestley Newsletter, 1977, No. 1, 3. J. Priestley to the Abbé R. J. Boscovich, 19 August 1778, in Schofield (note 3), p.167.
14. A. J. Ihde, "Priestley and Lavoisier", in L. Kieft (note 5), pp. 62-91. For Priestley's changing conception of dephlogisticated air and his regarding it as a compound of 'earth' and 'acid', see J. G. McEvoy, Ambix, 1978, 25, especially pp. 166-71.
15. A. J. Ihde (note 14), pp. 76-85.
16. J. H. Brooke, "Davy's Chemical Outlook: the Acid Test", in S. Forgan (ed.), "Science and the Sons of Genius: Studies on Humphry Davy", Science Reviews, London, 1980, pp. 121-75.
17. J. G. McEvoy, "Causes and Laws, Powers and Principles: the Metaphysical Foundations of Priestley's Concept of Phlogiston", unpublished paper read before the Priestley symposium held at the Wellcome Institute, London, 27 May 1983: "Priestley wished to emphasize the fact that ... the oxygen

theory and the phlogiston theory were equally at odds" with an "avowed commitment to the empiricist analysis of material substances. ... Although he preferred the phlogiston theory, Priestley did not defend traditional dogma in the Chemical Revolution so much as attack the newly established orthodoxy of his rivals".

18. J. Priestley to B. S. Barton, 7 July 1801, in Schofield (note 3), p. 308.
19. J. G. McEvoy, British Journal for the History of Science, 1979, 12, 1, especially p. 14.
20. S. Schaffer, "Priestley and the Politics of Spirit", unpublished paper read before the Priestley symposium, Wellcome Institute, London, 27 May 1983.
21. Cited by Gibbs (note 9), pp. 123-4.
22. Ibid., p. 168.
23. J. G. McEvoy, Ambix, 1978, 25, 1, especially pp. 13-18.
24. This is the main thesis of J. G. McEvoy and J. E. McGuire, "God and Nature: Priestley's Way of Rational Dissent", in R. McCormmach (ed.), Historical Studies in the Physical Sciences, 1975, 6, pp. 325-404. It is a pleasure to acknowledge my debt to this seminal study.
25. H. J. Abrahams, Ambix, 1964, 12, 44; especially, p. 59 footnote 19.
26. Ibid., p. 68.
27. Ibid.
28. J. G. McEvoy (note 23), especially pp. 94-101.
29. R. E. Schofield (note 3), p. 133.
30. M. Fitzpatrick, Enlightenment and Dissent, 1982, No. 1, 3.
31. See, for example, his letters to M. von Marum in Schofield (note 3), pp.246 and 251.
32. J. Priestley to J. Canton, 18 November 1771, ibid. (note 3), p. 92.
33. R. E. Schofield (note 3), p.ix.
34. J. G. McEvoy (note 23), p.7.
35. Cited by E. N. Hiebert, "The Integration of Revealed Religion and Scientific Materialism in the Thought of Joseph Priestley", in L. Kieft (ed.), (note 5) pp.27-61, 34.
36. Ibid., p.57.
37. J. Priestley, "The Doctrine of Phlogiston Established and that of the Composition of Water refuted", Northumberland, 1800, Preface; Hiebert (note 35 p. 58.
38. D. Orange, History Today, 1974, 24, 773.
39. According to Priestley's own account he took particular exception to the prospect of having to declare an experience of God's grace and having to "subscribe my assent to ten printed articles of the strictest Calvinist faith and repeat it every six months". C. Lawless, "The Dissenting Academies", in "Scientific Progress and Religious Dissent", Block 3 of

Open University Course, "Science and Belief from Copernicus to Darwin", AMST 283, Open University Press, Milton Keynes, 1974, p. 97.

40. J. Priestley, "Defences of the History of the Corruptions of Christianity in reply to Dr. Horsley and Mr. Babcock" (1786) in J. T. Rutt (note 10), Vol. 18, p. 37.
41. E. N. Hiebert (note 35), p. 32.
42. J. G. McEvoy and J. E. McGuire (note 24), p. 383.
43. C. Lawless (note 39), pp. 98-9.
44. Ibid.; also Hiebert (note 35), p. 32.
45. J. Priestley, "Disquisitions relating to Matter and Spirit" (1777), reprint edition, Arno Press, New York, 1975, p. 356.
46. J. Priestley, "An Appeal to the Serious and Candid Professors of Christianity", 4th. edition, London, 1772.
47. Ibid., pp. 5-7.
48. Ibid., p. 10.
49. Ibid., pp. 10-11.
50. Ibid., p. 18.
51. Ibid., p. 19.
52. Ibid., p. 21.
53. In a characteristic style of argument Priestley considered it in keeping with divine benevolence that the orthodox were rarely as immoral as their beliefs ought to make them! There were "principles implanted in our frame" by a "wise creator" who had "by no means left our moral conduct at the mercy of our opinions". J. Priestley, "Considerations on Differences of Opinion among Christians", London, 1769, pp. 22-4.
54. J. Priestley (note 46), pp. 8-10.
55. Ibid. Against the Calvinist emphasis on the primacy of sound doctrine, Priestley insisted that the criteria for evaluating religious maturity should not be concerned either with correct beliefs or with the right feelings. Love for God was ultimately to be judged by acts of virtue and obedience to the commandments. Ibid., pp. 21-4. To believe that salvation depended on holding present opinions was to close the door automatically on free inquiry and to live in dread of it. Priestley (note 53), pp. 14 and 31.
56. Ibid. (note 53), pp. 17-19.
57. Priestley (note 46), p. 21.
58. Priestley (note 53), pp. 17-18.
59. M. Fitzpatrick (note 30), p. 13.
60. Priestley (note 46), p. 12.
61. Priestley (note 11), "Letters to the Rev. Edward Burn ...", pp. ix-xi. He had come to see the established connection between church and state as "unnatural" since the one was a kingdom of this world, the other not.

62. M. Fitzpatrick (note 13).
63. This was certainly one of Priestley's major grievances. See, for example, note 11, p.xiv ("Letters ...) and pp. 44-5 ("Reflections ...).
64. S. Madan, "A Letter to Dr. Priestley in consequence of his Familiar Letters addressed to the Inhabitants of the town of Birmingham", Birmingham, 1790. Madan commended a "very shrewd resolution" which had been lately passed in the common hall at Leicester: "that all which the Dissenters urge respecting their own moderation is absurd, because no men give themselves bad characters". Priestley had felt bound to complain that Madan was representing the whole body of dissenters as armed with a king-killing dagger (pp. 16 and 25).
65. "Another foolish and unjust War, like that with America, which was chiefly urged by the clergy ... can hardly fail to bring their affairs to a crisis". Priestley, note 11 ("Letters ...), p. xiv.
66. Priestley, note 11 ("Reflections ...), pp. 41-2.
67. Ibid., p.39.
68. Ibid..
69. Priestley, note 11 ("Letters ...), p. xii.
70. Ibid.. The high value which Priestley put on his proselytic pen is emphasised by Fitzpatrick (note 30), p.25.
71. J. Macgowan, "Familiar Epistles to the Rev. Dr. Priestley", London 1771, p.14.
72. J. Stephens, Enlightenment and Dissent, 1982, No. 1, 43; M. Fitzpatrick (note 13).
73. J. Macgowan (note 71), p. 12.
74. F. W. Gibbs (note 9), p.7.
75. Ibid. p. 10.
76. C. Lawless (note 39), p.97.
77. As, for example, in (note 46), pp. 10-11.
78. Priestley, note 11 ("Letters ...), p. ix.
79. This argument for a connection between Priestley's assumptions about nature and his renunciation of a theological voluntarism is expounded in greater depth in J. G. McEvoy and J. E. McGuire (note 24), pp. 334-40. See also P. Heimann and J. E. McGuire, Historical Studies in the Physical Sciences, 1971, 3, 233.
80. For an introduction to the literature on the connection between a voluntarist theory of creation and the presuppositions of modern science see J. R. Milton, Arch. Europ. Sociol., 1981, 22, 173; and J. E. McGuire, Journal Hist. of Ideas, 1972, 33, 523.
81. J. Priestley, "The Doctrine of Divine Influence on the Human Mind Considered in a Sermon", Bath, 1779, pp. 12-13.

82. Ibid., p.14.
83. Priestley, note 11 ("Letters ..."), pp. vi-viii.
84. Ibid., p. 6.
85. Cited by Hiebert (note 35), p.50.
86. Ibid., pp. 51-3.
87. Priestley (note 46), pp. 4-6.
88. J. Priestley, "A Familiar Illustration of Certain Passages of Scripture", London, 1772.
89. Priestley (note 53), pp. 24-5.
90. J. Priestley, "The Importance and Extent of Free Inquiry in Matters of Religion", London, 1785, p.7.
91. Priestley, note 11 ("Reflections ..."), pp. 51-9.
92. Priestley (note 81), p. 9: "... the proper use of miracles has been to make more miracles unnecessary". These "unnecessary" miracles were the very ones assumed in the doctrine that God could directly influence the workings of the mind.
93. Ibid., p.25.
94. Cited by Orange (note 38), p. 781.
95. The parallel between Priestley's efforts and those of modern apologists on behalf of a demythologised Christianity has not gone unnoticed (cf. ibid., p. 774 and 781). For a modern equivalent one might turn to J. A. T. Robinson, "But that I can't believe", Collins, London, 1967.
96. From this perspective he would rebut the charge that Socinianism led straight into deism. On the contrary it was corrupted Christianity which led to deism and unbelief by a process of wholesale rejection. See his "Animadversions on some passages in Mr. White's Sermons at the Bampton Lecture" contained in (note 90), pp. 65-72.
97. Priestley, note 11 ("Reflections ...) pp. 48-9.
98. See Note 11 ("Letters ..."), pp. iii-vi. Undeterred, Priestley "rejoiced" that "so great a proportion of the human species are worshippers of the one true God ..." (note 96), p. 61
99. Priestley, note 11 ("Reflections ..."), p. 49.
100. R. E. Schofield, "The Professional Work of an Amateur Chemist", this volume.
101. E. N. Hiebert (note 35), pp. 52-3.
102. J. Priestley, "A History of the Corruptions of Christianity", reprinted from Rutt's edition, London, 1871, p. 302.
103. F. Manuel, "The Religion of Isaac Newton", Clarendon Press, Oxford, 1974, Chapter 3.
104. Priestley (note 102), p. 302.
105. Priestley (note 45), pp. 49-50.
106. Priestley (note 102), p. 302.
107. J. G. McEvoy and J. E. McGuire (note 24), p. 383. Priestley also ascribed

the doctrine of atonement to an over-reaction on the part of the reformers against the popish doctrine of salvation through merit. See (note 90), p. 12.
108. M. Canovan, Priestley-Price Newsletter, 1980, No. 4, 16, especially pp. 19-20.
109. Priestley (note 102), p. 305.
110. M. Canovan (note 108), p. 21. Because of this blindness on Priestley's part, Canovan can conclude that "although he had a highly developed sense of historical growth, which his Dissenting forebears had lacked, the eventual result of his work was to lead 'Rational Christianity' into a cul-de-sac" (p. 24).
111. Priestley (note 102), p. 310.
112. Priestley, note 11 ("Letters ..."), p. xiv.
113. Ibid.. Even the current rage against dissenters would have the effect, so Priestley thought, of making more dissenters agree with himself in opposing civil establishments of Christianity: "The greater their violence, the greater is our confidence of final success. Because it will excite public discussion, which is all that is necessary for our purpose" (p.xv).
114. Cited by J. G. McEvoy and J. E. McGuire (note 24), p. 380.
115. D. Orange (note 38), p. 781.
116. Priestley (note 45), pp. xxxviii and 16.
117. Cited by J. G. McEvoy and J. E. McGuire (note 24), p. 389.
118. On Priestley's debt to Hartley see ibid., pp. 348-57 and 380-2. As these authors point out, however, Hartley's view of the origin of an association of ideas in material vibrations in the brain did not entail the materialism of Priestley. Hartley himself retained a parallelism between mind and body rather than reduce the one to the other (p. 382).
119. A. Tapper, Enlightenment and Dissent, 1982, No. 1, 73; A. P. F. Sell, Priestley-Price Newsletter, 1979, No. 3, 41.
120. Priestley (note 45), pp. 41-2.
121. Priestley, note 11 ("Letters ..."), pp. 26-7.
122. Priestley (note 90), p. 22.
123. In using this phrase I mean no more by it than Priestley would have countenanced when he wrote that all that could be done to influence men's moral conduct was to present to their minds sufficient motives of hope or fear; and in Christianity, he continued, these were bound up with the joys and torments of a future life. Priestley (note 53), p. 15. From a broader perspective it is difficult to see in his intellectualist theology an opiate of the masses. J. J. Hoecker, Priestley-Price Newsletter, 1978, No. 2, 44, especially p. 62.
124. Such reformation would pave the way for the final perfecting of men in the after-life, when even the wicked might yet prove capable of improvement.

J: J. Hoecker (ibid.), pp. 55-6.
125. Priestley (note 45), p. 161.
126. A. Bicknell, "The Putrid Soul. A poetical epistle to Joseph Priestley on his Disquisitions Relating to Matter and Spirit", London, 1780, pp. 8-10.
127. Ibid., pp. 17-18.
128. J. Priestley, "A Free Discussion of the Doctrines of Materialism and Philosophical Necessity in a Correspondence between Dr. Price and Dr. Priestley", London 1778, p. 23. It is fair to agree, writes R. E. Schofield, that "if he materialized the spirit, he did so only by spiritualizing matter", Enlightenment and Dissent, 1983, No. 2, 69, especially p. 77.
129. R. Price in (note 128), p. 19.
130. Ibid., p. 73.
131. This same point has been cogently argued by Schofield (note 128). There is, however, the problem of deciding what kind of connections there were between Socinianism, Materialism and Philosophical Necessity, since, as McEvoy has observed, the mutual relations were not those of logical entailment (Ambix, 1979, 26, 16, especially p. 34). Priestley himself insisted that each of his principal doctrines stood on its own independent foundation (note 45, p. 356). Given that each was so controversial one might argue that he had good strategic reasons for saying so, not wishing to jeopardize a particular line of argument by fusing it with another that his readers could not accept. An example of this dissociation occurs in (note 81), pp. vii-viii where his critique of the doctrine of divine influence is de-coupled from his doctrine of necessity, even though the two were congruous. Similarly he contended that the non-preexistence of Christ could be proved independently of the materiality of man, E. N. Hiebert (note 35), p. 37.
132. F. W. Gibbs (note 9), pp. 99-100.
133. See, for example, McEvoy's critique of Schofield in Ambix, 1968, 15, 115.
134. Priestley's exact words were: "By what means we can ever come at the knowledge of the internal arrangement of the elementary parts of natural substances, I have not the least idea". Cited in ibid., p. 119.
135. R. E. Schofield (note 128), pp. 77-80.
136. Priestley (note 128), p. xxi.
137. Ibid., pp. xxiii-xxiv.
138. Priestley (note 81), p. 8.
139. Ibid., pp. 1-2.
140. Ibid., p. 27.
141. J. Priestley, "A Scripture Catechism", 2nd. edn., London, 1781.
142. " ... everything being what God foresaw and intended, and which must issue

of state to the exclusion of "the people" and "what has been the progress of science; of arts, of manufacturers, and commerce, by which the real welfare of nations is promoted". J. Priestley, "The proper objects of education in the present state of the world ...", London, 1791, p. 31. The historical cast of Priestley's mind allowed him to integrate scientific and theological elements in his doctrine of progress and human perfectibility. J. J. Hoecker (note 123), p. 45 and in Priestley-Price Newsletter, 1979, No. 3, 29. That human history had a direction was further reinforced by Priestley's biblical commitment to the Second Coming of Christ, for which the purification of Christianity and the reform of men were pre-requisites. In Priestley's vision of progress, however, Christ's second coming would precede, not succeed, the millennium. M. Fitzpatrick, "Priestley and the Millennium", unpublished paper read before the Priestley symposium, Wellcome Institute, London, 27 May 1983.

154. This point is developed more fully in J. J. Hoecker (note 123).
155. Priestley (note 81), p. xi.
156. J. Priestley to B. Lynde Oliver, 3 April 1800, in R. E. Schofield (note 3), p. 302.
157. Thus Priestley himself had been introduced to a course of scientific study whilst at Daventry, though there is the qualification that it was already rather old-fashioned. R. E. Schofield (note 5), p. 104.
158. From a sample of innovators in manufacturing, mining, transport, and agriculture for the year 1770, E. E. Hagen concluded that "The Nonconformists contributed about nine times as many entrepreneurs, relative to their total number in the population, as did the Anglicans". See G. K. Roberts in C. Lawless (note 39), pp. 108-9.
159. A. W. Thackray, American Historical Review, 1974, 79, 672; D. Orange, "Rational Dissent and Provincial Science: William Turner and the Newcastle Literary and Philosophical Society", in I. Inkster and J. Morrell, "Metropolis and Province: Science in British Culture 1780-1850", Hutchinson, London, 1983, p. 205.
160. M. Fitzpatrick (note 13), p. 15.
161. Cited in ibid., p. 14.
162. Thus he was anxious to disclaim the assertion that the reason of the individual was the sole umpire in matters of faith. Neither the doctrine of the unity of God nor the resurrection of man could have been discovered by unassisted reason. Priestley, note 11 ("Letters ..."), p. 26.
163. Despite his conviction that universal toleration would pave the way for unanimity on the subject of rational religious belief, he was at the same time realistic enough to appreciate that his doctrine of necessity "will probably never be within the clear comprehension of the vulgar, so that it will always be unpopular". Priestley (note 81), p. xi.

as he wishes", Priestley (note 128), p. xxii.

143. On the doctrine of necessity "whether those coincidences, which are ascribed to a particular providence, be brought about just at the time of the respective events, or were originally provided for in the general plan, the design is the very same", Priestley (note 81), p. vi.

144. Cited by McEvoy and McGuire (note 24), p. 393.

145. In this respect scientific investigation could be perceived as a religious duty, ibid., pp. 343-4 and 347-8.

146. J. Priestley, "Letters to a Philosophical Unbeliever", Bath, 1780, pp. 105-25. J. G. McEvoy and J. E. McGuire (note 24), pp. 357-62; J. J. Hoecker (note 123), pp. 58-9; J. H. Brooke, "Natural theology in Britain from Boyle to Paley", in "New Interactions between theology and natural science", constituting Block 4 of Open University Course "Science and Belief from Copernicus to Darwin", AMST 283, Open University Press, Milton Keynes, 1974

147. This achievement of economy via the category of "principles" was common in 18th century chemistry and did not require a theological justification. See, for example, E. M. Melhado, "Jacob Berzelius: The Emergence of His Chemical System", University of Wisconsin Press, Madison, 1981, Chapter 1. But in Priestley's case there is little doubt that his theism reinforced the quest for economy in scientific explanation, of which the ontology of "principles" was a striking example. For further discussion of this point see McEvoy (note 17) and McEvoy & McGuire (note 24), p. 397.

148. Priestley to J. Wedgwood, 8 January 1788, in R. E. Schofield (note 3), p. 249.

149. As in Schofield (ibid.), pp. 129, 156, 209, 249, 252, 271. Priestley's emphasis on phlogiston as a common principle meant that he did not "absolutely despair of the transmutation of metals" (ibid., p. 156). One might argue that his commitment to generic principles ought to have made him more not less, sympathetic to Lavoisier's oxygen qua principle of acidity. Had this been so the story would have been delightfully paradoxical, for Priestley would then have followed Lavoisier into error. He did go so far as to say that the principle of acidity was in dephlogisticated air (ibid., p. 256 and 325).

150. This and other examples are cited by McEvoy (note 23), p. 164.

151. Cited by F. W. Gibbs (note 9), p. 81.

152. J. G. McEvoy (note 23), pp. 100-1.

153. On Priestley's enthusiasm for applied science see R. E. Schofield, "The Lunar Society of Birmingham", Clarendon Press, Oxford, 1963. He had a keen eye for the uses of his airs: oxygen as a luxury; carbonated water as a possible remedy for scurvy; 'nitrous air' as a preservative ... In the same spirit he would complain of histories preoccupied with princes and ministers

Priestley and the Manipulation of Gases

By W. A. Campbell
DEPARTMENT OF INORGANIC CHEMISTRY, THE UNIVERSITY, NEWCASTLE UPON TYNE
NEI 7RU, U.K.

A gas occupies more space than the solid or liquid reactants from which it is generated. That artless sentence embodies all the tribulations of Van Helmont who, deriving the name 'gas' from 'chaos', believed that gases were wild spirits which must inevitably burst their confining vessels. It also sums up the manipulative problems of eighteenth century pneumatic chemistry: how to construct and manage apparatus which could accommodate changing volumes, and how to connect the constituent parts so that the restless gases could not escape. The practical difficulties were overcome to such an extent that during the closing decades of the eighteenth century five gaseous elements were discovered and some dozen gaseous compounds intensively investigated. Much of the credit for this belongs to Priestley.

General Laboratory Apparatus

In the second half of the eighteenth century the private research laboratory was still in the process of emerging from the kitchen. The Domus chymici described by Libavius in 1606, with its five laboratories, wine cellar, bathroom, gardens and sleeping accommodation for the technician, was never built. J.J. Becher's laboratory in Vienna was simply an instrument of propaganda for his economic theories, and Boyle's laboratory in London became, under Hanckewitz, the centre for the production of English phosphorus. Most laboratories were in fact associated with commercial chemistry including pharmacy, and for these a good deal of specialized apparatus was available; those who wished to pursue chemistry privately without thought of trading would make do with a domestic table furnished with bowls from the kitchen, bottles from the wine-cellar, and glasses from the butler's pantry.

In the commercial laboratory, furnaces occupied a dominant position. Thus Robert Dossie (1759) dismissed chemical vessels in five pages of his laboratory manual but required thirty-nine to describe furnaces.[1] These might include the athanor, a slow fire for lengthy distillations; the sand furnace for high temperatures; the reverberatory furnace for calcinations; the lamp furnace over which evaporations could be carried out, and the wind furnace blown by a bellows. The pioneers of pneumatic chemistry managed very well with a domestic grate and (later) a burning glass with which to focus the sun's rays.

Partly as a legacy from alchemy and partly due to pharmaceutical influences, a variety of distillation apparatus was available. By the middle of the eighteenth century three main types had emerged, the alembic (with or without a cooling worm), the retort, and the circulatory vessel for reflux distillations or protracted extractions. A round-bottomed flask bedded in sand with its neck tilted a few degrees below the horizontal was often used as an improvized retort; the furnace of 'thirty-two long necks' which Boyle possessed was an arrangement of this kind.

Priestley and the Manipulation of Gases

Depending on the ratio of height to diameter, the flask might be described as a phial, mattrass or balloon. Some flasks were described by the name of the wine bottle which they most nearly resembled, for example Florence or Frontignac. A flask with two necks, or with a tubulus at one side, was frequently called a receiver and played a prominent part in the pneumatic work of Woulfe, Pepys, Hassenfratz and Lavoisier.

Among apparatus of a general nature might be mentioned vessels for separating oil and water (the forerunner of the separating funnel, but without the tap) and for decanting a liquid from a solid; filtering and percolating devices; crucibles and cupels; pestles and mortars; and some assorted equipment such as jars, tongs, shovels and simple tools.

Against this common laboratory background, the study of gases evolved its own special apparatus, some of which pre-dated Priestley's work. Both Robert Boyle (1660) and John Mayow (1674) had collected gases against a moving boundary of liquid, actually the acid from which the gas was generated. The apparatus was merely a flask filled with acid and inverted in a bath of the same acid; the solid reactant was lodged in the neck of the flask.[2] Mayow also transferred gas from one bottle to another under water.

Stephen Hales (1727), vicar of Teddington, measured the quantity of gas evolved when such substances as coal, tobacco, saltpetre, sawdust, wheat, olive oil, amber and red lead were heated in a gun-barrel. He collected the gases over water, using an upturned flask suspended by string so that its mouth was below the surface of water in a dish.[3]

This was the first 'pneumatic trough'. Cavendish used a similarly inconvenient arrangement, though his idealized diagrams surely minimize the difficulties of suspending an inverted bottle of water by means of string. In fact, a better way to manage inverted flasks of water had already been found. In 1741 William Brownrigg, a Cumberland physician who had studied at Leyden, placed across a cistern of water a shelf perforated with holes of various sizes into which a flask might be fixed with wedges.[4]

Priestley modestly claimed that his pneumatic apparatus was 'nothing more than that of Dr. Hales, Dr. Brownrigg and Mr. Cavendish, diversified and made a little more simple'.[5] In fact his modifications were extremely important.

Discarding the unwieldy arrangements of his predecessors, Priestley fitted his trough with a submerged shelf on which the receiving vessel might stand; he also showed that jars were more convenient than flasks. Although his diagrams show a circular trough, the description reveals an oblong trough 26 inches long and 14 inches wide, made of stout wood with the joints well caulked with paint. The shelf was $1\frac{1}{2}$ inches thick in order that the underside of each hole might be widened into a funnel. His collecting jars were those made for batteries, of which he possessed a large number.

From the 1770s the pneumatic trough replaced the furnace as the salient feature of the chemical laboratory. Lavoisier's trough had a surface area of 14 square feet yet he complained that he was frequently pressed for space.

Flexible connectors

To connect the gas generator to the collecting vessel in the pneumatic trough, without creating an unworkable rigidity, some form of flexible piping was required. If only a slight easing of position was necessary, Hales' device of a length of lead piping (which he called a 'syphon') was adequate; and lead was more easily bent than was glass tubing. As late as the 1820s Berzelius continued to use lead tubes. For greater flexibility, a strip of soft leather was made up into a spiral tube. Priestley made use of such a connection in his work on artificial Pyrmont water, prescribing that the leather should be 'sewed with a waxed thread, in the manner used by shoemakers'. The leather tubes were sometimes fitted with brass mouthpieces.

When two glass tubes were to be connected in a glass-to-glass joint, gummed paper was often employed; often this was simply good quality writing paper smeared with starch paste. Both Robert Dossie (1759) and Charles Hopson (1789)[6] referred to this material and as late as 1827 Faraday was describing carefully how to make and use it.[7]

All three however preferred moist bladder, though the details of their practice differed. Dossie moistened the strips of bladder with white of egg or a strong solution of gum arabic. Hopson used bladder softened by soaking in water, tying the joint with pack thread. Faraday was aware that if the strips of bladder were soaked for some time in water and then 'applied as a bandage' no further adhesive or ligature was necessary. Nevertheless, until well into the 19th century laboratory manuals carried instructions on tying, often supplemented by diagrams of the kinds of knots appropriate to particular situations.

This is not to say that rubber connections were unknown. The earliest imports of rubber into Europe took the form of 'bottles' made by allowing the latex to dry on rough clay moulds. The necks of such bottles formed short rubber tubes. Very soon however attempts were made to soften the mass of rubber so that it might be moulded into any required form. To make tubing, boiling the rubber in water was usually sufficient. In an account of the chemistry of rubber, William Henry (1806) wrote: 'If long strips of caoutchouc are tied spirally round a glass or metal rod, and boiled for an hour or two, the edges cohere, and a hollow tube is formed'.[8]

After the discovery of Thomas Hancock's process for the production of sheet rubber it was found that tubes could be made simply by pressing together the newly cut edges. Such home-made tubes were sometimes difficult to remove from the glass which they joined; some workers therefore sprinkled a little flour over the glass ends, whilst others favoured the enclosure of a thin slip of paper.

Hancock had dissolved rubber in naphtha (available from the newly established coal gas industry) and had used the solution for waterproofing cloth. Thomas Thomson (1836) employed the same technique in making water- and gas-tight joints; he smeared one side of a ribbon with rubber solution, wrapped the ribbon tightly round the joint, and tied it with twine. Three years later Goodyear patented his 'dry heat' vulcanisation process, but good vulcanised tubing did not appear in Europe until the middle 1840s.

Lutes and luting

In alchemical practice, a glass retort which was to be heated in a furnace was coated with clay or some similar material to prevent cracking. It thus became effectively an earthenware vessel glazed on the inside. Priestley was fortunate, through his friendship with Josiah Wedgwood the potter, in being able to obtain retorts and tubes of porcelain and ware which could withstand prolonged heating.

By a natural extension, the clay which was formerly used for coating the retort also served to seal the retort to its receiver. In this capacity the clay was called a 'lute' and the practice arose of varying the composition of the lute according to the nature of the material to be distilled. In the second half of the 18th century, it was commonly held that the correct choice and application of the lute was the most important single operation in chemistry.

The most commonly recommended lute for general purposes was known as 'fat lute'. It was made by incorporating into linseed oil some such material as ground pipeclay or Windsor loam. It resembled glaziers' putty, and Faraday used the latter regularly. Common loam made into a paste with glue, or quicklime mixed with cheese curds or white of egg were prescribed for confining aqueous vapours; for spiritous distillates, common loam mixed with chopped flax and ox blood might be used; and for acid vapours pure clay beaten into a stiff dough with boiled linseed oil and mixed with chopped hair. Harder setting mixtures were: Plaster of Paris with glue and starch; red iron oxide, lime and clay: and iron filings with clay and milk. Amid all this complexity, Cavendish believed that he had found one mixture superior to all others, 'almond powder, made into a paste with glue and beat a good deal with a heavy hammer'.[9] On the other hand Priestley, a born sceptic, held that 'where the greatest possible accuracy is required, lutes are by no means to be trusted, since a variety of vapours, coming into contact with them, are considerably affected'.

Corks and stoppers

Although the use of cork as a resilient material for stopping the necks of vessels was known to the Romans (one of Horace's Odes mentions cork coated with pitch) its use did not become general until the late 17th century. The usual agent for stopping bottles was soft wood covered with melted resin. Although by Priestley's time corks were common enough laboratory objects, Lavoisier gave directions for fixing a retort to its receiver by introducing short pieces of soft wood. In fact, though corks were plentiful, good corks were scarce and even in 1857 Greville Williams could write: 'The greatest difficulty is sometimes experienced in obtaining really good corks............ Whenever a fresh quantity is obtained of the smaller sizes, the best should be selected and put aside in a box by themselves'.[10]

Before the invention of the cork-borer in 1837, two methods of passing glass tubing through corks found general favour. To avoid actually boring the cork, a semicircular notch was filed in one side of the cork for each tube that was to be accommodated; the whole junction, neck of vessel, cork and tubes, was then sealed with luting mixture. This method was used by Peter Woulfe (1767).[11] More usually the cork was pierced either by gouging out a hole with the sharpened tang of a file, or by burning with a red hot piece of iron; the hole was then finished off with a round file of suitable size. Again, the cork and tube would be luted together. Rubber stoppers appeared only after the development of satisfactory vulcanizing processes. Before the sizes of flasks became standard, it was possible to buy a long tapering stopper which could be cut up to provide closures for whatever flasks came to hand.

Priestley, whose dislike of lutes has been noted, made two contributions to the art of stopping flasks. He had a ground glass stopper pierced with small holes, which allowed him to control the transfer of a gas under water. Another glass stopper, perforated with one central hole, was drawn out to form a tube which could be bent to suit particular circumstances. The first device was made for Priestley by his pupil Benjamin Vaughan, and the second by the London instrument maker Parker.

Bladders as pneumatic vessels

When Carl Wilhelm Scheele prepared oxygen by heating saltpetre in a glass retort he collected the gas by tying an ox bladder over the neck of the retort; as the bladder became distended, the resulting pressure forced a blow-hole in the soft heated glass.[12] In another experiment, iron filings were placed in a bladder which was then squeezed to expel the air and tied over the neck of a bottle previously filled to the brim with acid. The bladder was shaken to eject some filings into the bottle of acid. When enough gas had collected the bladder was tied off and removed from the bottle. Scheele specialized in these techniques which allowed him to handle gases without the benefit of a pneumatic trough.

Priestley used a bladder as a device to regulate pressure. In his apparatus for saturating water with carbon dioxide (designed for anti-scurvy trials in the Navy) he interposed a pig's bladder between the gas generator and the vessel of water to be impregnated; this helped to take up any inequalities in the relative rates of generation and absorption. It was this bladder which caused Dr. Mervyn Nooth to complain that Priestley's soda water tasted of urine.[13]

The widespread use of bladders in pneumatic chemistry led to the appearance of various time-saving devices such as brass nozzles, connecting pieces and stopcocks. These were still offered for sale in the 1870s. As with corks, users were at pains to find a good bladder and to preserve it; occasional rubbing with oil was often recommended.

Conclusion

A hundred and fifty years separated Van Helmont's declaration of the uncontrollable nature of gases from William Murdock's illumination of Boulton & Watt's Soho Works by means of coal gas. In that time the wild spirits had been subdued and harnessed. The new gas industry relied heavily on the accumulated practice of the pneumatic chemists; the cast iron chamber in which the coal was pyrolised was even called, in a graceful gesture to the laboratory, a 'retort'. And the unlovely gasometer floating in its reservoir of water, what was this but a huge gas jar in a pneumatic trough?

References

[1] R. Dossie, "Institutes of Experimental Chemistry", J. Nourse, London, 1759, vol. 1, 120-158.

[2] J. Mayow, "Tractatus Quinque Medico-Physici" (1674), in Alembic Club Reprints No. 17, J.Thin, Edinburgh, 1907, 111.

[3] S. Hales, "Vegetable Staticks" (1731), reprinted by Oldbourne Book Co., London, 1961, Experiment LXXVII, 105.

[4] W. Brownrigg, Phil.Trans.Roy.Soc., 1765, 55, 218.

[5] J. Priestley, "Observations and Experiments on Different Kinds of Air", Pearson, Birmingham, 1790, vol. 1, 12.

[6] C.R. Hopson, "A General System of Chemistry, taken chiefly from the German of M. Wiegleb", Robinson, London,1789, 90.

[7] M. Faraday, "Chemical Manipulation", Phillips, London,1827, 477.

[8] W. Henry, "An Epitome of Chemistry", 4th ed., Johnson, Edinburgh, 1806, 274.

[9] H. Cavendish, Phil.Trans.Roy.Soc., 1766, 56, 143 in footnote.

[10] C.G. Williams, "Handbook of Chemical Manipulation", van Voorst, London, 1857, 439.

[11] P. Woulfe, Phil.Trans.Roy.Soc., 1767, 57, 517.

[12] C.W. Scheele, "Chemische Abhandlung von der Luft und dem Feuer" (1777) in Ostwald's Klassiker der exakten Wissenschaften, No. 58, Engelmann, Leipzig, 1894, 32.

[13] J. Priestley, "Directions for Impregnating Water with Fixed Air" in J. Elliot, "An Account of the Principal Mineral Waters in Great Britain", Johnson, London, 1781, 15.